Novel Molecules in Diabetes Melitus, Dyslipidemia and Cardiovascular Disease

Novel Molecules in Diabetes Melitus, Dyslipidemia and Cardiovascular Disease

Editors

Cosmin Mihai Vesa

Simona Gabriela Bungau

MDPI • Basel • Beijing • Wuhan • Barcelona • Belgrade • Manchester • Tokyo • Cluj • Tianjin

Editors
Cosmin Mihai Vesa
Preclinical Department
University of Oradea
Oradea
Romania

Simona Gabriela Bungau
Pharmacy Department
University of Oradea
Oradea
Romania

Editorial Office
MDPI
St. Alban-Anlage 66
4052 Basel, Switzerland

This is a reprint of articles from the Special Issue published online in the open access journal *International Journal of Molecular Sciences* (ISSN 1422-0067) (available at: www.mdpi.com/journal/ijms/special_issues/H149FMEW5G).

For citation purposes, cite each article independently as indicated on the article page online and as indicated below:

LastName, A.A.; LastName, B.B.; LastName, C.C. Article Title. *Journal Name* **Year**, *Volume Number*, Page Range.

ISBN 978-3-0365-7685-5 (Hbk)
ISBN 978-3-0365-7684-8 (PDF)

Contents

About the Editors

Cosmin Mihai Vesa

Cosmin Mihai Vesa became associate professor at University of Oradea, Faculty of Medicine and Pharmacy starting from 2022, having begun teaching in 2017 as assistant professor.

He teaches Physiology of the renal, respiratory, and cardiovascular system. He is medical doctor specialized in diabetes, nutrition and metabolic diseases and obtained his PhD diploma in 2019. His research interest include: the pathophysiology of diabetes mellitus complications, insulin resistance, metabolic syndrome, obesity, and novel drugs in the treatment of diabetes mellitus. He serves as guest editors in six MDPI Special Issues and is a reviewer at numerous journals.

Simona Gabriela Bungau

Simona Gabriela Bungau became a full professor in 2005, having a continuous teaching and research activity, started in 1992 and going through all the stages of a university teaching career. After two Bachelor's degrees (in Chemistry–physics and Pharmacy), she obtained her Doctorate (PhD) in Chemistry at "Babes-Bolyai" University in Cluj, Romania, in 2003. The particularly rich research activity was presented in over 400 Web of Science indexed publications, and over 10 books published in Romanian publishing houses and 2 in international publishing houses. The initial fields of interest were Chemistry, respectively Analytical Chemistry, then expanding to Pharmacology, Botany, Public health, Ecology, Scientific Research Methodology, and Sustainability/Environmental protection. Obtaining the qualification of PhD supervisor in the field of Pharmacy came naturally, in 2015, by presenting the Habilitation Thesis at "Carol Davila" University of Medicine and Pharmacy, Bucharest, Romania. Participating in over 20 international and national research projects, numerous Erasmus teaching programs, member of societies in her native country and abroad, Mrs. Bungau continues her activity for over 30 years at the same university, namely the University of Oradea, Romania.

Preface to "Novel Molecules in Diabetes Melitus, Dyslipidemia and Cardiovascular Disease"

The purpose of this reprint based on the Special issue "Molecules in Diabetes Melitus, Dyslipidemia and Cardiovascular Disease", *IJMS* journal, MDPI, is to present the impact in clinical practice as well as in medical research of novel molecules that have been introduced in the treatment of diabetes mellitus, dyslipidemia, and cardiovascular disease. The topic focuses on the improvements in disease outcomes as well as the numerous beneficial effects of use of certain medications such as GLP-1 agonists, dual GLP-1/ GIP agonists, and SGLT-2 inhibitors for diabetes mellitus treatment; PCSK9 inhibitors and small interfering RNA (siRNA) molecules for dyslipidemia; and antiaggregants, oral anticoagulants, scubitril/valsartan, and non-steroidal mineralocorticoid receptor agonists or dapagliflozin in cardiovascular diseases. We believe that the number of real-life studies and in vivo and in vitro research of the effects of these drugs is quite low at the moment, leading to a lack of understanding of the complex molecular effects and pleiotropic effects of these medications, as well as incomplete knowledge regarding the repurposing of these drugs in clinical practice.

With 11 papers published, we consider this reprint as a valuable reward for all of us, authors and editors.

Cosmin Mihai Vesa and Simona Gabriela Bungau
Editors

International Journal of *Molecular Sciences*

MDPI

Editorial

Novel Molecules in Diabetes Mellitus, Dyslipidemia and Cardiovascular Disease

Cosmin Mihai Vesa [1] and Simona Gabriela Bungau [2,*]

1 Department of Preclinical Disciplines, Faculty of Medicine and Pharmacy, University of Oradea, 410073 Oradea, Romania
2 Department of Pharmacy, Faculty of Medicine and Pharmacy, University of Oradea, 410087 Oradea, Romania
* Correspondence: sbungau@uoradea.ro

Citation: Vesa, C.M.; Bungau, S.G. Novel Molecules in Diabetes Mellitus, Dyslipidemia and Cardiovascular Disease. *Int. J. Mol. Sci.* **2023**, *24*, 4029. https://doi.org/10.3390/ijms24044029

Received: 30 January 2023
Revised: 13 February 2023
Accepted: 15 February 2023
Published: 17 February 2023

The purpose of this Special Issue is to present the impact in clinical practice as well as in medical research of novel molecules that have been introduced in the treatment of diabetes mellitus, dyslipidaemia, and cardiovascular disease. The topic focuses on the improvements in disease outcomes as well as the numerous beneficial effects of the use of certain medications such as glucagon-like peptide 1 (GLP-1) agonists, glucose-dependent insulinotropic polypeptide (GIP) agonists, and dual GLP-1/GIP agonists for the treatment of diabetes mellitus and obesity, as well as medication such as sacubitril/valsartan for the treatment of heart failure. Included in the topic is also the presentation of certain novel molecules such as derivatives of quinopimaric acid, useful in the inhibition of some enzymes involved in the glucose metabolism, or what is also presented is novel applications of the administration of older molecules (such as nicotinamide) for the inhibition of the key pathogenetic processes involved in diabetes complications. We believe that the number of real-life studies and in vivo and in vitro research of the effects of these drugs is, to date, rather low, leading to a lack of understanding of the complex molecular effects and pleiotropic effects of these medications, as well as incomplete knowledge regarding the repurposing of these drugs in clinical practice. Thus, this Special Issue is incredibly topical and important, and we look forward to your submissions which will help to clarify the issues surrounding such drugs.

Teryakova's team carried out a very interesting study where abietane-type derivatives with arylidene, heterocyclic, nitrile, and acetylene fragments were used to inhibit the enzyme, α-Glucosidase (α-GLy); the efficacy of these novel compounds was investigated in vitro and by molecular modelling approaches. The complex investigation, that included absorption, distribution, metabolism, excretion, and toxicity (ADMET) profiling or kinetic studies of the novel synthesized compounds, led to the conclusion that a lead diterpene indole with an alkyne substituent of 45 was identified as a competitive inhibitor for the enzyme α-GLy involved in the glucose metabolism, which provides an insight into novel antidiabetic drug development [1].

The team led by Zlacká demonstrated that phosphofructokinase 15 (PFK15), a synthetic compound, can act together with sunitinib and inhibit glycolysis in human umbilical vein endothelial (HUVEC) cells, therefore reducing angiogenesis, an important pathophysiological process in the apparition and progression of diabetic complications [2].

Lee's team had an important contribution with two impressive articles. In one study, they demonstrated that nicotinamide mononucleotide intraperitoneal injections in mice prevented retinal ischemia/reperfusion (I/R) injury, a mechanism that was explained by the activation of certain antioxidant pathways [3], while in the other study, the administration of nicotinamide mononucleotide in mice, where a surgical unilateral common carotid artery occlusion was performed, was associated with a reduced retinal dysfunction, preserved nicotinamide adenine dinucleotide (NAD^+) levels, normal glycolysis, and increased antioxidant activity mechanisms which can be important in clinical practice for the amelioration of diabetic retinopathy [4].

With respect to the above-mentioned data, Iovanovici et al. analysed the beneficial role of sacubitril/valsartan therapy in patients with heart failure with a reduced ejection fraction (HFrEF) and patients with heart failure with a preserved ejection fraction (HFpEF). The review presented the main clinical trials results, where it was demonstrated that the utilisation of sacubitril/valsartan therapy in HFrEF patients reduced the rehospitalization rate due to heart failure and improved patients' symptoms, while other trials demonstrated reversible cardiac remodelling and the drop of N-terminal proBNP (NT-proBNP) levels when the angiotensin receptor–neprilysin inhibitor (ARNI) is administered [5]. In HFpEF, sacubitril/valsartan was associated with reduced NT-proBNP levels, although there were not any significant results in cardiovascular mortality, the mortality of all causes, or an improvement in the NYHA class [5].

Another interesting review was provided by Forzano I. et. which analysed the benefits of a novel medication, tirzepatide, a molecule that acts by the dual agonism of GLP-1/GIP receptors. Tirezepatide's efficacy in reducing blood glucose, weight, and blood pressure values was confirmed in five clinical trials in patients with type 2 diabetes mellitus, while in non-diabetic obese patients, a weekly administration of the drug, in various concentrations, produced remarkable results in reducing body weight [6].

Fiorentino et al. analysed the benefits of statin use in children, a population where this medication is rarely used, and demonstrated numerous benefits such as a reduction in the atherosclerosis burden, with fewer side effects than in adults because of the presence of a very low number of comorbidities [7].

Numerous molecular pathways, such as those activated by GLP-1 agonists [8], still need to be investigated for reducing the impact that diabetes mellitus and cardiovascular disease have on human health, especially given the fact that the molecular understanding of a disease creates the possibility of targeted therapies and personalized care.

Conflicts of Interest: The authors declare no conflict of interest.

References

1. Tretyakova, E.; Smirnova, I.; Kazakova, O.; Nguyen, H.T.T.; Shevchenko, A.; Sokolova, E.; Babkov, D.; Spasov, A. New Molecules of Diterpene Origin with Inhibitory Properties toward α-Glucosidase. *Int. J. Mol. Sci.* **2022**, *23*, 13535. [CrossRef] [PubMed]
2. Zlacká, J.; Murár, M.; Addová, G.; Moravčík, R.; Boháč, A.; Zeman, M. Synthesis of Glycolysis Inhibitor PFK15 and Its Synergistic Action with an Approved Multikinase Antiangiogenic Drug on Human Endothelial Cell Migration and Proliferation. *Int. J. Mol. Sci.* **2022**, *23*, 14295. [CrossRef]
3. Lee, D.; Tomita, Y.; Miwa, Y.; Shinojima, A.; Ban, N.; Yamaguchi, S.; Nishioka, K.; Negishi, K.; Yoshino, J.; Kurihara, T. Nicotinamide Mononucleotide Prevents Retinal Dysfunction in a Mouse Model of Retinal Ischemia/Reperfusion Injury. *Int. J. Mol. Sci.* **2022**, *23*, 11228. [CrossRef]
4. Lee, D.; Tomita, Y.; Miwa, Y.; Jeong, H.; Shinojima, A.; Ban, N.; Yamaguchi, S.; Nishioka, K.; Negishi, K.; Yoshino, J.; et al. Nicotinamide Mononucleotide Protects against Retinal Dysfunction in a Murine Model of Carotid Artery Occlusion. *Int. J. Mol. Sci.* **2022**, *23*, 14711. [CrossRef] [PubMed]
5. Iovanovici, D.-C.; Bungau, S.G.; Vesa, C.M.; Moisi, M.; Babes, E.E.; Tit, D.M.; Horvath, T.; Behl, T.; Rus, M. Reviewing the Modern Therapeutical Options and the Outcomes of Sacubitril/Valsartan in Heart Failure. *Int. J. Mol. Sci.* **2022**, *23*, 11336. [CrossRef] [PubMed]
6. Forzano, I.; Varzideh, F.; Avvisato, R.; Jankauskas, S.S.; Mone, P.; Santulli, G. Tirzepatide: A Systematic Update. *Int. J. Mol. Sci.* **2022**, *23*, 14631. [CrossRef] [PubMed]
7. Fiorentino, R.; Chiarelli, F. Statins in Children, an Update. *Int. J. Mol. Sci.* **2023**, *24*, 1366. [CrossRef] [PubMed]
8. Iorga, R.A.; Bacalbașa, N.; Carsote, M.; Bratu, O.G.; Stănescu, A.M.A.; Bungău, S.; Pantiș, C.; Diaconu, C.C. Metabolic and cardiovascular benefits of GLP-1 agonists, besides the hypoglycemic effect (Review). *Exp. Ther. Med.* **2020**, *20*, 2396–2400. [CrossRef] [PubMed]

International Journal of *Molecular Sciences*

MDPI

Article

Clinical Study of Metabolic Parameters, Leptin and the SGLT2 Inhibitor Empagliflozin among Patients with Obesity and Type 2 Diabetes

Zsolt Szekeres [1], Barbara Sandor [1], Zita Bognar [2], Fadi H. J. Ramadan [2], Anita Palfi [1], Beata Bodis [3], Kalman Toth [4] and Eszter Szabados [1,*]

1 Division of Preventive Cardiology and Rehabilitation, 1st Department of Medicine, Medical School, University of Pecs, 7624 Pecs, Hungary
2 Department of Biochemistry and Medical Chemistry, University of Pecs, Medical School, 7624 Pecs, Hungary
3 Division of Endocrinology and Metabolism, 1st Department of Medicine, Medical School, University of Pecs, 7624 Pecs, Hungary
4 Division of Cardiology, 1st Department of Medicine, Medical School, University of Pecs, 7624 Pecs, Hungary
* Correspondence: szabados.eszter@pte.hu

Abstract: Obesity is a major public health problem worldwide, and it is associated with many diseases and abnormalities, most importantly, type 2 diabetes. The visceral adipose tissue produces an immense variety of adipokines. Leptin is the first identified adipokine which plays a crucial role in the regulation of food intake and metabolism. Sodium glucose co-transport 2 inhibitors are potent antihyperglycemic drugs with various beneficial systemic effects. We aimed to investigate the metabolic state and leptin level among patients with obesity and type 2 diabetes mellitus, and the effect of empagliflozin upon these parameters. We recruited 102 patients into our clinical study, then we performed anthropometric, laboratory, and immunoassay tests. Body mass index, body fat, visceral fat, urea nitrogen, creatinine, and leptin levels were significantly lower in the empagliflozin treated group when compared to obese and diabetic patients receiving conventional antidiabetic treatments. Interestingly, leptin was increased not only among obese patients but in type 2 diabetic patients as well. Body mass index, body fat, and visceral fat percentages were lower, and renal function was preserved in patients receiving empagliflozin treatment. In addition to the known beneficial effects of empagliflozin regarding the cardio-metabolic and renal systems, it may also influence leptin resistance.

Keywords: empagliflozin; type 2 diabetes mellitus; obesity; leptin; lipid metabolism

Citation: Szekeres, Z.; Sandor, B.; Bognar, Z.; Ramadan, F.H.J.; Palfi, A.; Bodis, B.; Toth, K.; Szabados, E. Clinical Study of Metabolic Parameters, Leptin and the SGLT2 Inhibitor Empagliflozin among Patients with Obesity and Type 2 Diabetes. *Int. J. Mol. Sci.* **2023**, *24*, 4405. https://doi.org/10.3390/ijms24054405

Academic Editors: Simona Gabriela Bungau and Vesa Cosmin

Received: 30 January 2023
Revised: 19 February 2023
Accepted: 21 February 2023
Published: 23 February 2023

1. Introduction

According to the latest data, nearly 2 billion adults (39% of the world's adult population) were estimated to be obese or overweight. If current trends continue, it is expected that 1 billion adults, nearly 20% of the world's population, will clinically be declared obese by 2025 [1]. Obesity is associated with many diseases and abnormalities, such as type 2 diabetes [2], dyslipidemia [3], cardiovascular diseases [4], hypertension [5], certain types of cancer [6,7], pneumological [6], nephrological [8], skeletal muscle [9], rheumatologic [10], dermatologic [11], and neuropsychologic [11] complications, and is it associated with premature mortality. Obesity, especially the dysfunctional visceral adipose tissue (VAT), is the main driver of many metabolic abnormalities including insulin resistance, hyperinsulinemia, glucose intolerance, atherogenic dyslipidemia (high triglyceride and apolipoprotein B levels, increased proportion of small, dense LDL [low-density lipoprotein] particles, low HDL [high-density lipoprotein] cholesterol levels, and small HDL particles), and is associated with a low-grade inflammation [6].

Leptin was the first identified adipokine in the 1990s known to suppress food intake through the suppression of appetite and mediate energy homeostasis including glucose and lipid metabolism [12]. The serum level of leptin is elevated paradoxically in obesity [13], and this high level of leptin may induce leptin resistance and result in altered glucose metabolism and insulin resistance [14]. Hyperleptinemia has also been associated with increased inflammation, oxidative stress, endothelial dysfunction, atherogenesis, and thrombosis [15]. Based on these effects, leptin is attributed to a significant role in the development of cardiovascular diseases. Additionally, patients with type 2 diabetes mellitus scored a higher percentage of hypertension, obesity, metabolic syndrome, and endothelial dysfunction if they had elevated leptin levels [16].

The link between obesity and type 2 diabetes mellitus [T2DM] has long been recognized and explains the high prevalence of type 2 diabetes mellitus. Type 2 diabetes mellitus is associated with many vascular complications. Microvascular complications include diabetic kidney disease, retinopathy, and neuropathy, whereas the macrovascular complications include coronary artery, cerebrovascular, and peripheral vascular diseases. The main goals of treatment in patients with T2DM are to achieve adequate glycemic control, reduce body weight and prevent vascular damage, and target organ damage [17]. Novel antidiabetic therapies such as sodium glucose co-transporter 2 (SGLT2) inhibitors provide a new approach to preventing or ameliorating the complications that insulin resistance and hyperglycemia create [18]. SGLT2 inhibitors are potent antihyperglycemic drugs, which inhibit glucose reabsorption in the proximal tubules of the kidney inducing glycosuria and improving blood glucose levels, and may reduce body weight through calorie loss. Numerous studies have shown they are associated with reduced cardiovascular morbidity and mortality, including vascular diseases and heart failure [19]. Furthermore, SGLT2 inhibitors have also demonstrated positive reno-metabolic effects [20]. In a cardiovascular outcome trial, the SGLT2 inhibitor empagliflozin proved superior to conventional antidiabetic therapy in reducing the rate of MACE, mortality, and hospitalization due to heart failure [21]. SGLT2 inhibitor therapy has been associated with a decrease in serum triglycerides, an increase in HDL cholesterol, and also a small increase in LDL cholesterol level was observed [20]. The presence of metabolic disturbances in obese patients results in oxidative stress [22]. Since obesity and insulin resistance is a major component of metabolic syndrome, it is strongly associated with oxidative stress [23]. The oxidative modification of lipoproteins can result in more atherogenic compounds, which may have a key role in the development of cardiovascular dysfunction in patients with diabetes mellitus [24,25].

The aim of our study was to investigate certain laboratory parameters such as lipids, inflammatory markers, blood glucose level, glycated hemoglobin [HbA1c] level, kidney function, leptin level, as well as body mass index [BMI], body fat and visceral fat percentage among patients afflicted with obesity and diabetes. We also investigated a subgroup of patients receiving empagliflozin treatment.

2. Results

2.1. *Body Mass Index, Body Fat, and Visceral Fat Were Significantly Lower in the Empagliflozin Treated Group*

BMI was significantly lower in the control group (C) when compared to the obese (O) ($p < 0.001$), to the obese and diabetic (OD) ($p < 0.001$), and to the empagliflozin treated (ODE) group ($p < 0.001$). It was also significantly lower in the diabetic (D) group when compared to the obese (O) ($p < 0.001$), to the obese and diabetic (OD) ($p < 0.001$), and to the empagliflozin-treated group ($p < 0.001$). BMI was significantly lower in the empagliflozin-treated group (ODE) when compared to the obese and diabetic (OD) group ($p < 0.001$). There was no significant difference between the other groups.

Body fat was significantly lower in the control group (C), when compared to the obese (O) ($p < 0.001$), and to the obese and diabetic (OD) ($p < 0.001$) groups. It was also significantly lower in the diabetic (D) group when compared to the obese (O) ($p = 0.001$) and to the obese and diabetic (OD) ($p = 0.001$) groups. Body fat was significantly lower in the empagliflozin-treated group (ODE) when compared to the obese and diabetic (OD) group ($p = 0.002$). There were no significant differences between the other groups.

Visceral fat was significantly lower in the control group (C) when compared to the obese (O) ($p < 0.001$), to the obese and diabetic (OD) ($p < 0.001$), and to the empagliflozin-treated group (ODE) ($p < 0.001$). It was also significantly lower in the diabetic (D) group when compared to the obese (O) ($p < 0.001$), to the obese and diabetic (OD) ($p < 0.001$), and to the empagliflozin-treated group ($p < 0.001$). Visceral fat was significantly lower in the empagliflozin-treated group (ODE) when compared to the obese and diabetic (OD) group ($p < 0.014$). There were no significant differences between the other groups (Table 1).

Table 1. Patients' general characteristics, n = number of patients, BMI = body mass index, kg = kilogram, m^2 = square meter, Body fat = body fat percentage, Visc. fat = visceral fat, HT = high blood pressure, DM = type 2 diabetes mellitus, CVD = cardiovascular disease, SGLT2i = Sodium glucose co-transporter 2 inhibitors.

Group	C (n = 20)	O (n = 20)	D (n = 19)	OD (n = 19)	ODE (n = 20)	Total (n = 98)
			Demographics and anthropometrics			
Age, years	65.95 ± 1.98	66.40 ± 2.23	74.58 ± 6.38	70.90 ± 1.74	65.2 ± 1.86	68.52 ± 0.90
Male sex, %	75.00	50.00	52.60	68.40	75.00	64.30
BMI, kg/m^2	26.01 ± 0.50	34.75 ± 0.85	26.50 ± 0.44	35.78 ± 0.91	31.61 ± 0.77	31.04 ± 0.52
Body fat, %	26.75 ± 6.73	38.44 ± 8.38	28.12 ± 6.62	37.24 ± 6.67	30.98 ± 6.06	32.93 ± 6.90
Visc. fat, %)	10.5 ± 0.56	16.65 ± 0.93	10.89 ± 0.58	19.01 ± 1.25	15.50 ± 0.67	14.51 ± 0.79
			Comorbidities			
HT, %	100.00	100.00	100.00	100.0	100.00	100.00
DM, %	0.00	0.00	100.00	100.00	100.00	59.20
CVD, %	70.40	69.40	78.60	84.70	64.30	73.48
			Medications			
Antihypertensive, %	100.00	100.00	100.00	100.00	100.00	100.00
Antidiabetics (other, than SGLT2i), %	0.00	0.00	100.00	100.00	100.00	59.20
Antihyperlipidemic, %	70.00	75.00	89.47	84.20	80.00	79.59

2.2. Hemoglobin Levels Were Significantly Higher among the Empagliflozin Treated Patients

Hemoglobin levels were significantly higher in the empagliflozin-treated group (ODE) when compared to the diabetic (D), and obese and diabetic (OD) groups ($p = 0.004$ and $p < 0.001$, respectively). There was no significant difference between the diabetic (D) and the obese and diabetic (OD) group ($p = 0.850$). The obese group (O) had a significantly higher hemoglobin when compared to the obese and diabetic group (OD) and a significantly lower level when compared to the empagliflozin-treated obese group (ODE) ($p = 0.033$ and $p = 0.007$ respectively) (Table 2).

2.3. Renal Parameters Were Significantly Higher in Diabetic Patients, Yet Were Reduced in the Empagliflozin Treated Group

Urea nitrogen level increases significantly with the appearance of diabetes in obesity (O vs. OD) ($p = 0.002$). In the empagliflozin-treated group (ODE), the urea nitrogen level was significantly lower when compared to the obese and diabetic (OD) group ($p = 0.008$) (Table 2).

Creatinine significantly increases with the appearance of diabetes in the obese groups (O vs. OD) ($p = 0.011$). In the empagliflozin-treated group (ODE), the creatinine level was significantly lower when compared to the obese and diabetic (OD) group ($p = 0.012$) (Table 2).

Table 2. Laboratory parameters in the different groups. C = control group, n = number of patients, O = obese group, D = diabetic group, OD = obese diabetic group, ODE = obese diabetic group treated with empagliflozin, g = gram, L = liter, mmol = millimole, mL = milliliter, CRP = C-reactive protein, mg = milligram, μmol = micromole, Total chol. = total cholesterol, HDL = High-density lipoprotein cholesterol, LDL = Low-density lipoprotein cholesterol, ng = nanogram.

Groups	C (n = 20)	O (n = 20)	D (n = 19)	OD (n = 19)	ODE (n = 20)
Hemoglobin, g/L	141.85 ± 20.16	139.85 ± 11.85	133.79 ± 18.20	126.32 ± 14.72	152.90 ± 10.56
HbA1c, %	5.48 ± 0.08	5.67 ± 0.93	6.72 ± 0.34	6.39 ± 0.15	7.68 ± 0.33
Blood glucose, mmol/L	5.48 ± 0.85	5.62 ± 1.17	6.79 ± 1.95	6.20 ± 1.53	7.01 ± 1.61
CRP, mg/L	1.93 ± 0.43	6.81 ± 2.08	3.83 ± 1.08	4.55 ± 1.61	3.94 ± 0.60
Urea nitrogen, mmol/L	6.30 ± 0.75	5.22 ± 0.32	6.34 ± 0.42	9.69 ± 0.28	5.71 ± 0.31
Creatinine, μmol/L	83 ± 3.99	81.55 ± 3.42	94.68 ± 3.75	120.26 ± 9.75	81.71 ± 3.87
Total chol., mmol/L	4.73 ± 0.25	4.44 ± 0.26	3.79 ± 0.24	4.12 ± 0.36	4.12 ± 0.28
HDL, mmol/L	1.33 ± 0.08	1.25 ± 0.06	1.11 ± 0.05	1.10 ± 0.08	1.06 ± 0.05
LDL, mmol/L	3.15 ± 0.25	2.58 ± 0.21	2.21 ± 0.22	2.38 ± 0.31	2.21 ± 0.23
Triglycerides, mmol/L	1.76 ± 0.38	1.77 ± 0.27	1.81 ± 0.19	1.78 ± 0.22	1.88 ± 0.16
Leptin, ng/mL	5.97 ± 0.70	19.42 ± 3.06	10.33 ± 2.21	29.86 ± 3.61	17.43 ± 2.99

2.4. Blood Glucose and HbA1c Levels Were Significantly Higher in Diabetic Patients, Yet There Was No Significant Difference between the Different Diabetic Groups

Blood glucose and HbA1c levels were significantly lower in the control group (C) when compared with the diabetic (D), the obese and diabetic (OD), and the empagliflozin-treated obese and diabetic groups ($p = 0.029$, $p = 0.005$, and $p < 0.001$, respectively). Blood glucose and HbA1c levels were significantly lower in the obese group when compared with the diabetic (D), the obese and diabetic (OD), and the empagliflozin-treated obese and diabetic groups ($p = 0.015$, $p = 0.008$, and $p < 0.001$, respectively). There were no significant differences between the other groups regarding blood glucose and HbA1c levels.

2.5. Leptin Levels Were Significantly Higher in Obese Patients, Yet Were Reduced in the Empagliflozin-Treated Group

Leptin levels were significantly higher with the appearance of obesity (O) ($p = 0.003$) even if obesity was present with diabetes (OD) ($p < 0.001$) when compared to the control (C) group. It was also significantly higher in diabetic patients (D) when compared with the control group (C) ($p = 0.029$). Obese and diabetic patients (OD) had a significantly higher level of leptin when compared to diabetic yet not obese (D) patients ($p = 0.001$). In the empagliflozin-treated group (ODE), the leptin level was significantly lower when compared to the obese and diabetic (OD) group ($p = 0.048$) (Table 2).

2.6. There Were No Significant Differences between the Other Measured Parameters

There was no significant difference in body muscle percentage, white blood cell count, red blood cell count, platelet count, fibrinogen levels, uric acid, triglyceride, sodium and potassium levels, and thyroid-stimulating hormone levels among the groups.

There was no significant difference in the cholesterol levels among the groups. It bears mentioning, cholesterol levels were strongly affected by the antihyperlipidemic agents.

The continuous variables did not differ from the normal distribution. Data are shown as means ± standard deviation.

3. Discussion

In our clinical study, we examined metabolic and inflammatory parameters, kidney function, and leptin levels among patients afflicted with hypertension, obesity, type 2 diabetes, and cardiovascular diseases. The aim of our study was to detect the severity of the metabolic state among these patients and to examine a subgroup of patients treated with empagliflozin. In our study, we found empagliflozin-treated obese, diabetic patients had significantly lower BMI, body fat, and visceral fat values as well as lower serum creatinine and leptin levels when compared to patients with obesity and type 2 diabetes treated with usual antidiabetics (such as biguanides and sulfonylureas). Leptin levels were already higher among patients with type 2 diabetes even with normal BMI, and were significantly higher in obese non-diabetic patients and were the highest in obese patients with type 2 diabetes. Furthermore, we discovered that increased visceral fat and leptin levels predicted diabetes similarly to HbA1c.

Excess visceral adiposity is a major risk factor for metabolic and cardiovascular disorders. It plays a crucial role in the development of a diabetogenic and atherogenic metabolic profile inducing insulin resistance and increased cardiometabolic risk [26]. In our study, BMI, body fat, and visceral fat percentage were the highest among patients with obesity and type 2 diabetes (Group OD). In the empagliflozin-treated obese, diabetic patients (Group ODE), BMI, body fat, and visceral fat were significantly lower when compared with obese and diabetic patients (OD) treated with usual antidiabetics (Table 1). In an animal study, empagliflozin suppressed weight gain by shifting energy metabolism towards fat utilization, elevated adenosine monophosphate-activated [AMP] protein kinase, and acetyl coenzyme A [acetyl-CoA] carboxylase phosphorylation in skeletal muscle. Furthermore, empagliflozin increased energy expenditure, heat production and browning, and attenuated obesity-induced inflammation and insulin resistance by polarizing M2 macrophages in white adipose tissue [WAT] and liver [27]. Thus, empagliflozin suppressed weight gain by enhancing fat utilization and browning and attenuated obesity-induced inflammation and insulin resistance.

White adipose tissue is an endocrine organ capable of producing and releasing numerous bioactive substances known as adipokines or adipocytokines. Dysregulated production of adipocytokines is involved in the development of obesity-related diseases. Leptin is one of the most examined adipokines. An increased leptin level is associated with insulin resistance and T2DM development [28]. In T2DM, a link has also been reported between high leptin concentrations and increased cardiovascular [CV risk], including the presence of microvascular complications and cardiac autonomic dysfunction [29]. Furthermore, obesity, hypertension, metabolic syndrome, and endothelial dysfunction are more frequent in T2DM patients with increased leptin levels [30]. In chronic heart disease (CHD) patients, elevated leptin levels were significantly associated with an increased risk of cardiac death, acute coronary syndrome, non-fatal MI, stroke, and hospitalization for congestive heart failure [31,32]. Similarly, higher leptin levels were significantly related to the number of stenotic coronary arteries and arterial stiffness in CHD patients [33]. The presence, severity, extent, and lesion complexity of coronary atherosclerosis have been associated with higher leptin levels in CHD patients [34]. Leptin may also affect cardiac remodeling, metabolism, and contractile function [35]. Other effects of leptin include activation of inflammatory responses, oxidative stress, thrombosis, and atherosclerosis, thereby resulting in endothelial dysfunction and atherosclerotic plaque [16].

In our study, the leptin level was already higher among patients with type 2 diabetes even with normal BMI (Group D), was significantly higher in obese non-diabetic patients (Group O), and was the highest in obese patients with type 2 diabetes (Group OD) when compared to the control group.

A link between increased plasma leptin concentrations and chronic kidney disease (CKD) has been reported, which is possibly due to reduced renal clearance [36]. Leptin concentrations gradually increased with the severity of CKD [37]. In CKD patients, plasma leptin levels have been inversely associated with glomerular filtration rate and directly associated with urinary albumin levels as well as age and obesity markers (BMI and waist circumference) [38]. Overall, hyperleptinemia has been linked to the presence, severity, and progression of CKD. In our study, creatinine levels were significantly higher with the appearance of diabetes and were the highest among obese patients with type 2 diabetes. Among the empagliflozin-treated obese and type 2 diabetic patients, the creatinine level was significantly lower eliciting improved renal function (Table 2).

We possess a vast amount of knowledge regarding the cardiovascular and renal effects of SGLT2 inhibitors [20,39–42]. In addition to their direct effect on glucose homeostasis, they have many other underlying mechanisms from which not all are fully understood. For instance, SGLT2 inhibitors may also act upon visceral adipose tissue. Dapagliflozin therapy was associated with a decreased circulating leptin level and an increased circulating adiponectin level among patients with type 2 diabetes, which, may contribute to the beneficial effects of SGLT2 inhibitors on metabolic homeostasis, such as improved insulin resistance and reduced cardiovascular risk [43–45]. Furthermore, dapagliflozin displayed significantly lower arterial stiffness in diabetic mice treated with dapagliflozin when compared to untreated diabetic mice [46]. The effects of empagliflozin on adipocytokines were examined in an animal study conducted on obese rats. Empagliflozin dose-dependently reduced body weight, body fat, adiponectin, and leptin following the 28-day treatment [39]. In our study, the leptin level was significantly lower in the empagliflozin-treated obese and type 2 diabetic patients (ODE) when compared to the obese, diabetic patients (OD) treated with other antidiabetics (Table 2). To the best of our knowledge, this is the first time the beneficial effect of empagliflozin on the leptin level has been demonstrated in a clinical setting.

HbA1c is a well-known screening and diagnostic tool in detecting diabetes. A score higher than 5.7 % value implies prediabetes, and consequently, higher than 6.5 % confirms diabetes. Our receiver operating characteristic [ROC] analysis has proven the recommended 5.7 % cut-off value effectively predicted altered glucose homeostasis with very high sensitivity and acceptable specificity. In the same analysis, leptin was found to be similar in the prediction of diabetes. This is congruent with previous observations stating elevated leptin levels are associated with insulin resistance and T2DM development [28].

The second ROC analysis with the composite endpoint diabetes and obesity showed, in addition to HgA1c, leptin, and visceral fat may have a role in the diagnosis of diabetes among obese adults. These findings emphasize patients with increased visceral fat, which is easily measured using a smart weight scale, are prime candidates to be screened for insulin resistance or diabetes with HbA1c and fasting glucose value.

Hemoglobin values were the highest in the empagliflozin-treated group, which, may imply a slight hemoconcentration, and may be related to the osmotic diuretic effect of empagliflozin treatment. It is worthwhile to draw the attention of patients to the need for adequate fluid intake during SGLT2 inhibitor treatment. Unexpectedly, HbA1c levels were the highest in the empagliflozin-treated group. Presumably, this is due to the fact that, in Hungary, SGLT2 inhibitor treatment can only be prescribed to patients with an HbA1c level above 7%. This also means this group is a more severe patient group in terms of diabetes, thus, the results obtained prove even more crucial.

There was no significant difference in C-reactive protein (CRP) levels among the examined groups; however, some differences were detected. The CRP level was the lowest in the non-obese, non-diabetic group (C). Although many factors can influence the CRP level, it may be important that it was higher among obese and diabetic patients, which may indicate a low level of inflammation and corresponds to previous observations [19]. Among patients receiving empagliflozin treatment (ODE), the CRP level was lower when compared to the obese and diabetic group (OD), which may reflect lower inflammation status, likely due to the empagliflozin treatment. It has been previously reported, that empagliflozin reduced renal inflammation and oxidative stress in spontaneously hypertensive rats [47] In the EMPA-CARD trial patients with type2 diabetes and coronary artery disease treated with empagliflozin had lower levels of interleukin 6, interleukin 1β and CRP levels compared to a placebo. There were elevations in superoxidase dismutase (SOD) activity, glutathione (GSHr), and total antioxidant capacity (TAC) with empagliflozin [48].

Notably, there was no significant difference in LDL cholesterol levels. This may be due to the fact in which LDL cholesterol levels were greatly influenced by antihyperlipidemic drugs. Previous literature data indicated a moderate increase in LDL level can be detected with SGLT2 inhibitor treatment. In our study, we did not observe higher LDL values in the empagliflozin-treated group when compared to the other groups. Additionally, in our study, CV disease incidence was provided primarily to describe the patient population. Although it was lower in the empagliflozin-treated group, it was not intended to examine this correlation.

The main strength of our study is that, to the best of our knowledge, this is the first examination that has demonstrated that empagliflozin treatment has a beneficial effect on serum leptin levels under clinical conditions. However, our study was conducted on a relatively small number of patients, so further studies on a larger patient population are needed to confirm our results.

4. Materials and Methods

4.1. Ethics

The study protocol was approved by the Regional Ethics Committee of Pecs (No. 7622—PTE 2019) and was conducted in accordance with the ethical principles stated in the Declaration of Helsinki. Written informed consent was obtained from all patients.

4.2. Patients

102 patients (35 female, 67 male) were enrolled in our study. Patients were recruited from different internal medicine and outpatient departments by various physicians. They voluntarily agreed to participate in our study in which they signed an informed consent letter. Subgroup analysis was performed based on different metabolic states. Patients who did not have type 2 diabetes and were not obese were assigned to group C (20 patients), declared as the control group. Obese patients without diabetes were assigned to group O (obese), (20 patients). Non-obese patients with type 2 diabetes were selected into group D (diabetic), (19 patients). Obese and diabetic patients were assigned into group OD (obese and diabetic), (19 patients). Obese, diabetic patients receiving empagliflozin therapy for at least 3 months were assigned to group ODE (20 patients). Patients were considered obese if their BMI was 30.0 kg/m^2 or higher. Antihypertensive, antidiabetic, and antihyperlipidemic therapies were recorded from the patient's history as well as their comorbidities, such as diabetes mellitus, hypertension, and cardiovascular diseases. Exclusion criteria include the following: previous SGLT2 inhibitor therapy for groups C, O, D, OD; active cancer disease; and refusing to sign the consent form. Four patients were excluded from the study for different reasons (low compliance, severe epileptic seizure, withdrawal of their consent, and urgent psychiatric ward admission).

Patients' general characteristics were as follows. The mean age for different groups was: 65.95 for group C, 66.40 for group O, 74.58 for group D, 70.90 for group OD, and 65.20 for group ODE. The distribution of sex (male to female percentage) in the groups was as follows: 75–25% for group C, 50–50% for group O, 52.60–47.40% for group D, 68.40–31.60% for group OD, and 75–25% for group ODE. Mean BMI values for different groups were as follows: 26.01 kg/m^2 for group C, 34.75 kg/m^2 for group O, 26.50 kg/m^2 for group D, 35.78 kg/m^2 for group OD, and 31.61 kg/m^2 for group ODE. All patients had high blood pressure in their medical history. All patients in the diabetic groups (D, OD, ODE) had identified type 2 diabetes mellitus in their medical history, whereas none were reported in the remaining groups (C, O). The percentage of patients with identified cardiovascular disease was 70.40% in group C, 69.40% in group O, 78.60% in group D, 64.30% in group OD, and 73.48% in group ODE. All patients received antihypertensive therapy. All diabetic patients (D, OD, ODE) received antidiabetic therapy, whereas none were administered in the non-diabetic groups (C, O). Empagliflozin was administered only in the ODE group. No other SGLT2 inhibitors were used in our study. The percentage of patients with antihyperlipidemic therapy was as follows: 70% for group C, 75% for group O, 89.47% for group D, 84.20% for group OD, and 80% for group ODE.

4.3. Study Design

102 patients were recruited into this clinical study. We assessed their body composition, followed by pre-prandial venous blood collected using a peripheral venous catheter in the cubital vein. The preparation and laboratory procedures were in full accordance with the recommendations of the laboratory kits. Laboratory tests were performed at the Department of Laboratory Medicine, University of Pecs, Pecs, Hungary. The leptin levels were determined using the immunoassay method (Human Leptin ELISA, Biovendor, Czech Republic) at the Department of Biochemistry and Medical Chemistry, University of Pecs, Pecs, Hungary.

4.4. Anthropometric Measurements

The patients' body composition was assessed using an Omron HBF-511 body composition scale (Omron HealthCare Co., Ltd., Kyoto, Japan). We measured weight, BMI, body fat percentage, and visceral fat percentage. Height was measured using a measuring tape.

4.5. Laboratory Tests

Pre-prandial laboratory tests were performed on every patient. These include complete blood count (red and white blood cell count, platelet count, hemoglobin level, hematocrit), fibrinogen, basic metabolic panel (pre-prandial glucose, sodium, potassium, calcium, blood urea nitrogen, and creatinine levels), lipid panel (total cholesterol, HDL cholesterol, LDL cholesterol, and triglyceride levels), liver panel (aspartate transaminase (AST), alanine transaminase (ALT), gamma-glutamyl transferase (GGT) levels), hemoglobin A1C level, and the thyroid stimulating hormone level.

4.6. Immunoassay Tests

Plasma leptin 1 levels were measured in duplicate using enzyme-linked immunosorbent assay (ELISA) kits (Cat. No. RD191001100). The blood samples were centrifuged at $2500\times g$ for 10 min. The recovered plasma was stored at -70 °C in aliquots until assayed. The tests were performed in full accordance with the recommendations of the manufacturer, with a detection limit of 0.08 and 0.2 ng/mL, respectively. (BioVendor GmbH., Brno, Czech Republic).

4.7. Statistical Analysis

IBM SPSS statistics, version 28.0.0. (SPSS, Chicago, IL, USA, 2022); software for statistical; was used to conduct descriptive analyses and to describe the sample. Data are shown as means $\pm$ standard deviation.

Differences in the continuous variables were evaluated using a one-way repeated ANOVA statistical test (Tamhane post-hoc test) following the administering of the Kolmogorov–Smirnov test to check the normality of the data distribution. The continuous variables did not differ from the normal distribution.

In the case of categorical variables, data are shown as percentages and incidence (absolute number compared to total number). Differences were evaluated by using chi-square test analyses.

Multivariate linear regression and stepwise analyses of the data were performed regarding the leptin values for HbA1c, LDL, triglyceride, creatinine, hemoglobin, and visceral fat.

Multiple regression analysis with various models including leptin, HbA1c, and visceral fat considering the principle of multicollinearity was performed to reveal which factors predict the occurrence of diabetes and obesity.

The diagnostic power of variables was assessed using the area under the curve (AUC) of the receiver operating characteristic (ROC) curve. The predicted probabilities were calculated from the variables produced by binary logistic regression analysis, in which $p \leq 0.05$ was considered statistically significant.

5. Conclusions

BMI, body fat, and visceral fat values as well as serum creatinine and leptin levels were improved with empagliflozin treatment. High leptin levels and leptin resistance in obesity are associated with insulin resistance, type 2 diabetes, increased risk of CV diseases, low-grade inflammation, and thrombosis. The markedly decreased circulating leptin levels observed in the empagliflozin-treated group may contribute to the known beneficial cardiovascular effects of empagliflozin treatment.

Author Contributions: Conceptualization, E.S. and A.P.; methodology, E.S., K.T.; software, Z.S., Z.B. and B.S.; validation, B.S., Z.B. and F.H.J.R.; formal analysis, Z.S., B.S.; investigation, Z.S., B.B., A.P., E.S.; resources, B.B. and E.S.; data curation, B.S.; writing—original draft preparation, Z.S.; writing—review and editing, E.S.; visualization, Z.S., B.S., E.S ; supervision, E.S. and K.T.; project administration, K.T., E.S.; funding acquisition, E.S. and Z.B. All authors have read and agreed to the published version of the manuscript.

Funding: The study was supported by KA-2019-29 project of the University of Pecs, Medical School, Hungary, the Excellence Program of the Ministry for Innovation and Technology; Hungarian Ministry for Innovation and Technology: TKP2021-EGA-17 by the János Bolyai Research Scholarship of the Hungarian Academy of Sciences and by the ÚNKP-22-5 New National Excellence Program of the Ministry for Innovation and Technology from the source of the National Research, Development and Innovation. The funder had no role in study design, data collection, and analysis, decision to publish, or preparation of the manuscript.

Institutional Review Board Statement: The study protocol was approved by the Regional Ethics Committee of Pecs (No. 7622—PTE 2019) and was conducted in accordance with the ethical principles stated in the Declaration of Helsinki.

Informed Consent Statement: A written informed consent was obtained from all patients.

Data Availability Statement: The data presented in this study are available on request from the corresponding author. The data are not publicly available due to ethical and privacy restrictions.

Conflicts of Interest: The authors declare no conflict of interest.

References

1. Loos, R.J.F.; Yeo, G.S.H. The genetics of obesity: From discovery to biology. *Nat. Rev. Genet.* **2022**, *23*, 120–133. [CrossRef]
2. Klein, S.; Gastaldelli, A.; Yki-Järvinen, H.; Scherer, P.E. Why does obesity cause diabetes? *Cell Metab.* **2022**, *34*, 11–20. [CrossRef]
3. Vekic, J.; Zeljkovic, A.; Stefanovic, A.; Jelic-Ivanovic, Z.; Spasojevic-Kalimanovska, V. Obesity and dyslipidemia. *Metabolism* **2019**, *92*, 71–81. [CrossRef]
4. Powell-Wiley, T.M.; Poirier, P.; Burke, L.E.; Després, J.-P.; Gordon-Larsen, P.; Lavie, C.J.; Lear, S.A.; Ndumele, C.E.; Neeland, I.J.; Sanders, P.; et al. Obesity and Cardiovascular Disease: A Scientific Statement From the American Heart Association. *Circulation* **2021**, *143*, e984–e1010. [CrossRef] [PubMed]
5. Mendoza, M.F.; Kachur, S.M.; Lavie, C.J. Hypertension in obesity. *Curr. Opin. Cardiol.* **2020**, *35*, 389–396. [CrossRef] [PubMed]
6. Piché, M.-E.; Tchernof, A.; Després, J.-P. Obesity Phenotypes, Diabetes, and Cardiovascular Diseases. *Circ. Res.* **2020**, *126*, 1477–1500. [CrossRef] [PubMed]
7. Ackerman, S.E.; Blackburn, O.A.; Marchildon, F.; Cohen, P. Insights into the Link between Obesity and Cancer. *Curr. Obes. Rep.* **2017**, *6*, 195–203. [CrossRef] [PubMed]
8. Kovesdy, C.P.; Furth, S.L.; Zoccali, C. Obesity and kidney disease: Hidden consequences of the epidemic. *J. Nephrol.* **2017**, *30*, 1–10. [CrossRef]
9. Urano, T.; Inoue, S. Recent genetic discoveries in osteoporosis, sarcopenia and obesity. *Endocr. J.* **2015**, *62*, 475–484. [CrossRef] [PubMed]
10. George, M.D.; Baker, J.F. The Obesity Epidemic and Consequences for Rheumatoid Arthritis Care. *Curr. Rheumatol. Rep.* **2016**, *18*, 6. [CrossRef]
11. Lause, M.; Kamboj, A.; Faith, E.F. Dermatologic manifestations of endocrine disorders. *Transl. Pediatr.* **2017**, *6*, 300–312. [CrossRef] [PubMed]
12. Farooqi, I.S.; O'Rahilly, S. Human disorders of leptin action. *J. Endocrinol.* **2014**, *223*, T63–T70. [CrossRef] [PubMed]
13. D'Souza, A.M.; Neumann, U.H.; Glavas, M.M.; Kieffer, T.J. The glucoregulatory actions of leptin. *Mol. Metab.* **2017**, *6*, 1052–1065. [CrossRef] [PubMed]
14. UK Prospective Diabetes Study (UKPDS) Group. Intensive Blood-Glucose Control with Sulfonylureas or Insulin Compared with Conventional Treatment and Risk of Complications in Patients with Type 2 Diabetes. *Endocrinologist* **1999**, *9*, 149. [CrossRef]

15. Ouchi, K.W.N.; Parker, J.L.; Lugus, J.J.; Walsh, K. Adipokines in inflammation and metabolic disease. *Nat. Rev. Immunol.* **2011**, *11*, 85–97. [CrossRef]
16. Katsiki, N.; Mikhailidis, D.P.; Banach, M. Leptin, cardiovascular diseases and type 2 diabetes mellitus review-article. *Acta Pharmacol. Sin.* **2018**, *39*, 1176–1188. [CrossRef]
17. Artasensi, A.; Pedretti, A.; Vistoli, G.; Fumagalli, L. Type 2 Diabetes Mellitus: A Review of Multi-Target Drugs. *Molecules* **2020**, *25*, 1987. [CrossRef]
18. Vesa, C.M.; Popa, L.; Popa, A.R.; Rus, M.; Zaha, A.A.; Bungau, S.; Tit, D.M.; Corb Aron, R.A.; Zaha, D.C. Current Data Regarding the Relationship between Type 2 Diabetes Mellitus and Cardiovascular Risk Factors. *Diagnostics* **2020**, *10*, 314. [CrossRef]
19. Musso, G.; Gambino, R.; Cassader, M.; Pagano, G. A novel approach to control hyperglycemia in type 2 diabetes: Sodium glucose co-transport (SGLT) inhibitors. Systematic review and meta-analysis of randomized trials. *Ann. Med.* **2012**, *44*, 375–393. [CrossRef]
20. Szekeres, Z.; Toth, K.; Szabados, E. The Effects of SGLT2 Inhibitors on Lipid Metabolism. *Metabolites* **2021**, *11*, 87. [CrossRef]
21. Fitchett, D.; Inzucchi, S.E.; Cannon, C.P.; McGuire, D.K.; Scirica, B.M.; Johansen, O.E.; Sambevski, S.; Kaspers, S.; Pfarr, E.; George, J.T.; et al. Empagliflozin Reduced Mortality and Hospitalization for Heart Failure Across the Spectrum of Cardiovascular Risk in the EMPA-REG OUTCOME Trial. *Circulation* **2019**, *139*, 1384–1395. [CrossRef]
22. Jakubiak, G.K.; Osadnik, K.; Lejawa, M.; Kasperczyk, S.; Osadnik, T.; Pawlas, N. Oxidative Stress in Association with Metabolic Health and Obesity in Young Adults. *Oxid. Med. Cell. Longev.* **2021**, *2021*, 9987352. [CrossRef] [PubMed]
23. Jakubiak, G.K.; Osadnik, K.; Lejawa, M.; Osadnik, T.; Golawski, M.; Lewandowski, P.; Pawlas, N. "Obesity and Insulin Resistance" Is the Component of the Metabolic Syndrome Most Strongly Associated with Oxidative Stress. *Antioxidants* **2022**, *11*, 79. [CrossRef]
24. Tsatsoulis, A.; Paschou, S.A. Metabolically Healthy Obesity: Criteria, Epidemiology, Controversies, and Consequences. *Curr. Obes. Rep.* **2020**, *9*, 109–120. [CrossRef] [PubMed]
25. Genovesi, S.; Antolini, L.; Orlando, A.; Gilardini, L.; Bertoli, S.; Giussani, M.; Invitti, C.; Nava, E.; Battaglino, M.G.; Leone, A.; et al. Cardiovascular Risk Factors Associated with the Metabolically Healthy Obese (MHO) Phenotype Compared to the Metabolically Unhealthy Obese (MUO) Phenotype in Children. *Front. Endocrinol.* **2020**, *11*, 27. [CrossRef] [PubMed]
26. Mahabadi, A.A.; Massaro, J.M.; Rosito, G.A.; Levy, D.; Murabito, J.M.; Wolf, P.A.; O'Donnell, C.J.; Fox, C.S.; Hoffmann, U. Association of pericardial fat, intrathoracic fat, and visceral abdominal fat with cardiovascular disease burden: The Framingham Heart Study. *Eur. Heart J.* **2009**, *30*, 850–856. [CrossRef]
27. Xu, L.; Nagata, N.; Nagashimada, M.; Zhuge, F.; Ni, Y.; Chen, G.; Mayoux, E.; Kaneko, S.; Ota, T. SGLT2 Inhibition by Empagliflozin Promotes Fat Utilization and Browning and Attenuates Inflammation and Insulin Resistance by Polarizing M2 Macrophages in Diet-induced Obese Mice. *Ebiomedicine* **2017**, *20*, 137–149. [CrossRef]
28. Andrade-Oliveira, V.; Câmara, N.O.S.; Moraes-Vieira, P.M. Adipokines as Drug Targets in Diabetes and Underlying Disturbances. *J. Diabetes Res.* **2015**, *2015*, 681612. [CrossRef]
29. Vavruch, C.; Länne, T.; Fredrikson, M.; Lindström, T.; Östgren, C.J.; Nystrom, F.H. Serum leptin levels are independently related to the incidence of ischemic heart disease in a prospective study of patients with type 2 diabetes. *Cardiovasc. Diabetol.* **2015**, *14*, 62. [CrossRef]
30. Morioka, T.; Emoto, M.; Yamazaki, Y.; Kawano, N.; Imamura, S.; Numaguchi, R.; Urata, H.; Motoyama, K.; Mori, K.; Fukumoto, S.; et al. Leptin is associated with vascular endothelial function in overweight patients with type 2 diabetes. *Cardiovasc. Diabetol.* **2014**, *13*, 10. [CrossRef]
31. Puurunen, V.-P.; Kiviniemi, A.; Lepojärvi, S.; Piira, O.-P.; Hedberg, P.; Junttila, J.; Ukkola, O.; Huikuri, H. Leptin predicts short-term major adverse cardiac events in patients with coronary artery disease. *Ann. Med.* **2017**, *49*, 448–454. [CrossRef] [PubMed]
32. Bickel, C.; Schnabel, R.B.; Zeller, T.; Lackner, K.J.; Rupprecht, H.J.; Blankenberg, S.; Sinning, C.; Westermann, D. Predictors of leptin concentration and association with cardiovascular risk in patients with coronary artery disease: Results from the AtheroGene study. *Biomarkers* **2016**, *22*, 210–218. [CrossRef] [PubMed]
33. Tsai, J.-P.; Wang, J.-H.; Chen, M.-L.; Yang, C.-F.; Chen, Y.-C.; Hsu, B.-G. Association of serum leptin levels with central arterial stiffness in coronary artery disease patients. *BMC Cardiovasc. Disord.* **2016**, *16*, 80. [CrossRef] [PubMed]
34. Hasan-Ali, H.; El-Mottaleb, N.A.A.; Hamed, H.B.; Abd-Elsayed, A. Serum adiponectin and leptin as predictors of the presence and degree of coronary atherosclerosis. *Coron. Artery Dis.* **2011**, *22*, 264–269. [CrossRef]
35. Feijóo-Bandín, S.; Portoles, M.; Roselló-Lletí, E.; Rivera, M.; Juanatey, J.R.G.; Lago, F. 20years of leptin: Role of leptin in cardiomyocyte physiology and physiopathology. *Life Sci.* **2015**, *140*, 10–18. [CrossRef]
36. Alix, P.M.; Guebre-Egziabher, F.; Soulage, C.O. Leptin as an uremic toxin: Deleterious role of leptin in chronic kidney disease. *Biochimie* **2014**, *105*, 12–21. [CrossRef]
37. Ambarkar, M. Adipokines and their Relation to Endothelial Dysfunction in Patients with Chronic Kidney Disease. *J. Clin. Diagn. Res.* **2016**, *10*, BC04–BC08. [CrossRef]
38. De Oliveira, R.B.; Liabeuf, S.; Okazaki, H.; Lenglet, A. The clinical impact of plasma leptin levels in a cohort of chronic kidney disease patients. *Clin. Kidney J.* **2012**, *6*, 63–70. [CrossRef]

39. Packer, M.; Zannad, F.; Butler, J.; Filippatos, G.; Ferreira, J.P.; Pocock, S.J.; Brueckmann, M.; Zeller, C.; Hauske, S.; Anker, S.D.; et al. Influence of endpoint definitions on the effect of empagliflozin on major renal outcomes in the EMPEROR-Preserved trial. *Eur. J. Heart Fail.* **2021**, *23*, 1798–1799. [CrossRef]
40. Packer, M.; Januzzi, J.L.; Ferreira, J.P.; Anker, S.D.; Butler, J.; Filippatos, G.; Pocock, S.J.; Brueckmann, M.; Jamal, W.; Cotton, D.; et al. Concentration-dependent clinical and prognostic importance of high-sensitivity cardiac troponin T in heart failure and a reduced ejection fraction and the influence of empagliflozin: The EMPEROR -Reduced trial. *Eur. J. Heart Fail.* **2021**, *23*, 1529–1538. [CrossRef]
41. Zinman, B.; Wanner, C.; Lachin, J.M.; Fitchett, D.; Bluhmki, E.; Hantel, S.; Mattheus, M.; Devins, T.; Johansen, O.E.; Woerle, H.J.; et al. Empagliflozin, Cardiovascular Outcomes, and Mortality in Type 2 Diabetes. *N Engl. J. Med.* **2015**, *373*, 2117–2128. [CrossRef] [PubMed]
42. Zinman, B.; Inzucchi, S.; Lachin, J.M.; Wanner, C.; Ferrari, R.; Fitchett, D.; Bluhmki, E.; Hantel, S.; Kempthorne-Rawson, J.; Newman, J.; et al. Rationale, design, and baseline characteristics of a randomized, placebo-controlled cardiovascular outcome trial of empagliflozin (EMPA-REG OUTCOME™). *Cardiovasc. Diabetol.* **2014**, *13*, 102. [CrossRef] [PubMed]
43. Wu, P.; Wen, W.; Li, J.; Xu, J.; Zhao, M.; Chen, H.; Sun, J. Systematic Review and Meta-Analysis of Randomized Controlled Trials on the Effect of SGLT2 Inhibitor on Blood Leptin and Adiponectin Level in Patients with Type 2 Diabetes. *Horm. Metab. Res.* **2019**, *51*, 487–494. [CrossRef] [PubMed]
44. Vickers, S.P.; Cheetham, S.; Headland, K.; Dickinson, K.; Grempler, R.; Mayoux, E.; Mark, M.; Klein, T. Combination of the sodium-glucose cotransporter-2 inhibitor empagliflozin with orlistat or sibutramine further improves the body-weight reduction and glucose homeostasis of obese rats fed a cafeteria diet. *Diabetes, Metab. Syndr. Obesity Targets Ther.* **2014**, *7*, 265–275. [CrossRef] [PubMed]
45. Bailey, C.J.; Iqbal, N.; T'Joen, C.; List, J.F. Dapagliflozin monotherapy in drug-naïve patients with diabetes: A randomized-controlled trial of low-dose range. *Diabetes, Obes. Metab.* **2012**, *14*, 951–959. [CrossRef]
46. Lee, D.M.; Battson, M.L.; Jarrell, D.K.; Hou, S.; Ecton, K.E.; Weir, T.L.; Gentile, C.L. SGLT2 inhibition via dapagliflozin improves generalized vascular dysfunction and alters the gut microbiota in type 2 diabetic mice. *Cardiovasc. Diabetol.* **2018**, *17*, 62. [CrossRef] [PubMed]
47. Malínska, H.; Hütti, M.; Miklánková, D.; Hojná, S.; Papaoušek, F.; Šilhavý, J.; Mlejnek, P.; Zicha, J.; Hrdlička, J.; Pravenec, M.; et al. Beneficial Effects of Empagliflozin Are Mediated by Reduced Renal Inflammation and Oxidative Stress in Spontaneously Hypertensive Rats Expressing Human C-Reactive Protein. *Biomedicines* **2022**, *10*, 2066. [CrossRef]
48. Gohan, S.; Reshadmanesh, T.; Khodabandehloo, H.; Karbalaee-Hasani, A.; Ahangar, H.; Arsang-Jang, S.; Ismail-Beigi, F.; Dadashi, M.; Ghanban, S.; Taheri, H.; et al. The effect of EMPAgliflozin on markers of inflammation in patients with concomitant type 2 diabetes mellitus and Coronary ARtery Disease: The EMPA-CARD randomized controlled trial. *Diabetol. Metab. Syndr.* **2022**, *13*, 170. [CrossRef]

International Journal of *Molecular Sciences*

MDPI

Article

Pre-Diabetes-Linked miRNA miR-193b-3p Targets PPARGC1A, Disrupts Metabolic Gene Expression Profile and Increases Lipid Accumulation in Hepatocytes: Relevance for MAFLD

Inês Guerra Mollet [1,2,*] and Maria Paula Macedo [1,3,*]

1 iNOVA4Health, NOVA Medical School (NMS), Faculdade de Ciências Médicas (FCM), Universidade NOVA de Lisboa, 1150-082 Lisboa, Portugal
2 UCIBIO-Requimte, Faculdade de Ciências e Tecnologia (FCT), Universidade NOVA de Lisboa, 2825-149 Caparica, Portugal
3 Associação Protectora dos Diabéticos de Portugal, Education Research Center (APDP-ERC), 1250-203 Lisbon, Portugal
* Correspondence: ines.mollet@nms.unl.pt (I.G.M.); paula.macedo@nms.unl.pt (M.P.M.)

Abstract: Distinct plasma microRNA profiles associate with different disease features and could be used to personalize diagnostics. Elevated plasma microRNA hsa-miR-193b-3p has been reported in patients with pre-diabetes where early asymptomatic liver dysmetabolism plays a crucial role. In this study, we propose the hypothesis that elevated plasma hsa-miR-193b-3p conditions hepatocyte metabolic functions contributing to fatty liver disease. We show that hsa-miR-193b-3p specifically targets the mRNA of its predicted target *PPARGC1A*/PGC1α and consistently reduces its expression in both normal and hyperglycemic conditions. *PPARGC1A*/PGC1α is a central co-activator of transcriptional cascades that regulate several interconnected pathways, including mitochondrial function together with glucose and lipid metabolism. Profiling gene expression of a metabolic panel in response to overexpression of microRNA hsa-miR-193b-3p revealed significant changes in the cellular metabolic gene expression profile, including lower expression of *MTTP*, *MLXIPL*/ChREBP, *CD36*, *YWHAZ* and *GPT*, and higher expression of *LDLR*, *ACOX1*, *TRIB1* and *PC*. Overexpression of hsa-miR-193b-3p under hyperglycemia also resulted in excess accumulation of intracellular lipid droplets in HepG2 cells. This study supports further research into potential use of microRNA hsa-miR-193b-3p as a possible clinically relevant plasma biomarker for metabolic-associated fatty liver disease (MAFLD) in dysglycemic context.

Keywords: metabolic syndrome; liver; hsa-miR-193b-3p; peroxisome proliferator activated receptor gamma coactivator 1 alpha; lipid metabolism

Citation: Mollet, I.G.; Macedo, M.P. Pre-Diabetes-Linked miRNA miR-193b-3p Targets PPARGC1A, Disrupts Metabolic Gene Expression Profile and Increases Lipid Accumulation in Hepatocytes: Relevance for MAFLD. *Int. J. Mol. Sci.* **2023**, *24*, 3875. https://doi.org/10.3390/ijms24043875

Academic Editors: Simona Gabriela Bungau and Vesa Cosmin

Received: 15 January 2023
Revised: 6 February 2023
Accepted: 11 February 2023
Published: 15 February 2023

1. Introduction

MicroRNAs are small, non-protein-coding, 20–22-nucleotide-long RNAs produced by cells of various tissues that can be actively released into the bloodstream in vesicles called exosomes and delivered to cells in the same or other tissues in the organism [1]. MicroRNAs target protein expression essentially by binding to the 3′ untranslated region (3′UTR) of messenger RNA (mRNA), thereby preventing translation of the mRNA into protein and directing the mRNA to nonsense-mediated decay [2]. We hypothesize that particular microRNAs detected in circulation in early disease states can be used as biomarkers of underlying dysmetabolism and also that the targets [3] of each microRNA can provide specific crucial organ-specific pathological information that could be used to inform early clinical intervention. Several circulating miRNAs have been reported as being altered in plasma from patients with metabolic disease, including pre-diabetes; however, the specific mechanisms through which they function remain unexplored [4].

Pre-diabetes, diagnosed as impaired fasting glucose (IFG: only fasting glycemia, between 100 and 125 mg/dL) and/or impaired glucose tolerance (IGT: glycemia 2 h after glucose challenge, between 140 and 199 mg/dL), is considered an early sign of metabolic syndrome [5]. NAFLD/MAFLD (nonalcoholic fatty liver disease/metabolic-associated fatty liver disease) [6] is a precursor of metabolic syndrome [7,8]. NAFLD/MAFLD is a condition characterized by a fatty liver, which can lead to complications, including cirrhosis, liver failure, liver cancer and cardiometabolic health problems that encompass dysmetabolism. Thus, analyzing the effects of pre-diabetes-associated circulating microRNA targets in liver cells is an appropriate premise to identify early intervention targets that may have potentially important clinical application in diagnosis, prevention and treatment, paving the way for precision medicine. Among these is microRNA hsa-miR-193b-3p, which is overexpressed in plasma of patients with impaired fasting glucose and impaired glucose tolerance and which tends to normalize upon exercise intervention in humans [4,9] but for which no data exist on a precise mechanism or relevant targeting.

In this study, we demonstrate that hsa-miR-193b-3p targets the expression of *PPARGC1A* mRNA and increases lipid accumulation in human hepatocyte-derived cells in hyperglycemic condition. *PPARGC1A* codes for PGC-1α, a central co-activator of transcriptional cascades that regulate several interconnected pathways including glucose metabolism [10], lipid metabolism [11,12], inflammation, mitochondrial function and oxidative stress [13]. PGC-1α functions by interacting with the nuclear peroxisome proliferator-activated receptors (PPARs), activating them to bind DNA in complex with retinoid X receptor (RXR) [14] on the promoter of numerous genes central to metabolic regulation, increasing transcription of specific genes and decreasing transcription of others. The peroxisome proliferator activated receptors (PPARs) are nuclear receptors that play key roles in the regulation of lipid and glucose metabolism, inflammation, cellular growth and differentiation. The receptors bind and are activated by a broad range of fatty acids and fatty acid derivatives serving as major transcriptional sensors of fatty acids [14]. In order to determine whether the expression of genes involved in metabolic pathways was also affected by overexpression of hsa-miR-193b-3p, we analyzed the expression of several metabolic master-switch genes, some of which are known to be transcriptionally regulated by *PPRAGC1A*/PGC1α, such as *GPT* [15]. Glucose and lipid metabolism depend on the ability of mitochondria to generate energy in cells; therefore, we also looked at genes involved in hepatic mitochondrial dysfunction, which plays a central role in the pathogenesis of insulin resistance in obesity, pre-diabetes and NAFLD/MAFLD [16].

We sought to investigate whether elevated levels of microRNA hsa-miR-193b-3p in hepatocytes could interfere with hepatic cell function in vitro under conditions that mimic pre-diabetes (hyperglycemia and hyperinsulinemia). We evaluated pre-diabetes associated microRNA hsa-miR-193b-3p predicted target *PPARGC1A* (PGC1α), hepatic lipid content and direct or indirect effects on expression of genes regulating cellular energy metabolism, in human hepatocyte-derived cells. The objective was to investigate which metabolic pathways microRNA hsa-miR-193b-3 interfered with so as to gauge the potential of hsa-miR-193b-3p as an early plasma biomarker of specific liver dysmetabolism to support precision medicine in early stages.

2. Results

2.1. MicroRNA hsa-miR-193b-3p Targets the 3'UTR of PPARGC1A mRNA and Downregulates PPARGC1A Expression in HepG2 Cells

To evaluate potential mRNA targets of microRNA hsa-miR-193b-3p, we performed a bioinformatic analysis of conserved predicted targets of hsa-miR-193b-3p from the online database TargetScan 7.1 [3] using the DAVID Bioinformatics Resources 6.8 [17,18] and clustering tools for functional Gene Ontology Annotations [19]. Of the 283 predicted conserved mRNA targets of microRNA hsa-miR-193b-3p, we found the top gene ontology cluster was composed of 43 genes involved in gene transcription. From this list, *PPRAGC1A* stood out as the gene with the most consistently conserved microRNA hsa-miR-193b-3p target site involved in energy metabolism. Peroxisome proliferator-activated receptor gamma coactivator 1-alpha (*PPARGC1A*) codes for PGC-1α, a transcriptional coactivator of numerous genes involved in energy metabolism. The endogenous level of hsa-miR-193b-3p in HepG2 cells was measured using real-time RT-qPCR. We observed identical Ct values of hsa-miR-193b-3p and the reference snoRNA RNU44 (Ct = 26.8 ± 0.4 and Ct = 26.7 ± 0.3, respectively), while snoRNA RNU48 had higher expression with Ct = 23.7 ± 0.5. Parrizas et al. 2015 [9] showed that higher levels of hsa-miR-193b-3p were observed in plasma of people with impaired fasting glucose by a difference of almost 6 dCt between lowest levels in non-diabetic controls and highest level in impaired fasting glucose, which corresponds to a 64-fold difference. To determine a physiologically relevant overexpression of hsa-miR-193b-3p to use in this study, we measured, by real-time RT-qPCR, the fold increase in hsa-miR-193b-3p using a 0.05 nM, 0.5 nM and 5 nM transfection of hsa-miR-193b-3p against the scrambled control microRNA and observed a fold increase of 1.9, 11.1 and 131, respectively. We therefore selected 5 nM hsa-miR-193b-3p as an appropriate physiologically relevant level of overexpression.

When hsa-miR-193b-3p was overexpressed in HepG2 cells for 72 h using a mimic, we observed robust downregulation of *PPARGC1A* mRNA expression at basal glucose concentration (5 mM glucose, Figure 1A), under hyperglycemia (20 mM glucose, Figure 1B), under hyperinsulinemia (5 mM glucose and final 24 h 10 nM insulin, Figure 1C) and under hyperglycemia/hyperinsulinemia (20 mM glucose and final 24 h 10 nM insulin, Figure 1D). We validated the predicted *PPARGC1A* 3′UTR target site for miR-193b-3p by cloning a "match", or a "mismatch" control (Figure 1E), into the 3′UTR of the firefly luciferase gene in the pmirGLO dual-luciferase miRNA target expression vector (Figure 1F), followed by dual-glow luciferase assay. Expression of firefly luciferase from the "match" clone was significantly downregulated in HepG2 cells when the hsa-miR-193b-3p mimic was overexpressed and cells were transfected with the pmirGLO match clone (Figure 1G). These results support our posit that microRNA hsa-miR-193b-3p directly targets and downregulates the human *PPARGC1A* mRNA by binding to its 3′UTR.

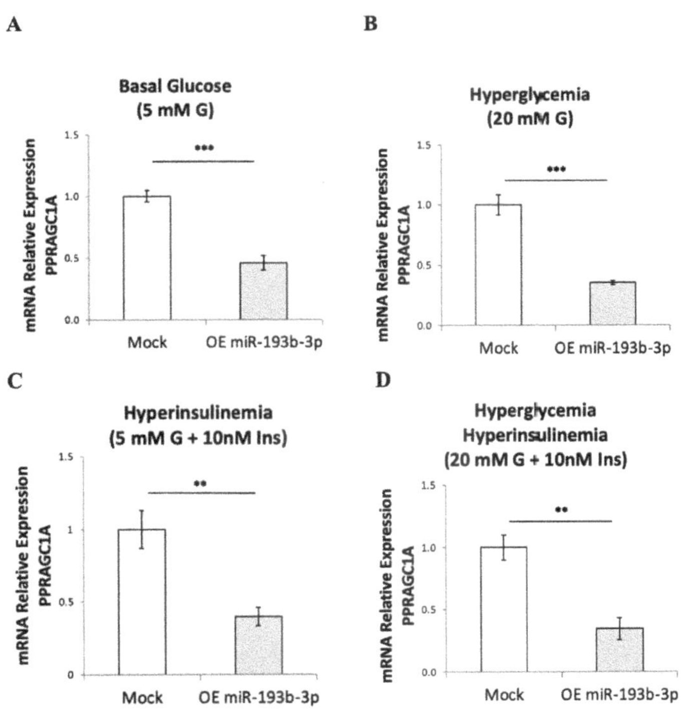

Figure 1. *Cont.*

E

```
Match cloning primer set of PPARGC1A 3'UTR hsa-miR-193b-3p target site
                 SacI                                            SalI
sense     5'-----CTAGCGGCCGCGTGTGTCTGAAGACTATCTTTGA-AGCCAGTATTTCTTTCCCG-----3'
antisense 3'-TCGAGATCGCCGGCGCACACAGACTTCTGATAGAAACT-TCGGTCATAAAGAAAGGGCAGCT-5'
                                     ||     ||||||  ||||||
                                   3'-TCGCCCTGAAACTCCCGGTCAA-5'   microRNA hsa-miR-193b-3p reverse complement

Mismatch cloning primer set of PPARGC1A 3'UTR hsa-miR-193b-3p target site
sense     5'-----CTAGCGGCCGCGTGTGTCTGAAGACTATGA--CT-AGCCAGTATTTCTTTCCCG-----3'
antisense 3'-TCGAGATCGCCGGCGCACACAGACTTCTGATACT--GA-TCGGTCATAAAGAAAGGGCAGCT-5'
                                     ||                 ||||||
                                   3'-TCGCCCTGAAACTCCCGGTCAA-5'   microRNA hsa-miR-193b-3p reverse complement
```

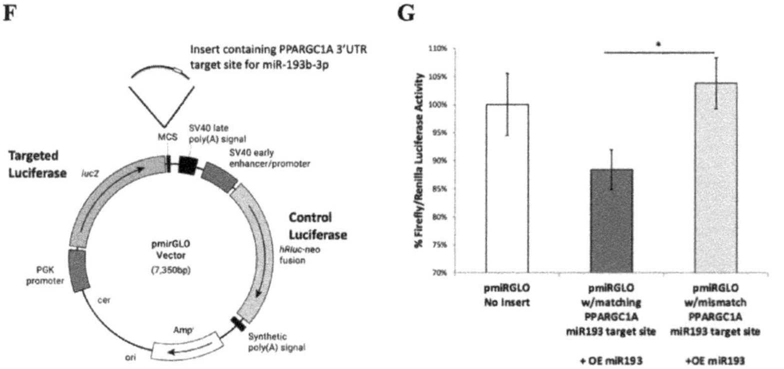

Figure 1. MicroRNA hsa-miR-193b-3p targets the 3′UTR of *PPARGC1A* mRNA and downregulates *PPARGC1A* expression in HepG2 cells. (**A–D**) mRNA expression, evaluated by real-time RT-qPCR, of *PPARGC1A* after a 72 h overexpression of hsa-miR-193b-3p in HepG2 cells in basal 5 mM glucose (**A**); hyperglycemia at 20 mM glucose (**B**); hyperinsulinemia with 5 mM glucose and final 24 h with 10 nM insulin (**C**); and hyperglycemia/hyperinsulinemia at 20 mM glucose and final 24 h with 10 nM insulin (**D**). (**E**) Match and mismatch primer pairs containing the *PPARGC1A* 3′UTR target site for hsa-miR-193b-3p that were cloned into pmirGLO miRNA target expression vector (**F**). (**G**) Luciferase luminescence measurement after overexpression of hsa-miR-193b-3p in HepG2 cells; pmirGLO plasmid was transfected into HepG2 cells 24 h before luciferase assay. Data are presented as mean ± SEM of n = 4. Mock—mock transfection, OE—overexpression of hsa-miR-193b-3p, G—glucose, Ins—insulin. Student's *t*-test *p*-values indicated by * < 0.05, ** < 0.01, *** < 0.001.

2.2. Intracellular Lipid Droplet Content Is Increased by Overexpression of microRNA hsa-miR-193b-3p in HepG2 Cells

PPARGC1A/PGC1α is of central importance for lipid regulation [11]. Fatty liver is associated with impaired activity of *PPARGC1A* [20]. Therefore, having verified that hsa-miR-193b-3p overexpression in HepG2 hepatoma cells reduced the expression of *PPARGC1A*, we proceeded to analyze the intracellular lipid content in these cells. We overexpressed hsa-miR-193b-3p mimic in HepG2 cells for 72 h, under normoglycemia (5 mM glucose) and hyperglycemia (20 mM glucose); visualized lipid droplet content using Oil Red O staining and brightfield microscopy (Figure 2); and performed quantitative image analysis. At hyperglycemia (20 mM glucose) we observed a significant increase in total area covered by lipid droplets (Figure 2G) and number of lipid droplets (Figure 2H); we also saw an increasing trend in lipid droplet size (Figure 2I), though not statistically significant. All three measurements also showed increasing trends at 5 mM. This result supports further research into the possibility of using increased plasma levels of microRNA hsa-miR-193b-3p as a biomarker for fatty liver disease.

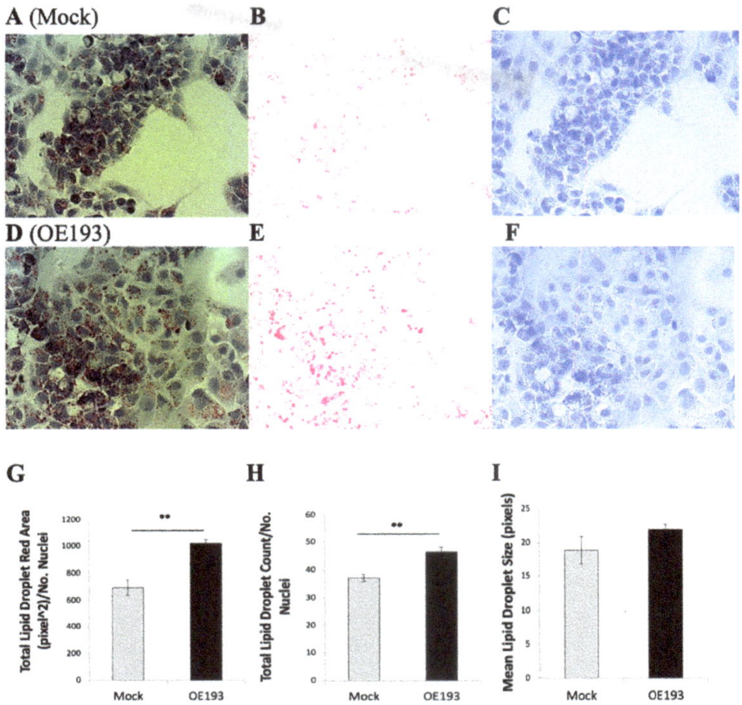

Figure 2. MicroRNA hsa-miR-193b-3p overexpression in HepG2 cells increases intracellular lipid droplet accumulation. HepG2 cells cultured at 20 mM glucose. Lipid droplets in HepG2 cells were visualized using Oil Red O staining. Brightfield 40× magnification images of Oil Red O staining in control mock transfection (**A**) with corresponding red (**B**) and blue (**C**) color deconvolution. Brightfield 40× magnification images of Oil Red O staining after microRNA hsa-miR193b-3p overexpression (**D**) with corresponding red (**E**) and blue (**F**) color deconvolution. Red color deconvolution used for red particle analysis; blue color deconvolution used for nuclei counting. Quantitative analysis of lipid droplet area (**G**), lipid droplet number (**H**) and lipid droplet size (**I**). Data are presented as mean ± SEM of n = 4; each n is average of ten fields per condition normalized to number of nuclei. Mock—mock transfection, OE—overexpression of hsa-miR-193b-3p. Student's t-test p-value ** < 0.01.

2.3. *Changes in Expression of Genes Involved in Lipid Metabolism May in Part Explain Intracellular Lipid Accumulation Following Overexpression of hsa-miR-193b-3p in HepG2 Cells*

Having observed that *PPARGC1A* expression is robustly downregulated and lipid droplets accumulate when microRNA hsa-miR-193b-3p is overexpressed in HepG2 cells under hyperglycemia–hyperinsulinemia, we proceeded to analyze the expression of other genes involved in lipid metabolism and processing that might be affected. The lower mRNA expression of microsomal triglyceride transfer protein (*MTTP*) (Figure 3), which is required for the secretion of plasma lipoproteins containing apolipoprotein B [21], such as VLDL, that is made in hepatocytes and secreted to deliver lipids, may be in part responsible for the increased lipid droplet accumulation observed. The increased expression of Tribbles pseudokinase 1 (*TRIB1*) (Figure 3) could also be responsible for the lipid droplet accumulation we observed [22]; indeed, Burkhardt et al. (2010) reported that hepatic overexpression of *TRIB1* in mice led to reduced VLDL secretion. The increased expression of low-density lipoprotein receptor (*LDLR*) (Figure 3) may increase LDL uptake in favor of increased lipid accumulation. No significant expression changes were observed in apolipoprotein B (*APOB*) (Figure 3).

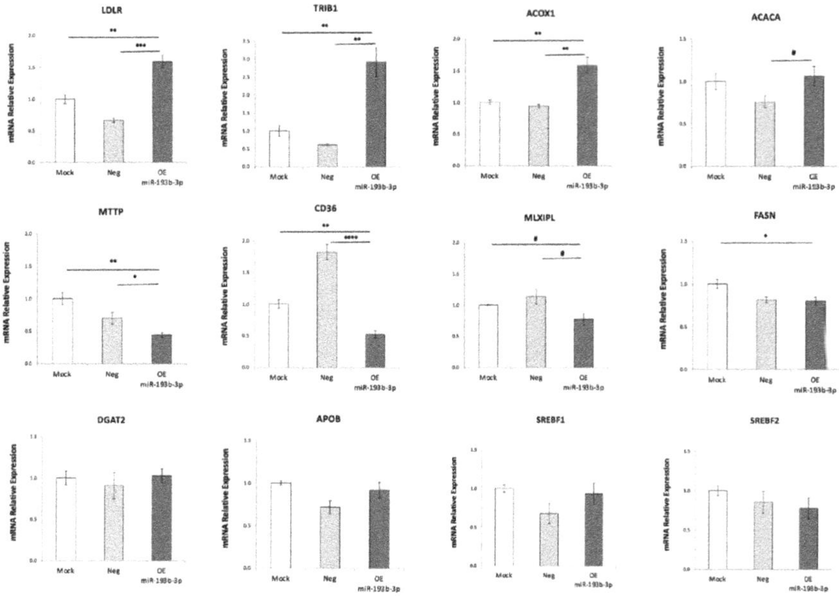

Figure 3. Overexpression of microRNA hsa-miR-193b-3p in HepG2 alters mRNA expression of fundamental genes involved in lipid processing. Lower mRNA expression of *MTTP* and higher expression of *TRIB1* suggest reduced VLDL secretion. Higher level of LDL receptor *LDLR*. No change in *APOB*. Lower mRNA level trend of *MLXIPL*/ChREBP, coordinating triglyceride synthesis. No change in *DGAT2*, *SREBF1* (n = 3) or *SREBF2* (n = 3). Lower expression of fatty acid synthase (*FASN*) (I). No change in *ACACA*. Increased *ACOX1* indicates increased fatty acid oxidation. Reduced fatty acid translocase *CD36*. Relative mRNA expression was evaluated by real-time RT-qPCR after 72 h overexpression of hsa-miR-193b-3p in HepG2 cells under hyperglycemia/hyperinsulinemia (20 mM glucose, with 10 nM insulin during last 24 h). Data are presented as mean ± SEM of n = 4, unless otherwise indicated. Mock—mock transfection, OE—overexpression of hsa-miR-193b-3p. Student's *t*-test *p*-values indicated by # < 0.1, * < 0.05, ** < 0.01, *** < 0.001 or **** < 0.0001.

Expression of carbohydrate-responsive element-binding protein (ChREBP), also known as MLX-interacting protein-like (*MLXIPL*), showed a reduced trend of 22% (vs. mock control) and 32% (vs. negative control) (Figure 3) when hsa-miR-193b-3p was overexpressed. This transcription factor binds and activates carbohydrate response element (ChoRE) motifs on the promoters of triglyceride synthesis genes in a glucose-dependent manner. Thus, we expect to see reduced expression of genes involved in triglyceride synthesis such as *DGAT2* [23], which catalyzes the final reaction in triglyceride synthesis. However, no changes in expression of *DGAT2* (Figure 3) were observed. No expression changes were observed in either of two major transcriptional regulators, sterol regulatory element binding transcription factors 1 and 2 coded by *SREBF1* (Figure 3) and *SREBF2* (Figure 3), which, in concert with *MLXIPL*/ChREBP, regulate hepatic lipid metabolism by inducing transcription of lipogenic enzymes that direct lipogenesis in the liver [24]. We also observed a 20% reduction in expression of fatty acid synthase (*FASN*) (vs. mock) (Figure 3), indicating reduced fatty acid synthesis, which was expected since this gene is regulated by *MLXIPL*/ChREBP. However, no change was observed in acetyl-CoA carboxylase (*ACACA*) (Figure 3), the enzyme which catalyzes the carboxylation of acetyl-CoA to malonyl-CoA, the rate-limiting step in fatty acid synthesis. This lack of downregulation of *ACACA* is important, because although we observe increased expression of Acyl-CoA oxidase 1 (*ACOX1*) (Figure 3), indicating increased fatty acid oxidation in the mitochondria, the malonyl-CoA produced by *ACACA* in the cytoplasm can inhibit transfer of cytoplasmic fatty acids into the mitochondria for beta oxidation contributing to increased lipids in the cytoplasm. Unexpectedly, we observed a 50% reduction in the fatty acid translocase *CD36* (Figure 3), which has been described as positively correlated with fatty liver [25]. Thus, the changes we observed in mRNA levels of genes involved in lipid metabolism do not all provide a clear rationale to explain lipid accumulation in hepatocytes during overexpression of hsa-miR-139b-3p.

2.4. Changes Favoring Insulin Signaling, Reduced Glycolysis, and Reduced Mitochondrial Biogenesis Are Observed When microRNA hsa-miR-193b-3p Is Overexpressed in HepG2 Cells

In addition to lipid metabolism, *PPARGC1A*/PGC1α also regulates insulin signaling, hepatic glucose metabolism and mitochondrial turnover [20,26,27]. Therefore, we investigated whether expression of genes involved in these mechanisms were altered. We observed downregulation of *YWHAZ* (Figure 4). *YWHAZ* belongs to the 14-3-3 family of proteins that mediate signal transduction; it binds to insulin receptor substrate-1 (IRS-1) and is thought to interrupt the association between the insulin receptor and IRS-1, thus interfering with insulin signaling [28]. Our result indicates decreased interference with insulin signaling by this particular mechanism. No change was observed in Tribbles homolog 3 (*TRIB3*) (Figure 4) that has been linked to insulin resistance and hepatic production of glucose in the liver [29,30].

PPARGC1A/PGC1α is known to regulate key hepatic gluconeogenic enzymes leading to increased glucose output [27], and we saw a significant increase in expression of pyruvate carboxylase (*PC*) (Figure 4), the mitochondrial matrix enzyme that regulates fuel partitioning toward gluconeogenesis in hepatocytes. We saw reduced expression trend of the plasma membrane bidirectional glucose transporter *SLC2A2*/GLUT2 (Figure 4), indicating both reduced uptake and release of glucose. The reduced expression of liver phosphofructokinase (*PFKL*) (Figure 4) and pyruvate kinase (*PKLR*) (Figure 4) indicate reduced glycolysis. Reduced expression of the pyruvate dehydrogenase E1 subunit alpha 1 (*PDHA1*) (Figure 4), a central component of the pyruvate dehydrogenase complex that is the primary link between glycolysis and the TCA cycle, indicated lower conversion of pyruvate to acetyl-CoA in mitochondria. This is further supported by a five-fold increased expression of the pyruvate dehydrogenase inhibitor mitochondrial kinase *PDK4* (Figure 4). Reduced expression of glutamic-pyruvic transaminase *GPT* (Figure 4) suggests lower conversion of alanine to pyruvate. Based on all this, we can predict a drop in ATP production.

We also observed reduced expression of mitochondrial transcription factor A (*TFAM*) (Figure 4) that is required for mitochondrial biogenesis through maintenance of mitochondrial DNA and replication of normal levels of mitochondrial DNA; reduced expression trend of mitofusin-2, *MFN2* (Figure 4), which regulates mitochondrial fusion dynamics; and reduced expression trend of isocitrate dehydrogenase 3 catalytic subunit alpha (*IDH3A*) (Figure 4), which catalyzes the rate-limiting step of the TCA cycle in the mitochondrial matrix.

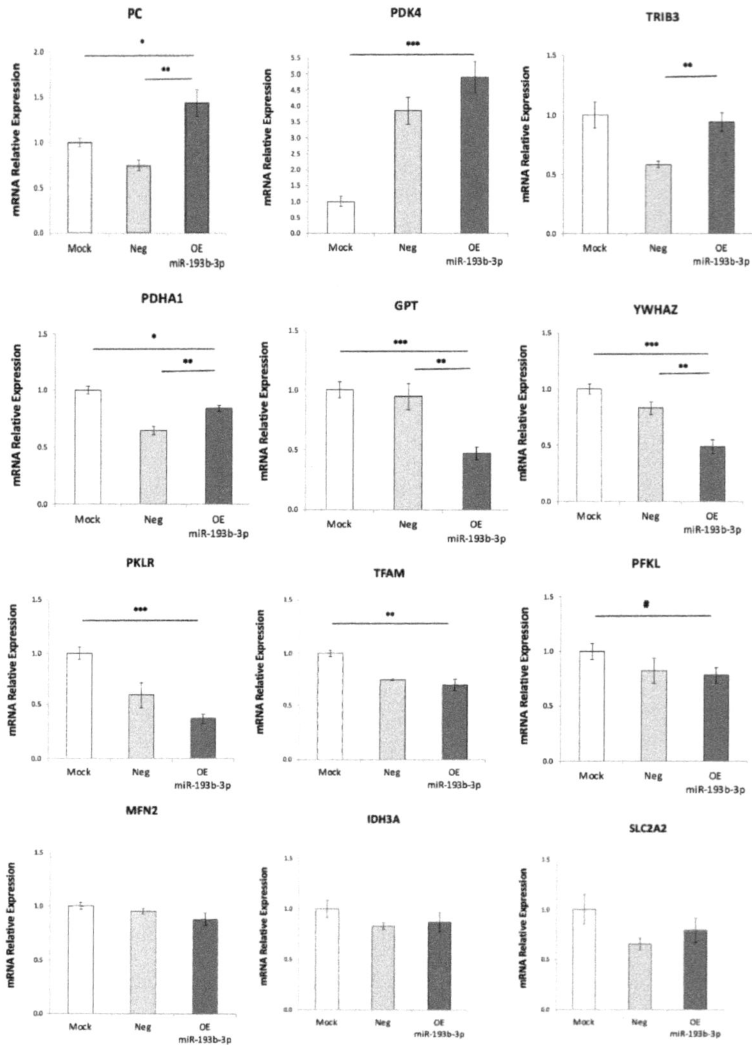

Figure 4. Changes in expression of genes regulating insulin signaling, glucose metabolism, pyruvate availability and mitochondrial biogenesis when hsa-miR-193b-3p is overexpressed in HepG2 cells. Downregulation of *YWHAZ* indicates increased insulin signaling. Increased pyruvate carboxylase (*PC*). Reduced trend in glucose transporter *SLC2A2*/GLUT2 indicates reduced glucose transport. Reduced levels of phosphofructokinase (*PFKL*) and pyruvate kinase (*PKLR*) indicate reduced glycolysis. Reduced levels of pyruvate dehydrogenase *PDHA1* and increase in its inhibiting kinase *PDK4* indicate lower conversion of pyruvate to acetyl-CoA. Reduced glutamic–pyruvic transaminase *GPT* indicates lower conversion of alanine to pyruvate. Reduced levels of mitochondrial transcription factor A (*TFAM*), and reduced trend in mitofusin 2 (*MFN2*) and isocitrate dehydrogenase 3 catalytic subunit alpha (*IDH3A*) indicate reduced levels of mitochondrial biogenesis, fusing dynamics and function, respectively. Relative mRNA expression was evaluated by real-time RT-qPCR after 72 h overexpression of hsa-miR-193b-3p in HepG2 cells under hyperglycemia/hyperinsulinemia (20 mM glucose, with 10 nM insulin during last 24 h). Data are presented as mean ± SEM of n = 4. Mock—mock transfection, OE—overexpression of hsa-miR-193b-3p. Student's *t*-test *p*-values indicated by # < 0.1, * < 0.05, ** < 0.01, *** < 0.001.

3. Discussion

This study is based on recent research showing that circulating microRNAs detected in plasma can enter cells in various tissues and there interfere with gene expression [31], and that, in humans, altered levels of circulating microRNAs are detected in many disease states including metabolic diseases [4]. In humans, higher than normal levels of the microRNA hsa-miR-193b-3p have been detected in plasma of patients with impaired fasting glucose and impaired glucose tolerance, two parameters that are used to diagnose pre-diabetes, an early sign of metabolic syndrome [5]. Nonalcoholic fatty liver disease, also referred to as metabolic-associated fatty liver disease (NAFLD/MAFLD) [6–8], characterized by excess accumulation of lipids in the liver (known as steatosis) is considered a precursor of metabolic syndrome. In this study, we have focused on identifying and validating a direct target of hsa-miR-193b-3p and evaluating changes in expression of genes involved in cellular energy metabolism after overexpression of microRNA hsa-miR-193b-3p in the hepatocyte cell line HepG2, which is considered an appropriate cell model [32]. This strategy is in line with recent attempts to try to understand metabolism in type 2 diabetes beyond glycemia [33].

Our results show that microRNA hsa-miR-193b-3p directly targets *PPRAGC1A*/PGC1α, reducing its expression, and that it causes excess intracellular lipid droplet accumulation in HepG2 cells. We also show that the accumulation of lipids observed may be caused in part by altered expression of genes involved in lipid processing, which may cause reduced VLDL secretion, as well as reduced glycolysis and reduced mitochondrial activity shifting the metabolic balance toward lipogenesis. We observed a 60% reduction in *MTTP* required for VLDL secretion, and a 3-fold increased expression of *TRIB1*, which reduces secretion of VLDL. Indeed, hepatic overexpression of Trib1 in mice reduces secretion of VLDL from the liver into circulation [34]. Although no change in *APOB* was observed, the decreased expression trend of *MLXIPL*/ChREBP suggests a reduced rate of triglyceride synthesis that could lead to insufficient *APOB* lipidation and secretion.

Insulin signaling normally directs de novo lipogenesis in the liver [35], among several other pathways. The observed downregulation of *YWHAZ*, known to bind insulin receptor substrate-1 (IRS-1) [28], predicts reduced interference of *YWHAZ* with IRS-1 at this point in the insulin signaling cascade. However, linking this observation to increased lipogenesis requires investigating additional downstream steps in the insulin signaling cascade towards lipogenesis.

When insulin and glucose are associated in liver cells, as in our experimental conditions, the stimulatory effect of glucose on *SLC2A2*/GLUT2 gene expression is predominant [36]. However, when hsa-miR-193b-3p was overexpressed, we observed a decreasing trend in *SLC2A2*/GLUT2 expression, suggesting lower glucose uptake, which may contribute to hyperglycemia, a feature associated with the plasma increase in hsa-miR-193b-3p in humans. We also observed lower glycolysis (reduced *PFKL* and *PKLR*), producing less pyruvate; less pyruvate is also expected coming from amino acid metabolism given the reduced *GPT* expression. Inhibition of conversion of pyruvate to acetyl-CoA for the TCA cycle is predicted via reduced *PDHA1* and increased *PDK4* expression; all of this points to a decrease in energy combustion and lower mitochondrial-generated ATP. In addition, we observed lower expression of *TFAM* pointing to less mitochondrial biogenesis, expected to contribute to a decrease in energy combustion. In line with our results, mitochondrial dysfunction has been shown to play a central role in the pathogenesis of metabolic diseases and associated complications [37], and reduced energy combustion through reduced glycolysis has been described as critical to excess lipid storage in the liver [38]. Energy combustion in the liver is modulated by PPARα-regulated fatty acid beta-oxidation. Interference with PPARα lipid sensing can reduce energy burning and result in accumulation of lipids in hepatocytes. Fatty liver is also associated with impaired activity of *PPARGC1A* (PGC1α) and reduced mitochondrial biogenesis in mice [20]. Although we saw increased expression of *ACOX1*, indicating increased fatty acid beta-oxidation in mitochondria, this could be a compensation for reduced mitochondrial biogenesis. We saw no change in acetyl-CoA carboxylase (*ACACA*), the enzyme which catalyzes the carboxylation of acetyl-CoA to malonyl-CoA, the rate-limiting step in fatty acid synthesis; this is relevant, as the malonyl-CoA produced by ongoing fatty acid synthesis will inhibit fatty acid transport into mitochondria for beta-oxidation.

Hepatic disruption of fatty acid translocase *CD36* in JAK2L livers has been shown to lower triglyceride (TG), diacylglycerol (DAG) and cholesterol ester (CE) content, significantly improving steatosis [25]. Therefore, one might speculate that our unexpected observation of a 50% reduction in the fatty acid translocase *CD36* alongside lipid accumulation in HepG2 cells under hyperglycemia/hyperinsulinemia might be a compensatory mechanism.

PPRAGC1A/PGC-1α interacts with the PPAR nuclear receptors [14], on the promoter of numerous genes involved in metabolic regulation. However, the gene expression changes that we observed when microRNA hsa-miR-193b-3p was overexpressed under hyperglycemia/hyperinsulinemia in the HepG2 cell model may result from either direct transcriptional regulation via *PPRAGC1A*/PGC-1α or other indirect regulation. In future work, one might consider rescuing the downregulation of *PPRAGC1A*/PGC-1α observed by overexpressing *PPRAGC1A*/PGC-1α along with hsa-miR-193b-3p in this cell model. As there are a further 283 predicted conserved mRNA targets of microRNA hsa-miR-193b-3p, including 43 genes involved in gene transcription, other predicted gene targets of hsa-miR-193b-3p must also be investigated. With regards to the specific targeting of *PPARGC1A*/PGC-1α by hsa-miR-193b-3p, and given the central importance of PGC1α in inflammation, oxidative stress and energy in cells other than hepatocytes, much work remains to be done on these pathways and on other cell types and in vivo. Future work on this subject should also include investigating the effects of inhibiting hsa-miR-193b-3p on hepatocyte lipid content.

Given that *PPRAGC1A*/PGC-1α interacts with the nuclear receptor PPAR-γ [14], our results also warrant bearing in mind the drug class of thiazolidinediones, which are potent PPAR-γ agonists, that can be used in the treatment of type 2 diabetes to improve hepatic sensitivity to insulin [39,40] and improve fibrosis in nonalcoholic steatohepatitis [40,41]. In this context, further work should differentiate effects which are mediated directly through hsa-miR-193b-3p targeting of *PPRAGC1A*/PGC-1α and PPAR transcriptional regulation, and which might be indirect.

Our results show that elevated levels of microRNA hsa-miR-193b-3p in HepG2 hepatocytes cause a cascade of gene-expression changes and increased lipid accumulation in these cells, and specific direct targeting of *PPARGC1A*/PGC1α mRNA. As levels of plasma hsa-miR-193b-3p may be elevated in pre-diabetes in humans [9], these results support the hypothesis of using hsa-miR-193b-3p as an early diagnostic biomarker of liver dysmetabolism in pre-diabetes. The potential clinical relevance of these results is highlighted by recent research showing that improvement of fatty liver disease reduces the risk of type 2 diabetes [42]. Thus, plasma levels of hsa-miR-193b-3p could conceivably be used to flag pre-diabetic patients for early intervention to target asymptomatic fatty liver, thereby preventing aggravation of pre-diabetes and the development of type 2 diabetes.

4. Materials and Methods

Cell model. The HegG2 hepatoma cell line [32] was maintained in Dulbecco's modified Eagle's medium (DMEM, 25 mM high glucose (21969-035, Gibco, Life Technologies, Carlsbad, CA, USA), 4 mM glutamine (2503-149, Gibco, Life Technologies, Carlsbad, CA, USA), 10% heat inactivated fetal bovine serum (S0615, Biochrom, Cambridge, UK), 100 IU/mL penicillin and 100 ug/mL streptomycin (15140-122, Life Technologies, Carlsbad, CA, USA. Experiments at basal 5 mM glucose were performed with DMEM 4 mM glutamine, 1 mM sodium pyruvate (SH30021.FS, Thermo Scientific), 10% heat-inactivated fetal bovine serum, 100 IU/mL penicillin and 100 ug/mL streptomycin.

Overexpression of miR-193b-3p in HepG2 cells. microRNA miR-193b-3p overexpression in HepG2 cells was performed over 72 h by reverse transfection on the day of plating, followed by forward transfection the next day, with an additional day of growth, before RNA extraction, plasmid transfection or fixing cells for imaging. MicroRNA mimics used were hsa-miR-193b-3p miRCURY LNA™ microRNA Mimic (472852-001, Exiqon, Hovedstaden, Denmark) and Negative Control 4 miRCURY LNA™ microRNA Mimic (479903-001, Exiqon, Hovedstaden, Denmark). Transfection of microRNA mimic and control into HegG2 cells was performed at 5 nM final concentration using Lipofectamine® RNAiMAX Transfection Reagent (13778-100, Life Technologies, Carlsbad, CA, USA) with 20 min lipofectamine/mimic complex formation in Opti-MEM® I Reduced Serum Medium (31985-062, Life Technologies, Carlsbad, CA, USA) and cells incubated in DMEM culture as described above with no penicillin or streptomycin added. Mock transfection was performed with lipofectamine alone.

RNA preparation and relative gene expression. Total RNA was prepared by organic extraction using Trizol Reagent (15596-018, Life Technologies, Carlsbad, CA, USA) and chloroform (Sigma, St Louis, MO, USA). RNA was precipitated in absolute ethanol, and dissolved in DNase/RNaseFree Distilled Sterile Water (10977-035, Gibco, Billings, MT, USA). RNA concentration was determined using Nanodrop 2000 (Thermo Fisher Scientific, Waltham, MA, USA). Relative gene expression was determined by two-step reverse-transcription real-time quantitative PCR (real-time RT-qPCR) with cDNA prepared from total extracted RNA using a High-Capacity cDNA Reverse Transcription Kit (4368814, Applied Biosystems, Waltham, MA, USA) on a PCR MyCycler (Biorad, CA, USA) (25 °C 10 min, 37 °C 120 min, 85 °C 5 min). For microRNA expression, the protocol was modified according to the manufacturer's instructions with added RNAse Inhibitor (N8080119, Life Technologies, Carlsbad, CA, USA) and pooled stem loop primer mix of hsa-miR-193b-3p, RNU44 and RNU48 (0.3 uL each stem-loop primer in 15 uL reaction volume: stem loop primers (TaqMan® MicroRNA Assay, hsa-miR-193b-3p, Assay ID 002366, 002366/4427975, Life Technologies, Carlsbad, CA, USA). The two internal reference genes used were TaqMan® MicroRNA Control Assays RNU44 (Assay ID 001094, 001094/4427975, Life Technologies, Carlsbad, CA, USA) and RNU48 (Assay ID 001006, 001006/4427975, Life Technologies, Carlsbad, CA, USA). TaqMan Universal Master Mix II, with UNG (4440042, Life Technologies, Carlsbad, CA, USA), was used to determine relative expression of the miRNA miR-193b-3p using TaqMan MicroRNA Assays qPCR primers; the geometric mean of both references was used to calculate $2^{-ddCt.}$ For protein-coding genes, forward and reverse primers for qPCR for each gene of interest were designed across constitutive exons using Primer Quest (Integrated DNA Technologies, IDT, https://www.idtdna.com/pages/tools/primerquest, accessed on 24 September 2018). The list of primers used is presented in Table 1. The custom-designed primers were ordered from Invitrogen (Waltham, MA, USA). Relative gene expression was determined from cDNA using NZYTaq 2× Green Master Mix (MB03903, NZYtech, Lisbon, Portugal) on MicroAmp Optical 96-well reaction plates (N8010560, Applied Biosystems, Waltham, MA, USA). Amplification and Ct values for two technical replicates of each sample were obtained on qPCR LightCycler and integrated software (Roche, Boston, MA, USA). Amplicon specificity was verified by a high-resolution melting curve using LightCycler integrated software. Relative mRNA expression was determined using the relative gene expression data using real-time quantitative PCR and the 2^{-ddCt} method [43] using *TBP* primers as reference gene.

Table 1. Primers designed for real-time RT-qPCR.

Gene (Exon Boundary)	Forward Primer	Reverse Primer
ACACA (31_32)	TTTGTCAGGATCTTTGATGAAGTG	TCATAAAGAGACGTGTGACCTG
ACOX1 (13_14)	GATGTGACACTTGGCTCTGT	TTCGTGGACCTCTGCTTTG
APOB (20_21)	TTCTGTCAGCGCAACCTATG	CTTCGCACCTTCTGCTTGA
CD36 (4_5)	GTCCTTATACGTACAGAGTTCGTT	CAGCCTCTGTTCCAACTGATAG
DGAT2 (6_7)	CAAAGAATGGGAGTGGCAATG	CAGGTCAGCTCCATGACG
FASN (3_4)	GTGGACGGAGGCATCAAC	TGTAGCCCACGAGTGTCT
GPT (4_5)	CCATCGTGACGGTGCTG	GCCGAGTAGAGTGGGTACT
IDH3A (10_11)	GCTCAGTGCCGTGATGAT	GTCAAGCTCTTTCCGTCCTT
LDLR (15_16)	CGTAAGGACACAGCACACA	GCCCAGAGCTTGGTGAG
MFN2 (17_18)	AAGTCCAGCAGGAACTGTC	ATTTCCTGCTCCAGGTTCTC
MLXIPL (15_16)	TTTGACCAGATGCGAGACAT	GATGCTGAACACCCAGAACT
MTTP (16_17)	CATTCTCAGGAACTTCAGTTACAATC	ACTCACGATACCACAAGCTAAA
PC (12_13)	CTGTGGACACCCAGTTCATC	TGACATGGCCGAGGTAGT
PDHA1 (2_3)	GAAATTAAGAAATGTGACCTTCACC	CAGTCTGCATCATCCTGTAGTA
PDK4 (7_8)	TCCAGACCAACCAATTCACATC	GCCCGCATTGCATTCTTAAATAG
PFKL (13_14)	ATCTCCCATGGACACACAGTAT	TACTTCTTGCACCTGACCCT
PKLR (10_11)	CTTTACCGTGAACCTCCAGAAG	CACGGAGCTTTCCACTTTCA
PPARGC1A (8_9)	GCAGTAGATCCTCTTCAAGATCC	AACGTGATCTCACATACAAGGG
SLC2A2 (8_9)	GACGGCTGGTATCAGCAAA	CTCCACAAGGAATACAGAGACAG
SREBF1 (5_6)	CACTGAGGCAAAGCTGAATAAAT	TAGGTTCTCCTGCTTGAGTTTC
SREBF2 (2_3)	CTGCAACAACAGACGGTAATG	GCTGAAGGACTTGAAAGCTAGTA
TBP (4_5) Reference	TCCACAGTGAATCTTGGTTGT	AGCAAACCGCTTGGGATTA
TFAM (2_3_4)	GCTCAGAACCCAGATGCAAA	TGCCACTCCGCCCTATAA
TRIB1 (2_3)	AGGAGAGAACCCAGCTTAGA	TGGGCAGCCATGTTTGT
TRIB3 (3_4)	GACCGTGAGAGGAAGAAGC	CTTGTCCCACAGGGAATCAT
YWHAZ (5_6)	GAAGCCATTGCTGAACTTGATAC	TCCACAATGTCAAGTTGTCTCT

miRNA target plasmid construct and luciferase assay. To validate the miR-193 b-3p target site on the mRNA of *PPARGC1A* (PGC1a gene), the pmirGLO dual-luciferase miRNA target expression vector (E1330, Promega, Madison, WI, USA) was used to clone the predicted *PPARGC1A* target site for microRNA miR-193b-3p along with a mismatch control (Figure 1E,F). The custom primers for cloning were ordered from Invitrogen. Primer pairs were cloned by double digestion using *Sac*I and *Sal*I restriction enzymes (R0156, R0156, CutSmart Buffer, New England Biolabs, Ipswich, MA, USA) and ligation with T4 DNA Ligase (EL0014, Thermo Fisher Scientific, Waltham, MA, USA) and FastAP thermosensitive alkaline phosphatase (EF0654, Thermo Fisher Scientific, Waltham, MA, USA). Selection of clones was made with *Xho*I/*Bam*HI double digest (R0136, R0146, CutSmart Buffer, New England Biolabs, Ipswich, MA, USA), with the *Xho*I site being on the excised multiple cloning site and *Bam*HI being present on the core plasmid. Digestion of positive clones produces a single fragment as opposed to two fragments. Clones were visualized after electrophoresis on 2% agarose using GreenSafe Premium (MB13201, NZYTech, Lisbon, Portugal) on Chemidoc Touch (Biorad, CA, USA). Plasmids were reproduced by chemical transformation and culture of DH5alpha *E. coli* (inhouse stock) in LB medium (1% tryptone, 0.5% yeast extract, 1% NaCl, pH 7.0). Colonies were grown on 1.5% agar/LB (20767.232, VWR Chemicals, Radnor, PA, USA) with ampicillin selection (20767.232, VWR Chemicals, Radnor, PA, USA). Plasmid DNA was extracted and purified using a Plasmid DNA Mini Kit I (D6943-01, E.Z.N.A., Norcross, GA, USA). Plasmid clones were sequenced to verify correct cloned insert with custom sequencing primer 5′-CGAACTGGAGAGCATCCTG-3′ using StabVida SANGER sequencing services (https://www.stabvida.com/sanger-sequencing-service, accessed on 17 January 2020). The selected pmirGLO dual-luciferase plasmid clone containing the hsa-miR-193b-3p target site from *PPARGC1A* was transfected into HepG2 cells grown at 5 mM glucose using the DharmaFECT kb DNA transfection reagent lipofectamine (T-2006-01, Dharmacon, Lafayette, CO, USA). Expression of firefly luciferase was normalized to internal renilla luciferase using a Dual-Glo® Luciferase Assay System

(E2920, Promega, Madison, WI, USA), with chemiluminescent data obtained in relative light units (RLU) measured using SpectraMax i3x (Molecular Devices, San Jose, CA, USA).

Lipid droplet quantification. To identify intracellular lipid content, HepG2 cells were cultured at 5 mM glucose or 20 mM glucose on 13 mm glass coverslips in 24-well plates. At the end of the experiment, cells were fixed with 4% paraformaldehyde in phosphate-buffered saline (PBS) at room temperature for 15 min, then washed with PBS before staining with the Oil Red O method [44] using a kit (010303, Diapath, Italy) according to the manufacturer's instructions. Nuclei were stained blue with hematoxylin. Coverslips were mounted on glass slides for microscopy. Stained cells were analyzed on a Zeiss Z2 fluorescent microscope with ZEN Pro 2012 software (Zeiss, Oberkochen, Germany. For each condition, ten brightfield color images were captured at 40× magnification with an Axiocam 105 color camera. Image quantitative analysis of lipid droplets was performed on each of ten images per condition with Fiji (ImageJ) software [45] using Color Deconvolution (v1.7, FastRed FastBlue DAB plugin); red color was used for binary/watershed particle analysis to obtain parameters of total red area (pixels^2), average particle size (pixels) and number of particles; blue color was used for manual nuclei counting using Fiji software; number of nuclei per image was used for normalization of red data per image. Results per condition are the average of ten images.

Statistical Analysis. Statistical analyses were performed using Excel functions. Data are presented as mean ± standard error of the mean (SEM). Statistical significance was evaluated using two-tailed unpaired two-sample equal variance Student's *t*-test, with 0.05 threshold chosen for statistical significance; *p*-values indicated by # < 0.1, * < 0.05, ** < 0.01, *** < 0.001 and **** < 0.0001.

Author Contributions: Conceptualization, I.G.M. and M.P.M.; methodology, I.G.M. and M.P.M.; validation, I.G.M.; formal analysis, I.G.M.; investigation, I.G.M. and M.P.M.; resources, M.P.M.; data curation, I.G.M.; writing—original draft preparation, I.G.M.; writing—review and editing, I.G.M. and M.P.M.; visualization, I.G.M.; supervision, I.G.M. and M.P.M.; project administration, M.P.M.; funding acquisition, M.P.M. All authors have read and agreed to the published version of the manuscript.

Funding: This research was funded by [Fundação para a Ciência e Tecnologia" FCT] grant number [PTDC/BIM-MET/2115/2014, PTDC/MEC-MET/29314/2017 and UIDB/Multi/04462/2020], and [European Commission Marie Skłodowska-Curie Actions H2020] grant number [Grant No. 734719].

Institutional Review Board Statement: Not applicable.

Informed Consent Statement: Not applicable.

Data Availability Statement: All data are contained within the article. All raw data pertaining to the manuscript can be shared upon request to the corresponding author.

Acknowledgments: The authors acknowledge Sandra Tenreiro for sequencing support, and the NMS Microscopy and Histology facilities with support from Ana Farinho.

Conflicts of Interest: The authors declare no conflict of interest.

References

1. Zhang, J.; Li, S.; Li, L.; Li, M.; Guo, C.; Yao, J.; Mi, S. Exosome and Exosomal MicroRNA: Trafficking, Sorting, and Function. *Genom. Proteom. Bioinform.* **2015**, *13*, 17–24. [CrossRef]
2. Lai, E.C. Micro RNAs Are Complementary to 3′ UTR Sequence Motifs That Mediate Negative Post-Transcriptional Regulation. *Nat. Genet.* **2002**, *30*, 363–364. [CrossRef]
3. Agarwal, V.; Bell, G.W.; Nam, J.-W.; Bartel, D.P. Predicting Effective MicroRNA Target Sites in Mammalian MRNAs. *eLife* **2015**, *4*, e05005. [CrossRef]
4. Párrizas, M.; Novials, A. Circulating MicroRNAs as Biomarkers for Metabolic Disease. *Best Pract. Res. Clin. Endocrinol. Metab.* **2016**, *30*, 591–601. [CrossRef]
5. Mayans, L. Metabolic Syndrome: Insulin Resistance and Prediabetes. *FP Essent* **2015**, *435*, 11–16.
6. Eslam, M.; Sanyal, A.J.; George, J. International Consensus Panel MAFLD: A Consensus-Driven Proposed Nomenclature for Metabolic Associated Fatty Liver Disease. *Gastroenterology* **2020**, *158*, 1999–2014.e1. [CrossRef]
7. Kotronen, A.; Yki-Järvinen, H. Fatty Liver: A Novel Component of the Metabolic Syndrome. *Arterioscler. Thromb. Vasc. Biol.* **2008**, *28*, 27–38. [CrossRef] [PubMed]

8. Lonardo, A.; Ballestri, S.; Marchesini, G.; Angulo, P.; Loria, P. Nonalcoholic Fatty Liver Disease: A Precursor of the Metabolic Syndrome. *Dig. Liver Dis.* **2015**, *47*, 181–190. [CrossRef] [PubMed]
9. Párrizas, M.; Brugnara, L.; Esteban, Y.; González-Franquesa, A.; Canivell, S.; Murillo, S.; Gordillo-Bastidas, E.; Cussó, R.; Cadefau, J.A.; García-Roves, P.M.; et al. Circulating MiR-192 and MiR-193b Are Markers of Prediabetes and Are Modulated by an Exercise Intervention. *J. Clin. Endocrinol. Metab.* **2015**, *100*, E407–E415. [CrossRef]
10. Wu, H.; Deng, X.; Shi, Y.; Su, Y.; Wei, J.; Duan, H. PGC-1α, Glucose Metabolism and Type 2 Diabetes Mellitus. *J. Endocrinol.* **2016**, *229*, R99–R115. [CrossRef] [PubMed]
11. Cheng, C.-F.; Ku, H.-C.; Lin, H. PGC-1α as a Pivotal Factor in Lipid and Metabolic Regulation. *Int. J. Mol. Sci.* **2018**, *19*, 3447. [CrossRef] [PubMed]
12. Liang, H.; Ward, W.F. PGC-1α: A Key Regulator of Energy Metabolism. *Adv. Physiol. Educ.* **2006**, *30*, 145–151. [CrossRef] [PubMed]
13. Rius-Pérez, S.; Torres-Cuevas, I.; Millán, I.; Ortega, Á.L.; Pérez, S. PGC-1α, Inflammation, and Oxidative Stress: An Integrative View in Metabolism. *Oxid. Med. Cell. Longev.* **2020**, *2020*, 1452696. [CrossRef] [PubMed]
14. la Cour Poulsen, L.; Siersbæk, M.; Mandrup, S. PPARs: Fatty Acid Sensors Controlling Metabolism. *Semin. Cell Dev. Biol.* **2012**, *23*, 631–639. [CrossRef] [PubMed]
15. Charos, A.E.; Reed, B.D.; Raha, D.; Szekely, A.M.; Weissman, S.M.; Snyder, M. A Highly Integrated and Complex PPARGC1A Transcription Factor Binding Network in HepG2 Cells. *Genome Res.* **2012**, *22*, 1668–1679. [CrossRef] [PubMed]
16. Sanyal, A.J.; Campbell-Sargent, C.; Mirshahi, F.; Rizzo, W.B.; Contos, M.J.; Sterling, R.K.; Luketic, V.A.; Shiffman, M.L.; Clore, J.N. Nonalcoholic Steatohepatitis: Association of Insulin Resistance and Mitochondrial Abnormalities. *Gastroenterology* **2001**, *120*, 1183–1192. [CrossRef]
17. Huang, D.W.; Sherman, B.T.; Lempicki, R.A. Systematic and Integrative Analysis of Large Gene Lists Using DAVID Bioinformatics Resources. *Nat. Protoc.* **2009**, *4*, 44–57. [CrossRef]
18. Huang, D.W.; Sherman, B.T.; Lempicki, R.A. Bioinformatics Enrichment Tools: Paths toward the Comprehensive Functional Analysis of Large Gene Lists. *Nucleic Acids Res.* **2009**, *37*, 1–13. [CrossRef]
19. Camon, E.; Barrell, D.; Brooksbank, C.; Magrane, M.; Apweiler, R. The Gene Ontology Annotation (GOA) Project—Application of GO in SWISS-PROT, TrEMBL and InterPro. *Comp. Funct. Genom.* **2003**, *4*, 71–74. [CrossRef]
20. Aharoni-Simon, M.; Hann-Obercyger, M.; Pen, S.; Madar, Z.; Tirosh, O. Fatty Liver Is Associated with Impaired Activity of PPARγ-Coactivator 1α (PGC1α) and Mitochondrial Biogenesis in Mice. *Lab. Investig.* **2011**, *91*, 1018–1028. [CrossRef]
21. Khatun, I.; Walsh, M.T.; Hussain, M.M. Loss of Both Phospholipid and Triglyceride Transfer Activities of Microsomal Triglyceride Transfer Protein in Abetalipoproteinemia. *J. Lipid Res.* **2013**, *54*, 1541–1549. [CrossRef] [PubMed]
22. Burkhardt, R.; Toh, S.-A.; Lagor, W.R.; Birkeland, A.; Levin, M.; Li, X.; Robblee, M.; Fedorov, V.D.; Yamamoto, M.; Satoh, T.; et al. Trib1 Is a Lipid- and Myocardial Infarction-Associated Gene That Regulates Hepatic Lipogenesis and VLDL Production in Mice. *J. Clin. Investig.* **2010**, *120*, 4410–4414. [CrossRef]
23. Shin, E.; Bae, J.-S.; Han, J.-Y.; Lee, J.; Jeong, Y.-S.; Lee, H.-J.; Ahn, Y.-H.; Cha, J.-Y. Hepatic DGAT2 Gene Expression Is Regulated by the Synergistic Action of ChREBP and SP1 in HepG2 Cells. *Anim. Cells Syst.* **2016**, *20*, 7–14. [CrossRef]
24. Xu, X.; So, J.-S.; Park, J.-G.; Lee, A.-H. Transcriptional Control of Hepatic Lipid Metabolism by SREBP and ChREBP. *Semin. Liver Dis.* **2013**, *33*, 301–311. [CrossRef] [PubMed]
25. Wilson, C.G.; Tran, J.L.; Erion, D.M.; Vera, N.B.; Febbraio, M.; Weiss, E.J. Hepatocyte-Specific Disruption of CD36 Attenuates Fatty Liver and Improves Insulin Sensitivity in HFD-Fed Mice. *Endocrinology* **2016**, *157*, 570–585. [CrossRef] [PubMed]
26. Besse-Patin, A.; Jeromson, S.; Levesque-Damphousse, P.; Secco, B.; Laplante, M.; Estall, J.L. PGC1A Regulates the IRS1:IRS2 Ratio during Fasting to Influence Hepatic Metabolism Downstream of Insulin. *Proc. Natl. Acad. Sci. USA* **2019**, *116*, 4285–4290. [CrossRef]
27. Yoon, J.C.; Puigserver, P.; Chen, G.; Donovan, J.; Wu, Z.; Rhee, J.; Adelmant, G.; Stafford, J.; Kahn, C.R.; Granner, D.K.; et al. Control of Hepatic Gluconeogenesis through the Transcriptional Coactivator PGC-1. *Nature* **2001**, *413*, 131–138. [CrossRef]
28. Ogihara, T.; Isobe, T.; Ichimura, T.; Taoka, M.; Funaki, M.; Sakoda, H.; Onishi, Y.; Inukai, K.; Anai, M.; Fukushima, Y.; et al. 14-3-3 Protein Binds to Insulin Receptor Substrate-1, One of the Binding Sites of Which Is in the Phosphotyrosine Binding Domain. *J. Biol. Chem.* **1997**, *272*, 25267–25274. [CrossRef]
29. Du, K.; Herzig, S.; Kulkarni, R.N.; Montminy, M. TRB3: A Tribbles Homolog That Inhibits Akt/PKB Activation by Insulin in Liver. *Science* **2003**, *300*, 1574–1577. [CrossRef]
30. Lima, A.F.; Ropelle, E.R.; Pauli, J.R.; Cintra, D.E.; Frederico, M.J.S.; Pinho, R.A.; Velloso, L.A.; De Souza, C.T. Acute Exercise Reduces Insulin Resistance-Induced TRB3 Expression and Amelioration of the Hepatic Production of Glucose in the Liver of Diabetic Mice. *J. Cell. Physiol.* **2009**, *221*, 92–97. [CrossRef]
31. Thomou, T.; Mori, M.A.; Dreyfuss, J.M.; Konishi, M.; Sakaguchi, M.; Wolfrum, C.; Rao, T.N.; Winnay, J.N.; Garcia-Martin, R.; Grinspoon, S.K.; et al. Adipose-Derived Circulating MiRNAs Regulate Gene Expression in Other Tissues. *Nature* **2017**, *542*, 450–455. [CrossRef] [PubMed]
32. Sefried, S.; Häring, H.-U.; Weigert, C.; Eckstein, S.S. Suitability of Hepatocyte Cell Lines HepG2, AML12 and THLE-2 for Investigation of Insulin Signalling and Hepatokine Gene Expression. *Open Biol.* **2018**, *8*, 180147. [CrossRef] [PubMed]

33. Pina, A.F.; Patarrão, R.S.; Ribeiro, R.T.; Penha-Gonçalves, C.; Raposo, J.F.; Gardete-Correia, L.; Duarte, R.; MBoavida, J.; LMedina, J.; Henriques, R.; et al. Metabolic Footprint, towards Understanding Type 2 Diabetes beyond Glycemia. *J. Clin. Med.* **2020**, *9*, 2588. [CrossRef] [PubMed]
34. Nagiec, M.M.; Skepner, A.P.; Negri, J.; Eichhorn, M.; Kuperwasser, N.; Comer, E.; Muncipinto, G.; Subramanian, A.; Clish, C.; Musunuru, K.; et al. Modulators of Hepatic Lipoprotein Metabolism Identified in a Search for Small-Molecule Inducers of Tribbles Pseudokinase 1 Expression. *PLoS ONE* **2015**, *10*, e0120295. [CrossRef]
35. Wong, R.H.F.; Sul, H.S. Insulin Signaling in Fatty Acid and Fat Synthesis: A Transcriptional Perspective. *Curr. Opin. Pharm.* **2010**, *10*, 684–691. [CrossRef]
36. Postic, C.; Burcelin, R.; Rencurel, F.; Pegorier, J.P.; Loizeau, M.; Girard, J.; Leturque, A. Evidence for a Transient Inhibitory Effect of Insulin on GLUT2 Expression in the Liver: Studies in Vivo and in Vitro. *Biochem. J.* **1993**, *293 Pt 1*, 119–124. [CrossRef]
37. Kim, J.-A.; Wei, Y.; Sowers, J.R. Role of Mitochondrial Dysfunction in Insulin Resistance. *Circ. Res.* **2008**, *102*, 401–414 [CrossRef]
38. Reddy, J.K.; Rao, M.S. Lipid Metabolism and Liver Inflammation. II. Fatty Liver Disease and Fatty Acid Oxidation. *Am. J. Physiol. Gastrointest. Liver Physiol.* **2006**, *290*, G852–G858. [CrossRef]
39. Miyazaki, Y.; Mahankali, A.; Matsuda, M.; Mahankali, S.; Hardies, J.; Cusi, K.; Mandarino, L.J.; DeFronzo, R.A. Effect of Pioglitazone on Abdominal Fat Distribution and Insulin Sensitivity in Type 2 Diabetic Patients. *J. Clin. Endocrinol. Metab.* **2002**, *87*, 2784–2791. [CrossRef]
40. Martins, F.O.; Delgado, T.C.; Viegas, J.; Gaspar, J.M.; Scott, D.K.; O'Doherty, R.M.; Paula Macedo, M.; Jones, J.G. Mechanisms by Which the Thiazolidinedione Troglitazone Protects against Sucrose-Induced Hepatic Fat Accumulation and Hyperinsulinaemia: Thiazolidinediones Enhance Insulin Clearance. *Br. J. Pharmacol.* **2016**, *173*, 267–278. [CrossRef]
41. Musso, G.; Cassader, M.; Paschetta, E.; Gambino, R. Thiazolidinediones and Advanced Liver Fibrosis in Nonalcoholic Steatohepatitis: A Meta-Analysis. *JAMA Intern. Med.* **2017**, *177*, 633. [CrossRef] [PubMed]
42. Cho, H.J.; Hwang, S.; Park, J.I.; Yang, M.J.; Hwang, J.C.; Yoo, B.M.; Lee, K.M.; Shin, S.J.; Lee, K.J.; Kim, J.H.; et al. Improvement of Nonalcoholic Fatty Liver Disease Reduces the Risk of Type 2 Diabetes Mellitus. *Gut Liver* **2019**, *13*, 440–449. [CrossRef] [PubMed]
43. Livak, K.J.; Schmittgen, T.D. Analysis of Relative Gene Expression Data Using Real-Time Quantitative PCR and the 2−ΔΔCT Method. *Methods* **2001**, *25*, 402–408. [CrossRef] [PubMed]
44. McVean, D.E.; Patrick, R.L.; Witchett, C.E. An Aqueous Oil Red O Fixative Stain for Histological Preparations. *Am. J. Clin. Pathol.* **1965**, *43*, 291–293. [CrossRef] [PubMed]
45. Schindelin, J.; Arganda-Carreras, I.; Frise, E.; Kaynig, V.; Longair, M.; Pietzsch, T.; Preibisch, S.; Rueden, C.; Saalfeld, S.; Schmid, B.; et al. Fiji: An Open-Source Platform for Biological-Image Analysis. *Nat. Methods* **2012**, *9*, 676–682. [CrossRef] [PubMed]

International Journal of
Molecular Sciences

MDPI

Review

Statins in Children, an Update

Riccardo Fiorentino and Francesco Chiarelli *

Department of Paediatrics, University of Chieti, Via dei Vestini, 66100 Chieti, Italy
* Correspondence: chiarelli@unich.it

Abstract: Since lipid abnormalities tend to progress from childhood to adulthood, it is necessary to early identify and treat children and adolescents with dyslipidemia. This is important in order to reduce the cardiovascular risk, delay the development of fatty streaks, slow the progression of atherosclerosis and reverse atherosclerotic plaques. Together with therapeutic lifestyle changes, statins are the most common lipid-lowering drugs. By inhibiting the endogenous cholesterol synthesis in the liver, statins increase the catabolism of LDL-C, reduce VLDL-C, IDL-C and TG and modestly increase HDL-C. Regardless of their lipid-lowering effect, statins have also pleiotropic effects. Statins have increasingly been prescribed in children and adolescents and mounting evidence suggests their beneficial role. As with adults, in children, several studies have demonstrated that statin therapy is efficient at lowering lipid levels and reducing CIMT progression and cumulative estimated atherosclerotic burden in children. Statins are generally very well-tolerated in both adults and children and adverse events are quite uncommon. When evaluating the need and the timing for statin treatment, the presence of several factors (secondary causes, familial history, additional risk factors) should also be considered. Before initiating statins, it is imperative for clinical practitioners to consult patients and families and, as with any new medication therapy, to monitor patients taking statins. Despite being safe and effective, many children with lipid disorders are not on statin therapy and are not receiving the full potential benefit of adequate lipid-lowering therapies. It is therefore important that clinicians become familiar with statins.

Keywords: dyslipidemia; statin; children; adolescents

Citation: Fiorentino, R.; Chiarelli, F. Statins in Children, an Update. *Int. J. Mol. Sci.* **2023**, *24*, 1366. https://doi.org/10.3390/ijms24021366

Academic Editors: Simona Gabriela Bungau and Vesa Cosmin

Received: 4 December 2022
Revised: 7 January 2023
Accepted: 9 January 2023
Published: 10 January 2023

1. Introduction

Atherosclerosis is a chronic inflammatory condition, characterized by the build-up of plaques inside the arteries and the progressive thickening of the arterial wall [1]. Despite a long initial asymptomatic phase, it is known that atherosclerosis is a potentially serious disease. The progression of the insidious atherosclerotic process may lead to cerebrovascular disease, coronary events and other atherosclerotic cardiovascular diseases (ASCVD), which are currently the leading cause of morbidity, disability and premature death worldwide [2,3]. The atherosclerotic process is mainly driven by lipids and initiates with the deposition of lipids on the inner lining of arteries. Lipid accumulation over time leads to an endothelial dysfunction and activates pro-inflammatory processes, which induce the formation of fatty streaks (the earliest atherosclerotic lesions) and their gradual evolution into atherosclerotic plaques [4,5]. Although clinical outcomes of atherosclerosis are usually observed in adults, several studies have confirmed that the pathological process begins much earlier [6]. Landmark studies showed that fatty streaks were found in the coronary arteries of children as young as 2 years of age [7]. More recently, it has been described that children with traditional cardiovascular risk factors had increased carotid intima–media thickness (CIMT), which is a surrogate biomarker of atherosclerosis [8]. In addition, recent studies have show that coronary intimal thickening may begin in utero [9,10]. Therefore, these results suggest the precocious origins of adult diseases and allows us to consider atherosclerosis as a significant problem in children. Several cardiovascular risk factors and conditions contribute to atherosclerosis in children: among them, childhood dyslipidemia plays a key role [11]. Indeed, it is well known that lipids are essential for cellular organization, stability and functionality as well as for steroid hormones, vitamins and bile acids synthesis [12]. However, lipid and lipoprotein abnormalities can be detrimental to the human body and are linked to the initiation and progression of atherosclerosis in children [13]. In particular, higher levels and longer exposures to increased low-density lipoprotein (LDL-C) are the main factors that exacerbate the atherosclerotic progression [14]. Since lipid abnormalities tend to progress from childhood to adulthood, it is necessary to early identify and treat children and adolescents with dyslipidemia. This is important in order to reduce the cardiovascular risk, delay the development of fatty streaks, slow the progression of atherosclerosis and reverse atherosclerotic plaques, whenever possible. Together with therapeutic lifestyle changes, statins are the most common lipid-lowering drugs and represent the cornerstone of primary and secondary prevention of cardiovascular diseases in adults [15,16]. Statins were increasingly prescribed in children and adolescents and mounting evidence suggested their beneficial role in children and adolescents. This narrative review aims to highlight the efficacy, safety and rationale for the use of statins in pediatrics. Moreover, we summarize the indications for clinical use of statin therapy, focusing on initiation and monitoring of treatment in children and adolescents.

2. Cholesterol Metabolism and Mechanism of Action for Statins

All cholesterol in the human body comes from two major sources: it can be either obtained through diet (20% of total cholesterol) or synthetized de novo (80% of total cholesterol). Although all cells are capable of synthesizing cholesterol, the major cholesterol production occurs in the liver [17]. Cholesterol synthesis is a complex process; several enzymes are involved, and the biosynthetic pathway is tightly regulated at several points [18]. Cholesterol production starts with the formation of mevalonate from acetate; subsequently, mevalonate is transformed to squalene (the biochemical precursor of all steroids) in the endoplasmic reticulum. Squalene is further converted to lanosterol and finally transformed into cholesterol [12,18]. Newly synthesized cholesterol then must leave the endoplasmic reticulum to exert its functions: through different mechanisms, it moves to the membranes and is transported to peripheral tissues. However, due to its insolubility in plasma, cholesterol is required to be packaged with lipoproteins: lipoproteins are spherical macromolecules, consisting of a hydrophobic core (containing cholesterol esters and triglycerides) and a hydrophilic coat (containing free cholesterol, phospholipids and apolipoproteins) [19]. There are several lipoproteins and different lipoprotein pathways. Via the exogenous pathway, dietary lipids are incorporated in chylomicrons, transported from the intestine to peripheral tissues and then taken up by the liver; via the endogenous pathway, lipids are transported from the liver to peripheral tissues [20]. The endogenous pathway begins in the liver with the production of very-low-density lipoproteins (VLDL-C), which contain cholesterol and a large amount of triglycerides (TG). In the circulation, the triglycerides carried in VLDL-C are hydrolyzed by lipoprotein lipase (LPL), resulting in the formation of intermediate-density lipoproteins (IDL-C) and the further conversion of IDL-C to low-density lipoprotein (LDL-C). LDL-C (the major cholesterol-rich lipoproteins) then reach peripheral cells and are cleared by circulation through the interaction with LDL-C receptor, which is present in many tissues and most abundant in the liver [19,20]. Lastly, through the reverse transport pathway (reverse cholesterol transport) the organism removes the excess fats from peripheral cells and transports them to the liver: the process is mediated by high-density lipoproteins (HDL-C), which exhibit various anti-atherogenic and cardio-protective effects [21]. Statins, also known as hydroxymethylglutaryl-CoA (HMG-CoA) reductase inhibitors, are a class of lipid-lowering drugs that work by inhibiting the endogenous cholesterol synthesis in the liver. In particular, statins are potent competitive inhibitors of HMG-CoA reductase [22]. HMG-CoA reductase is involved in the final step of the mevalonate pathway and catalyzes the reaction where HMG-CoA is reduced to mevalonate (NADP-dependent synthesis of mevalonate). This enzymatic reaction is not only an irreversible and tightly regulated step, but also the rate-limiting one of cholesterol biosynthesis [23]. On the basis of their structure, there are two different types of statins: type 1 statins (pravastatin, simvastatin and lovastatin) are products of natural origin, whereas type 2 statins (atorvastatin, fluvastatin, pitavastatin, rosuvastatin) are synthetic products that are characterized by attached fluorophenyl groups and larger hydrophobic regions [18]. The capacity of statins to inhibit HMG-CoA reductase is due to the presence of an HMG-like moiety that binds to the HMG binding site. Moreover, the hydrophobic region of type 2 statins contributes to block the binding of HMG-CoA to the active site [18,24]. By competitively blocking the HMG-CoA reductase, statins alter the endogenous pathway and reduce the intrahepatic cholesterol amount: consequently, this promotes the hepatic upregulation of the LDL-C receptor (increased synthesis and expression) and increases the catabolism of circulating LDL-C. In addition, statins reduce VLDL-C, IDL-C and TG and modestly increase HDL-C (improving the reverse transport pathway) [25]. Lastly, regardless of their lipid-lowering effect, statins have pleiotropic effects on endothelial function, inflammation, coagulation and oxidative stress (Figure 1) [26,27].

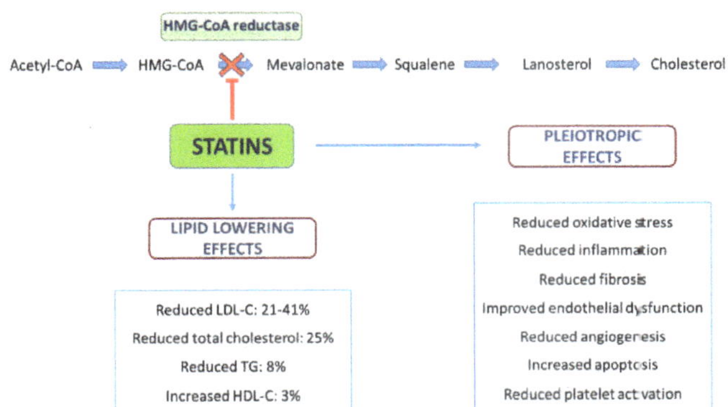

Figure 1. Mechanisms and effects of statins.

3. Efficacy

It is well known that statins are effective in adults: in addition to their variable lipid-lowering effect, statins are beneficial in improving the lipid profile, conferring cardiovascular protection (in terms of primary and secondary prevention) and reducing cardiovascular morbidity and mortality [22,23]. Over time, statins have increasingly been used in children and adolescents and several studies and meta-analyses have evaluated the efficacy of statins with children and adolescents [28–30]. Although most studies have been conducted on patients with pediatric familial hypercholesterolemia (FH), they highlighted the beneficial effect of different statins in youths with dyslipidemia [31,32]. As with adults, early studies demonstrated that statin therapy is efficient at lowering lipid levels in children: despite a variable efficacy through the trials, all statins led to a significant reduction in LDL-C levels when compared to placebo [14,33,34]. The differences in the mean lipid-lowering effect were probably due to different baseline LDL-C values, statin dosages and formulations [29]. In fact, it is important to remember that the lipid-lowering effect of statins changes when the dose of HMG-CoA reductase inhibitors increases (dose-dependent effect) and when a high-potency class of statins is used. Because of the longest terminal half-life, rosuvastatin, pitavastatin and atorvastatin are the most potent statins; on the other hand, lovastatin, pravastatin and fluvastatin are the least potent; simvastatin is a moderately potent statin [34]. In accordance with recent studies, HMG-CoA reductase inhibitors in children led to a relative reduction in LDL-C values by approximately 21–41% (mean relative decrease of 32%) [28,35]; moreover, thanks to statin treatment, the majority of children might achieve the desired LDL-C goals [30,36]. In addition, a recent meta-analysis demonstrated that statins were helpful in reducing total cholesterol and TG (by 25% and

8% respectively) and increasing HDL-C (by 3%), when compared with placebo [30]. As mentioned above, the use of statins in adults significantly reduces cardiovascular morbidity and this also applies to patients with FH; in adults with FH, statins may significantly reduce the ASCVD rates and delay the median age of onset of cardiovascular events in comparison with the pre-statin era [14,37]. In children, the results seem very similar: by performing a 20-year follow-up study, a recent study aimed to compare the incidence of cardiovascular events and death in patients with FH who initiated statins in childhood and in their affected parents who initiated statins much later in life. Very interestingly, after 20 years, the initiation of statin therapy at an age of 8–18 years resulted in the significant reduction of ASCVD rates and death from cardiovascular events. Moreover, the cumulative cardiovascular disease-free survival was significantly higher among patients who received statins [35]. Although these findings are particularly surprising, it is important to note that the study is only observational and other studies are needed to confirm the results. Nowadays, however, randomized control trials designed to measure the effects of statins during childhood on cardiovascular outcomes are lacking; since cardiovascular events rarely occur in children, there are probably many challenges associated with the conception of event-driven trials. Therefore, several studies have tried to evaluate the effects of HMG-CoA reductase inhibitors on surrogate markers of atherosclerosis. In particular, the CIMT represented the most evaluated noninvasive measure of early atherosclerosis [38]. In a landmark trial (2004), children with FH were randomized to pravastatin or placebo: over a two-year treatment period, the CIMT was significantly greater in those receiving a placebo. Moreover, trends toward CIMT progression and CIMT regression were found in the placebo and the statin-treated group, respectively [39]. The placebo-controlled phase was then followed by an open-label extension: some authors demonstrated that the earlier initiation of pravastatin was associated with a lower CIMT following approximately 4 years of treatment [40]. Most recently, a 10-year follow-up study found that, before starting the treatment, children with FH showed a greater CIMT when compared with non-FH controls; however, when statin therapy was initiated, important improvements in CIMT (reduced CIMT, slowed progression) were observed: very interestingly, the CIMT progression rates were similar between statin-treated children and their unaffected siblings [41]. These findings were then confirmed in other studies, 20 years after the initiation of statin therapy [35]. In view of these observations, the efficacy of statins in pediatrics seems clear; children respond well to HMG-CoA reductase inhibitors and reduce their cumulative estimated atherosclerotic burden: when initiated early (from age 10 years), statins result in delaying the burden from age 35 to 53 years [42,43].

4. Safety

Statins are generally very well-tolerated in both adults and children and adverse events are quite uncommon [44]. Although adverse events may be possible, they are often nonserious reversible events, which do not require adjustment in therapy. Moreover, in children receiving statins, the majority of adverse events are as frequent as in those receiving a placebo and they are only drug-related in few cases (e.g., elevated creatine phosphokinase levels after physical activity) [14,34]. Numerous randomized control trials, Cochrane reviews and subsequent meta-analyses consistently established the good short-, medium- and long-term safety of statins in children, especially in those at greater cardiovascular risk (e.g., FH) [28–30,41,44–46]. Currently, the longest study evaluating the safety profile was performed in children with FH in a 20-year follow-up [35]. Adverse effects primarily include liver toxicity, muscle-related adverse events and abnormalities in laboratory measurements [47]; in addition, statins may have teratogenic effects and may interfere with other drugs.

Liver toxicity: The risk of liver toxicity appears to be particularly low. As with adults, the risk of serious hepatotoxicity is extremely low and no pediatric cases were reported [33,44]. On the other hand, transient elevations in liver enzymes (alanine aminotransferase and aspartate aminotransferase) may occur more frequently (up to 5% of statin-treated children); however, while increased, transaminases did not exceed three times the upper limit of normal or normalized without discontinuation of statin treatment [34]. Interestingly, recent studies did not note any differences in the risk of liver enzymes abnormalities between treatment and placebo groups [41,44].

Muscle-related adverse events: In children receiving statins, the risk of muscle-related adverse events is also particularly low. Myopathy is very uncommon (<0.1%) and no cases of pediatric rhabdomyolysis have been attributed to statins to date [28,44]. Myalgia appears to occur more frequently, although the real extent of the problem remains controversial [14]. In blinded randomized control trials, myalgia was often not associated with elevations in muscular enzymes (creatine phosphokinase); moreover, the frequency of myalgia was lower than that reported in clinical practice, suggesting a "nocebo effect" [48,49]. As with regard to creatine phosphokinase levels, no significant changes (10 times above the upper limit of normal) were found when starting statin therapy and no differences were noted between statin-treated children and a placebo group [28,50]. Although the fear of muscle-related adverse events is a common reason for delaying or not initiating statin treatment, it is important to know that patients often remain asymptomatic and, if muscular symptoms occur, they usually resolve spontaneously without therapy discontinuation [34,50].

Growth, development and cognition: Over time, several concerns have emerged regarding the risk of interfering with the cholesterol synthesis pathway. As mentioned above, in children, cholesterol is essential for cellular functionality, steroid hormones and vitamin synthesis and for adequate growth and development (especially for normal brain development) [51]. Several studies have been published on this issue and most of them agree that statins have no effect on growth and development. When compared to a placebo, there was no convincing evidence that statins may alter growth velocity, cognitive function and educational level, sexual maturation (assessed by Tanner staging and age at menarche), hormone concentrations (e.g., estradiol, testosterone), menstrual cycle length or erectile function [14,28,34,41]. Concerns of statins affecting growth and development were also proven unfounded when statins were used in homozygous FH children as early as 2 years of age [25].

Diabetes mellitus: The risk of developing type 2 diabetes mellitus in children receiving statin treatment is not well clarified, but seems to be lower than in adults (0.2% per year approximately). It is likely that adults have a higher risk of new type 2 diabetes mellitus because of older age, a more aggressive therapy and more frequent comorbidity and concomitant drug use [44]. Interestingly, in a 20-year follow-up study, 1 of the 184 children receiving statins (0.5%) and 2 of the 77 unaffected siblings (2.6%) developed diabetes [35].

Other side-effects: As with adults, mild adverse effects may include hypersensitivity reactions (rash, urticarial), gastrointestinal (dyspepsia, abdominal pain, constipation, diarrhea) and neurological symptoms (headache, dizziness, asthenia). Nevertheless, there is no support of a relationship between statin use and peripheral neuropathy [52]. Almost all of these symptoms resolve with continued use of the statin [14].

Teratogenicity: In accordance with the most recent guidelines for childhood dyslipidemia, statins should be avoided during the preconception period, pregnancy and lactation due to their possible teratogenic effects [53]. However, it is important to note that a recent study systematically reviewed the existing studies on this topic, including almost 1.3 million participants. Very interestingly, the authors concluded that statins were not associated with increased birth defects and that there was no clear evidence that statins were teratogenic [54]. Further studies are therefore needed to investigate this issue.

Drug interactions: As a substrate of cytochrome P450, statins are associated with multiple drug interactions. Competing substrates are able to increase the systemic plasma concentrations of statins and their lipid-lowering effects, as well as the frequency of adverse events [55]. Because of their extensive first-pass hepatic metabolism, drug interaction effects are highest for simvastatin and lovastatin [14]; on the other hand, not being metabolized by a cytochrome P450 isoenzyme, pravastatin is the preferred choice for patients at greater risk of drug interactions [44]. Competing substrates that should be avoided include antifungal azoles, macrolides, cyclosporine, protease inhibitors and grapefruit; instead, dose adjustments are often needed for amiodarone and calcium channel-blockers. Moreover, when receiving statins, patients should also avoid natural remedies containing varying amounts of natural monacolin K (e.g., red yeast rice) [56].

5. Indications and Recommendations

Seven commercially available statins are currently available for children and adolescents, being indicated at varying ages and dosages [32]. Table 1 provides a summary of the approved statins in children and adolescents (Table 1).

Table 1. Commercially available statins for children and adolescents.

Drug	Potency	Pediatric FDA Approvals	Age	Dose	Mean LDL-C % Reduction	Supporting Studies
Atorvastatin	High-potency	Heterozygous FH	10–17 years	10–20 mg/day	40%	McCrindle et al. [57]
Fluvastatin	Low-potency	Heterozygous FH	10–16 years	20–80 mg/day	34%	Van der Graaf et al. [58]
Lovastatin	Low-potency	Heterozygous FH	10–17 years	10–40 mg/day	17–37%	Lambert et al. [59] Stein et al. [60] Clauss et al. [61]
Pitavastatin	High-potency	Heterozygous FH	≥8 years	1–4 mg/day	23–39%	Braamskamp M.J. et al. [62]
Pravastatin	Low-potency	Heterozygous FH	8–18 years	20 mg (8–13 years) 40 mg (14–18 years)	23–33%	Knipscheer et al. [63]
Rosuvastatin	High-potency	Heterozygous FH	8–17 years	5–20 mg/day	38–50%	Avis et al. [64]
		Homozygous FH	≥7 years			
Simvastatin	Moderate-potency	Heterozygous FH	10–17 years	10–40 mg/day	31–41%	Couture et al. [65] de Jongh et al. [66]

Rosuvastatin was recently approved for use in children from 7 years old (initially it was authorized for use in children from 10 years old); pitavastatin and pravastatin are indicated in children as young as 8 years old; atorvastatin, fluvastatin, lovastat in and simvastatin may be used in children ≥ 10 years old [31,67]. At present, statins are approved by the Food and Drug Administration (FDA) for children with FH and represent the first-line lipid-lowering drug (statins supplanted bile acid sequestrants) [68]. Patients with FH, due to the underlying genetic disorder, have persistent LDL-C elevations and are more prone to accelerated atherosclerosis and premature ASCVD (high-risk condition) [69]. It is therefore important to early recognize and effectively manage these patients, despite being asymptomatic [6]. Several guidelines and statements provided recommendations for statin treatment in pediatric populations [70–72]. According to The National Heart, Lung, and Blood Institute (NHLBI) guidelines and American Heart Association (AHA) Scientific Statement, the average of at least two separate fasting lipid profiles (obtained at more than two weeks, but no more than three months apart) is used to determine which patients are potential candidates for statin treatment [33,69]. Moreover, when evaluating the need and the timing for drug treatment, the presence of secondary causes, familial history of premature ASCVD, additional high-risk factors and risk-modifying conditions should also be considered (Figure 2).

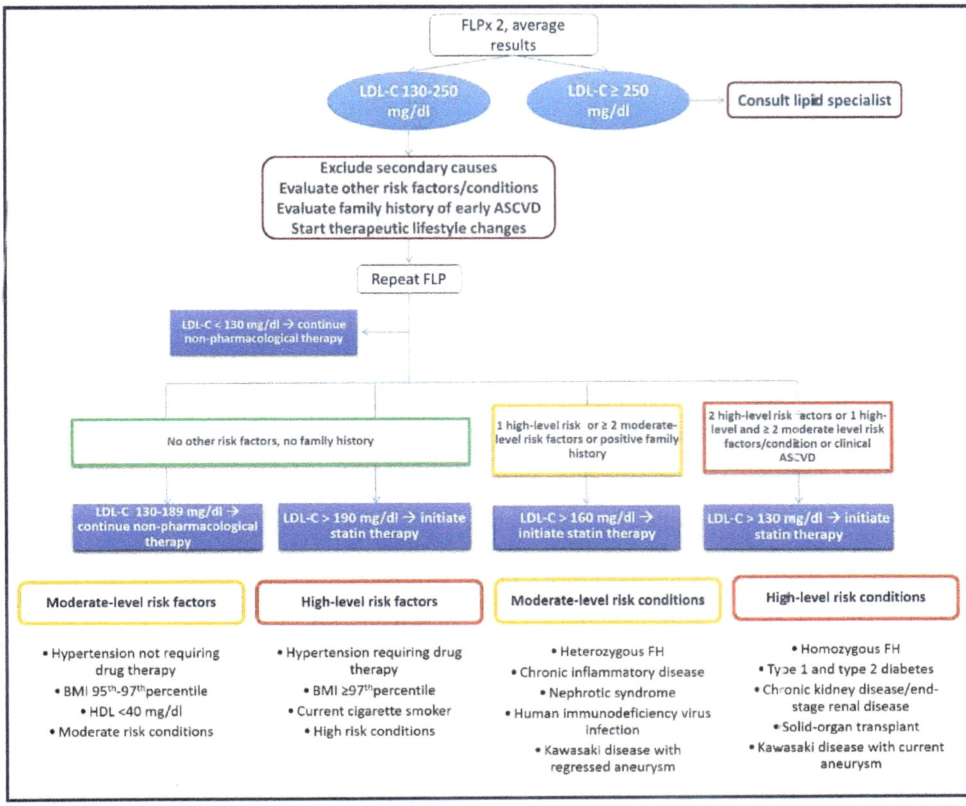

Figure 2. Recommendations for statin treatment in pediatric populations.

In children with heterozygous FH, guidelines recommend starting statin therapy from the age of 8–10 years: statins should be initiated after secondary causes of dyslipidemia have been excluded and in such cases of persistently elevated LDL-C concentrations despite 3–6 months of lifestyle modifications [63,73,74]. In heterozygous FH children with more severe forms of LDL-C abnormalities, statin treatment may also be started concurrently with therapeutic lifestyle changes [25,31]. On the other hand, in children with homozygous FH, the lipid-lowering therapy should be initiated at diagnosis and often in infancy, although the FDA granted approval for statins from the age of 7 years [63]. Given the most severe lipid abnormalities and the notably increased cardiovascular risk, these patients necessitate an earlier treatment in order to change the natural history of the disease, normalize the atherosclerotic burden in adulthood, and to improve the prognosis [14]. In fact, it is important to remember that the severity of atherosclerosis depends on the lifetime exposure to elevated LDL-C concentrations [1]. In homozygous FH, statins represent the pillar of the treatment despite not being sufficient alone for achieving the LDL-C goals (only 20% of children reach the target) [35]. As with regard to the therapeutic targets, it is noteworthy that there are no evidence-based LDL-C goals for pediatrics: however, when statin therapy is initiated, commonly used targets are:

- LDL-C values < 130 mg/dL in moderate-risk patients (heterozygous FH)
- LDL-C values < 100 mg/dL or a 50% reduction in LDL-C from baseline values in high-risk patients (homozygous FH) [75,76].

6. Prescribing Statin Use in Youth

When statin therapy should be initiated, the decision of a particular statin is mainly a matter of healthcare provider preference, but the choice should also take into account the child's age, baseline LDL-C values and LDL-C goals [31]. Moreover, the family and patient's preferences should also be considered, especially if a family member is already taking a particular statin formulation [14]. Statins should be administered once a day, at the lowest available dose: no dose adjustments based on body weight are required [33]. Because most cholesterol synthesis occurs overnight, HMG-CoA reductase inhibitors are typically prescribed with the evening meal or at night [44]. However, this is not always necessary: statins may be taken with or without meals and some of them (e.g., rosuvastatin, atorvastatin), having a long terminal half-life (approximately 19 and 14 h respectively), may be taken at any time [77]. Before initiating statins, it is imperative for clinical practitioners to consult patients and families in order to help to address their hesitancies and concerns regarding the benefits, risks and impacts (e.g., impact on quality of life) of statin therapy. Clinicians should explain the potential need for a lifetime drug therapy and the importance of adopting healthy habits as a synergistic lipid-lowering mechanism [25,78]. Given the possible teratogenicity of statins, all sexually active females must be counselled on having an appropriate contraception [79]; moreover, in view of their pharmacokinetics, a detailed review of potential drug interactions should always be addressed when prescribing HMG-CoA reductase inhibitors [80]. Considering that there are over 30 million patients with FH worldwide (20–25% of whom are younger than 19 years old) and most of them are not diagnosed and/or treated, it is important to promote and increase the knowledge on lipid disorders and their treatment [81,82]. A recent European study, comprising approximately 3.000 heterozygous FH children, showed surprising results: first, there were large between-country differences in the mean age of diagnosis of FH (e.g., 3 years in Greece, 11 years in Belgium). In addition, a significant proportion of the children who were potential candidates for lipid-lowering treatment were not on statin therapy in Europe; in fact, across the European countries the proportion was again very different. This was probably associated with the very different diagnostic strategies and policies used in the different countries [83]. In view of these observations, it is important that clinicians know the current recommendations and be familiar with the characteristics, the dosage and the formulations of at least one of the statins [1].

7. Surveillance

As with any new medication therapy, it is important for clinicians to monitor patients taking statins [84]. In particular, the efficacy of treatment, the adherence to therapy and the onset of adverse effects should all be carefully monitored. The most up-to-date guidelines suggest repeating a fasting lipid-profile 4–8 weeks after statin initiation; if adequate LDL-C reduction is observed, a fasting lipid profile should be repeated every 3–6 months (in the first year) and longitudinally every 6–12 months (Figure 3) [33]. If sufficient LDL-C reductions are not achieved, the adherence to therapeutic lifestyle changes and to statin treatment must be assessed [85]. While the adoption of healthy habits may be challenging, the adherence to statin therapy is usually good: according to some studies, 84% of patients took 80% or more of the prescribed medications [35]. When drug compliance is adequate and the tolerance is good, it is recommended to increase the statin dose by one increment (doubling the dose): no dose adjustments based on body weight are needed [14]. However, it is important to known that doubling the statin dose has only modest effects on LDL-C concentrations (resulting in further reduction of approximately 6%) [25]. Moreover, the increased dose may result in more adverse events, although the maximal dose of potent statins for children is between 25% and 50% of the one approved for adults [32]. Only when the combination of maximally dosed statins and lifestyle changes results in an inadequate LDL-C reduction, the addition of a second lipid-lowering therapy (e.g., ezetimibe) should be considered, in accordance with a lipid specialist [69]. It is important to check a fasting lipid profile at each dose and therapy modification or when it is clinically indicated. Current guidelines do recommend assessing transaminases and creatine phosphokinase levels at baseline and to constantly monitor the onset of new symptoms. After initiating statin therapy, it is also suggested to monitor liver enzymes and creatine phosphokinase at 1–2 months and to assess liver enzymes at 3–6 months and periodically thereafter: routine monitoring of creatine phosphokinase is not required because it may result in incidental increases of muscular enzymes that may be related to physical activity [14,33]. Nevertheless, if myopathy symptoms occur, creatine phosphokinase should be measured immediately. When slight increases in liver enzymes or creatine phosphokinase occurs, the discontinuation of statin treatment is not necessary; however, when transaminases and creatine phosphokinase levels increase over 3-fold and over 10-fold, respectively, a prompt discontinuation of statins and the exclusion of other causes of muscular and liver dysfunction is suggested [44]. When children are on statin therapy, it is also advised to monitor growth, development and sexual maturation and it may be reasonable to periodically monitor fasting glucose concentrations (especially in patients at risk of diabetes) [85]. As mentioned above, adolescent females should be counselled on the potential teratogenicity of statins: they should be informed about the importance of contraception and discontinuing statin therapy prior to planned conception (at least 3 months before), pregnancy and lactation [63]. At each visit, it is also recommended to review the potential drug interactions of statins. HMG-CoA reductase inhibitors are typically safe and statin intolerance is uncommon in children; even though side effects occur, it is not contraindicated to reinitiate the previous statin therapy (once laboratory abnormalities and symptoms are resolved). Alternatively, if LDL-C targets were achieved, it is also possible to administer a lower dose of the same statin, the same dose on alternate days or another statin formulation (with different pharmacokinetics) before using a non-statin therapy [14,44].

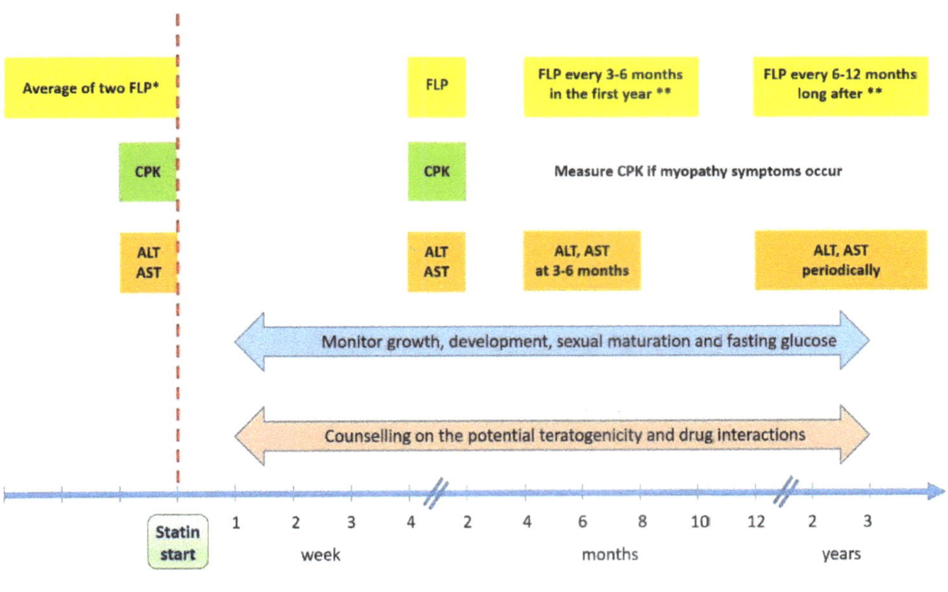

* Obtain fasting lipid profiles (FLP) at least 2 weeks but no more than 3 months apart.
**If Target LDL-C achieved
Creatine phosphokinase (CPK); Alanine aminotransferase (ALT); Aspartate aminotransferase (AST)

Figure 3. Surveillance scheme in children and adolescents taking statins.

8. Limitations

As mentioned above, due to their possible teratogenic effects, statins should be avoided during the preconception period and pregnancy; it is recommended to cease statin therapy for at least three months prior to becoming pregnant or as soon as pregnancy is recognized. Female patients should also avoid statins during lactation; if they require statin therapy, they should receive direction to immediately stop breastfeeding [14]. Statin contraindications also include children with hypersensitivity to any of its components and patients with active liver disease [44]. Statins should be prescribed with caution in patients with concurrent administration of interfering drugs, predisposing factors for myopathy and patients with chronic kidney disease. In the latter case, dose modification is required and only atorvastatin and simvastatin can be used (being not metabolized in the kidneys) [55].

9. Conclusions

Mounting evidence suggests that cardiovascular risk factors, such as dyslipidemia, may be present early on in life. In view of this, and considering that children with lipid abnormalities are more likely to become adults with dyslipidemia, it is extremely important to identify and manage children with lipid abnormalities early (especially those with more severe forms of dyslipidemia). However, in order to achieve these goals, it is important to widely promote the knowledge in this field. The treatment of childhood dyslipidemia has undergone significant changes in recent years: statins are now considered the first-line pharmacologic therapy and the cornerstone of FH treatment during childhood. Statins have increasingly been used in children and adolescents, including those at risk for premature ASCVD (e.g., FH): as in adults, statins effectively improve the lipid profile (mainly by lowering LDL-C), slow the progression of the atherosclerotic process and reduce atherosclerotic burden in adulthood. Current data suggest that the use of statins during childhood may potentially normalize the future cardiovascular risk despite not reaching LDL-C goals. As a consequence, these findings highlight the potential importance of the pleiotropic effects of statins and the utility of precocious treatment when atherosclerosis is reversible. HMG-CoA reductase inhibitors are safe and well-tolerated; as opposed to adults, less aggressive therapy is initiated and less frequent comorbidities are present. Therefore, the risk of adverse events is lower in children than adults. Despite being safe and effective, many children with lipid disorders are not on statin therapy and are not receiving the full potential benefit of adequate lipid-lowering therapies. It is therefore important that clinicians become familiar with statins: this also applies to primary care providers, who are often the first to diagnose childhood dyslipidemia. Lastly, given that most studies were in pediatric FH, more studies are needed to evaluate statin use in children with other risk conditions, such as obesity and diabetes. The increasing prevalence of obesity and diabetes altering the landscape of childhood dyslipidemia combined with dyslipidemia of obesity currently comprise the most frequent phenotype of lipid abnormalities.

Author Contributions: Writing—Original Draft Preparation, R.F.; Writing—Review and Editing, R.F. and F.C.; Supervision, F.C. All authors have read and agreed to the published version of the manuscript.

Funding: This research received no external funding.

Institutional Review Board Statement: Not applicable.

Informed Consent Statement: Not applicable.

Data Availability Statement: Not applicable.

Conflicts of Interest: The authors declare no conflict of interest.

References

1. Pahwa, R.; Jialal, I. Atherosclerosis. 2021 Sep 28. In *StatPearls [Internet]*; StatPearls Publishing: Treasure Island, FL, USA, 2021.
2. Benjamin, E.J.; Virani, S.S.; Callaway, C.W.; Chamberlain, A.M.; Chang, A.R.; Cheng, S.; Chiuve, S.E.; Cushman, M.; Delling, F.N.; Deo, R.; et al. Heart Disease and Stroke Statistics-2018 Update: A Report From the American Heart Association. *Circulation* **2018**, *137*, e67–e492, Erratum in *Circulation* **2018**, *137*, e493. [CrossRef] [PubMed]
3. Gatto, L.; Prati, F. Subclinical atherosclerosis: How and when to treat it? *Eur. Heart J. Suppl.* **2020**, *22*, E87–E90. [CrossRef] [PubMed]
4. Ellulu, M.S.; Patimah, I.; Khaza'Ai, H.; Rahmat, A.; Abed, Y.; Ali, F. Atherosclerotic cardiovascular disease: A review of initiators and protective factors. *Inflammopharmacology* **2016**, *24*, 1–10. [CrossRef] [PubMed]
5. Linton, M.F.; Yancey, P.G.; Davies, S.S.; Jerome, W.G.; Linton, E.F.; Song, W.L.; Doran, A.C.; Vickers, K.C. The Role of Lipids and Lipoproteins in Atherosclerosis. In *Endotext [Internet]*; Feingold, K.R., Anawalt, B., Boyce, A., Chrousos, G., de Herder, W.W., Dhatariya, K., Dungan, K., Hershman, J.M., Hofland, J., Kalra, S., et al., Eds.; MDText.com, Inc.: South Dartmouth, MA, USA, 2000.
6. Burlutskaya, A.V.; Tril, V.E.; Polischuk, L.V.; Pokrovskii, V.M. Dyslipidemia in pediatrician's practice. *Rev. Cardiovasc. Med.* **2021**, *22*, 817–834. [CrossRef]
7. Baroncini, L.A.V.; Sylvestre, L.D.C.; Filho, R.P. Assessment of Intima-Media Thickness in Healthy Children Aged 1 to 15 Years. *Arq. Bras. Cardiol.* **2016**, *106*, 327–332. [CrossRef] [PubMed]
8. Gooty, V.D.; Sinaiko, A.R.; Ryder, J.; Dengel, D.R.; Jacobs, D.R.; Steinberger, J. Association Between Carotid Intima Media Thickness, Age, and Cardiovascular Risk Factors in Children and Adolescents. *Metab. Syndr. Relat. Disord.* **2018**, *16*, 122–126. [CrossRef]
9. Briana, D.D.; Malamitsi-Puchner, A. Coronary Intimal Thickening Begins in Fetuses: Proof of Concept for the "Fetal Origins of Adult Disease" Hypothesis. *Angiology* **2019**, *71*, 89. [CrossRef]
10. Guerri-Guttenberg, R.; Castilla, R.; Cao, G.; Azzato, F.; Ambrosio, G.; Milei, J. Coronary Intimal Thickening Begins in Fetuses and Progresses in Pediatric Population and Adolescents to Atherosclerosis. *Angiology* **2019**, *71*, 62–69. [CrossRef]
11. Candelino, M.; Tagi, V.M.; Chiarelli, F. Cardiovascular risk in children: A burden for future generations. *Ital. J. Pediatr.* **2022**, *48*, 57. [CrossRef]
12. Röhrl, C.; Stangl, H. Cholesterol metabolism—Physiological regulation and pathophysiological deregulation by the endoplasmic reticulum. *Wien. Med. Wochenschr.* **2018**, *168*, 280–285. [CrossRef]
13. McMahan, C.A.; Gidding, S.S.; Malcom, G.T.; Tracy, R.E.; Strong, J.P.; McGill, H.C. Pathobiological Determinants of Atherosclerosis in Youth Research Group: Pathobiological Determinants of Atherosclerosis in Youth Risk Scores Are Associated with Early and Advanced Atherosclerosis. *Pediatrics* **2006**, *118*, 1447–1455. [CrossRef] [PubMed]
14. Khoury, M.; McCrindle, B.W. The Rationale, Indications, Safety, and Use of Statins in the Pediatric Population. *Can. J. Cardiol.* **2020**, *36*, 1372–1383. [CrossRef]
15. Grundy, S.M.; Stone, N.J.; Bailey, A.L.; Beam, C.; Birtcher, K.K.; Blumenthal, R.S.; Braun, L.T.; de Ferranti, S.; Faiella-Tommasino, J.; Forman, D.E.; et al. 2018 AHA/ACC/AACVPR/AAPA/ABC/ACPM/ADA/AGS/APhA/ASPC/NLA/PCNA Guideline on the Management of Blood Cholesterol: Executive Summary. *J. Am. Coll. Cardiol.* **2018**, *73*, 3168–3209, Erratum in *J. Am. Coll. Cardiol.* **2019**, *73*, 3234–3237. [CrossRef]
16. Arnett, D.K.; Blumenthal, R.S.; Albert, M.A.; Buroker, A.B.; Goldberger, Z.D.; Hahn, E.J.; Himmelfarb, C.D.; Khera, A.; Lloyd-Jones, D.; McEvoy, J.W.; et al. 2019 ACC/AHA Guideline on the primary prevention of cardiovascular disease: A report of the american college of cardiology/American heart association task force on clinical practice guidelines. *Circulation* **2019**, *140*, e596–e646, Erratum in *Circulation* **2019**, *140*, e649–e650; Erratum in *Circulation* **2020**, *141*, e60; Erratum in *Circulation* **2020**, *141*, e774. [CrossRef] [PubMed]
17. Trapani, L. Regulation and deregulation of cholesterol homeostasis: The liver as a metabolic "power station". *World J. Hepatol.* **2012**, *4*, 184–190. [CrossRef] [PubMed]
18. Cerqueira, N.M.F.S.A.; Oliveira, E.F.; Gesto, D.S.; Santos-Martins, D.; Moreira, C.; Moorthy, H.N.; Ramos, M.J.; Fernandes, P.A. Cholesterol Biosynthesis: A Mechanistic Overview. *Biochemistry* **2016**, *55*, 5483–5506. [CrossRef] [PubMed]
19. Feingold, K.R. Introduction to Lipids and Lipoproteins. In *Endotext [Internet]*; Feingold, K.R., Anawalt, B., Boyce, A., Chrousos, G., de Herder, W.W., Dhatariya, K., Dungan, K., Hershman, J.M., Hofland, J., Kalra, S., et al., Eds.; MDText.com, Inc.: South Dartmouth, MA, USA, 2000.
20. Lent-Schochet, D.; Jialal, I. Biochemistry, Lipoprotein Metabolism. In *StatPearls [Internet]*; StatPearls Publishing: Treasure Island, FL, USA, 2022.
21. Ouimet, M.; Barrett, T.J.; Fisher, E.A. HDL and Reverse Cholesterol Transport. *Circ. Res.* **2019**, *124*, 1505–1518. [CrossRef] [PubMed]
22. Pinal-Fernandez, I.; Casal-Dominguez, M.; Mammen, A.L. Statins: Pros and cons. *Med. Clínica (Engl. Ed.)* **2017**, *150*, 398–402. [CrossRef]
23. Collins, R.; Reith, C.; Emberson, J.; Armitage, J.; Baigent, C.; Blackwell, L.; Blumenthal, R.; Danesh, J.; Smith, G.D.; DeMets, D.; et al. Interpretation of the evidence for the efficacy and safety of statin therapy. *Lancet* **2016**, *388*, 2532–2561, Erratum in *Lancet* **2017**, *398*, 602. [CrossRef]
24. Istvan, E. Statin inhibition of HMG-CoA reductase: A 3-dimensional view. *Atheroscler. Suppl.* **2003**, *4*, 3–8. [CrossRef]

25. Elkins, C.; Fruh, S.; Jones, L.; Bydalek, K. Clinical Practice Recommendations for Pediatric Dyslipidemia. *J. Pediatr. Health Care* **2019**, *33*, 494–504. [CrossRef] [PubMed]
26. Ahmadi, M.; Amiri, S.; Pecic, S.; Machaj, F.; Rosik, J.; Łos, M.J.; Alizadeh, J.; Mahdian, R.; da Silva Rosa, S.C.; Schaafsma, D.; et al. Pleiotropic effects of statins: A focus on cancer. *Biochim. Biophys. Acta (BBA) Mol. Basis Dis.* **2020**, *1866*, 165968. [CrossRef]
27. Zhang, Q.; Dong, J.; Yu, Z. Pleiotropic use of Statins as non-lipid-lowering drugs. *Int. J. Biol. Sci.* **2020**, *16*, 2704–2711. [CrossRef]
28. Vuorio, A.; Kuoppala, J.; Kovanen, P.T.; Humphries, S.E.; Tonstad, S.; Wiegman, A.; Drogari, E.; Ramaswami, U. Statins for children with familial hypercholesterolemia. *Cochrane Database Syst. Rev.* **2017**, *7*, CD006401. [CrossRef]
29. Radaelli, G.; Sausen, G.; Cesa, C.C.; Santos, F.D.S.; Portal, V.L.; Neyeloff, J.; Pellanda, L.C. Statin Treatments and Dosages in Children with Familial Hypercholesterolemia: Meta-Analysis. *Arq. Bras. Cardiol.* **2018**, *111*, 810–821. [CrossRef]
30. Anagnostis, P.; Vaitsi, K.; Kleitsioti, P.; Mantsiou, C.; Pavlogiannis, K.; Athyros, V.G.; Mikhailidis, D.P.; Goulis, D.G. Efficacy and safety of statin use in children and adolescents with familial hypercholesterolaemia: A systematic review and meta-analysis of randomized-controlled trials. *Endocrine* **2020**, *69*, 249–261. [CrossRef]
31. Fiorentino, R.; Chiarelli, F. Treatment of Dyslipidaemia in Children. *Biomedicines* **2021**, *9*, 1078. [CrossRef]
32. Ferrari, F.; Martins, V.M.; Viviane, Z. Rochac Lipid Clinic Heart Institute (Incor), University of São Paulo Medical School Hospital, São Paulo, Brazil; Santos, R.D. Advances with lipid-lowering drugs for pediatric patients with familial hypercholesterolemia. *Expert Opin. Pharmacother.* **2020**, *22*, 483–495. [CrossRef]
33. Expert Panel on Integrated Guidelines for Cardiovascular Health and Risk Reduction in Children and Adolescents; National Heart, Lung, and Blood Institute. Expert panel on integrated guidelines for cardiovascular health and risk reduction in children and adolescents: Summary report. *Pediatrics* **2011**, *128* (Suppl. 5), S213–S256. [CrossRef] [PubMed]
34. Miller, M.L.; Wright, C.C.; Browne, B. Lipid-lowering medications for children and adolescents. *J. Clin. Lipidol.* **2015**, *9*, S67–S76. [CrossRef] [PubMed]
35. Luirink, I.K.; Wiegman, A.; Kusters, D.M.; Hof, M.H.; Groothoff, J.W.; de Groot, E.; Kastelein, J.J.; Hutten, B.A. 20-Year Follow-up of Statins in Children with Familial Hypercholesterolemia. *N. Engl. J. Med.* **2019**, *381*, 1547–1556. [CrossRef] [PubMed]
36. Mendelson, M.M.; Regh, T.; Chan, J.; Baker, A.; Ryan, H.H.; Palumbo, N.; Johnson, P.K.; Griggs, S.; Boghani, M.; Desai, N.K.; et al. Correlates of Achieving Statin Therapy Goals in Children and Adolescents with Dyslipidemia. *J. Pediatr.* **2016**, *178*, 149–155.e9. [CrossRef]
37. Gidding, S.S.; Champagne, M.A.; De Ferranti, S.D.; Defesche, J.; Ito, M.K.; Knowles, J.W.; McCrindle, B.; Raal, F.; Rader, D.; Santos, R.D.; et al. The Agenda for Familial Hypercholesterolemia. *Circulation* **2015**, *132*, 2167–2192, Erratum in *Circulation* **2015**, *132*, e397. [CrossRef] [PubMed]
38. Willeit, P.; Tschiderer, L.; Allara, E.; Reuber, K.; Seekircher, L.; Gao, L.; Liao, X.; Lonn, E.; Gerstein, H.C.; Yusuf, S.; et al. Carotid Intima-Media Thickness Progression as Surrogate Marker for Cardiovascular Risk. *Circulation* **2020**, *142*, 621–642. [CrossRef]
39. Wiegman, A.; Hutten, B.A.; De Groot, E.; Rodenburg, J.; Bakker, H.D.; Büller, H.R.; Sijbrands, E.; Kastelein, J.J.P. Efficacy and Safety of Statin Therapy in Children with Familial Hypercholesterolemia. *JAMA* **2004**, *292*, 331–337. [CrossRef] [PubMed]
40. Rodenburg, J.; Vissers, M.N.; Wiegman, A.; van Trotsenburg, A.S.P.; Van Der Graaf, A.; De Groot, E.; Wijburg, F.A.; Kastelein, J.J.P.; Hutten, B.A. Statin Treatment in Children with Familial Hypercholesterolemia. *Circulation* **2007**, *116*, 664–668. [CrossRef]
41. Kusters, D.M.; Avis, H.J.; De Groot, E.; Wijburg, F.A.; Kastelein, J.J.P.; Wiegman, A.; Hutten, B.A. Ten-Year Follow-up After Initiation of Statin Therapy in Children with Familial Hypercholesterolemia. *JAMA* **2014**, *312*, 1055–1057. [CrossRef]
42. Wiegman, A.; Gidding, S.S.; Watts, G.F.; Chapman, M.J.; Ginsberg, H.N.; Cuchel, M.; Ose, L.; Averna, M.; Boileau, C.; Borén, J.; et al. Familial hypercholesterolaemia in children and adolescents: Gaining decades of life by optimizing detection and treatment. *Eur. Heart J.* **2015**, *36*, 2425–2437. [CrossRef]
43. Vuorio, A.; Docherty, K.F.; Humphries, S.E.; Kuoppala, J.; Kovanen, P.T. Statin treatment of children with familial hypercholesterolemia—Trying to balance incomplete evidence of long-term safety and clinical accountability: Are we approaching a consensus? *Atherosclerosis* **2012**, *226*, 315–320. [CrossRef]
44. Newman, C.B.; Preiss, D.; Tobert, J.A.; Jacobson, T.A.; Page, I.R.L.; Goldstein, L.B.; Chin, C.; Tannock, L.R.; Miller, M.; Raghuveer, G.; et al. Statin Safety and Associated Adverse Events: A Scientific Statement from the American Heart Association. *Arter. Thromb. Vasc. Biol.* **2019**, *39*, e38–e81, Erratum in *Arter. Thromb. Vasc. Biol.* **2019**, *39*, e158. [CrossRef]
45. Mamann, N.; Lemale, J.; Karsenty, A.; Dubern, B.; Girardet, J.-P.; Tounian, P. Intermediate-Term Efficacy and Tolerance of Statins in Children. *J. Pediatr.* **2019**, *210*, 161–165. [CrossRef]
46. Kavey, R.-E.W.; Manlhiot, C.; Runeckles, K.; Collins, T.; Gidding, S.S.; Demczko, M.; Clauss, S.; Harahsheh, A.S.; Mietus-Syder, M.; Khoury, M.; et al. Effectiveness and Safety of Statin Therapy in Children: A Real-World Clinical Practice Experience. *CJC Open* **2020**, *2*, 473–482. [CrossRef]
47. Wiegman, A. Lipid Screening, Action, and Follow-up in Children and Adolescents. *Curr. Cardiol. Rep.* **2018**, *20*, 80. [CrossRef]
48. Desai, N.K.; Mendelson, M.; Baker, A.; Ryan, H.H.; Griggs, S.; Boghani, M.; Yellen, E.; Buckley, L.; Gillman, M.W.; Zachariah, J.P.; et al. Hepatotoxicity of Statins as Determined by Serum Alanine Aminotransferase in a Pediatric Cohort with Dyslipidemia. *J. Craniofacial Surg.* **2019**, *68*, 175–181. [CrossRef]
49. Gupta, A.; Thompson, D.; Whitehouse, A.; Collier, T.; Dahlof, B.; Poulter, N.; Collins, R.; Sever, P. Adverse events associated with unblinded, but not with blinded, statin therapy in the Anglo-Scandinavian Cardiac Outcomes Trial—Lipid-Lowering Arm (ASCOT-LLA): A randomised double-blind placebo-controlled trial and its non-randomised non-blind extension phase. *Lancet* **2017**, *389*, 2473–2481. [CrossRef] [PubMed]

50. Johnson, P.K.; Mendelson, M.; Baker, A.; Ryan, H.H.; Warren, S.; Graham, D.; Griggs, S.S.; Desai, N.K.; Yellen, E.; Buckley, L.; et al. Statin-Associated Myopathy in a Pediatric Preventive Cardiology Practice. *J. Pediatr.* **2017**, *185*, 94–98. [CrossRef] [PubMed]
51. Corkins, M.R.; Daniels, S.R.; de Ferranti, S.D.; Golden, N.H.; Kim, J.H.; Magge, S.N.; Schwarzenberg, S.J. Nutrition in Children and Adolescents. *Med. Clin. N. Am.* **2016**, *100*, 1217–1235. [CrossRef] [PubMed]
52. Eiland, L.S.; Luttrell, P.K. Use of Statins for Dyslipidemia in the Pediatric Population. *J. Pediatr. Pharmacol. Ther.* **2010**, *15*, 160–172. [CrossRef] [PubMed]
53. Cohen, H.; The Mighty Medic Satellite Research Group for Pediatric Dyslipidemia; Stefanutti, C. Current Approach to the Diagnosis and Treatment of Heterozygote and Homozygous FH Children and Adolescents. *Curr. Atheroscler. Rep.* **2021**, *23*, 30. [CrossRef] [PubMed]
54. Vahedian-Azimi, A.; Makvandi, S.; Banach, M.; Reiner, Ž.; Sahebkar, A. Fetal toxicity associated with statins: A systematic review and meta-analysis. *Atherosclerosis* **2021**, *327*, 59–67. [CrossRef] [PubMed]
55. Maliachova, O. Familial Hypercholesterolemia in Children and Adolescents: Diagnosis and Treatment. *Curr. Pharm. Des.* **2019**, *24*, 3672–3677. [CrossRef] [PubMed]
56. Farkouh, A.; Baumgärtel, C. Mini-review: Medication safety of red yeast rice products. *Int J. Gen. Med.* **2019**, *12*, 167–171. [CrossRef]
57. McCrindle, B.W.; Ose, L.; Marais, A. Efficacy and safety of atorvastatin in children and adolescents with familial hypercholesterolemia or severe hyperlipidemia: A multicenter, randomized, placebo-controlled trial. *J. Pediatr.* **2003**, *143*, 74–80. [CrossRef]
58. Van Der Graaf, A.; Nierman, M.C.; Firth, J.C.; Wolmarans, K.H.; Marais, A.D.; De Groot, E. Efficacy and safety of fluvastatin in children and adolescents with heterozygous familial hypercholesterolaemia. *Acta Paediatr.* **2006**, *95*, 1461–1466. [CrossRef] [PubMed]
59. Lambert, M.; Lupien, P.-J.; Gagné, C.; Lévy, E.; Blaichman, S.; Langlois, S.; Hayden, M.; Rose, V.; Clarke, J.T.R.; Wolfe, B.M.J.; et al. Treatment of Familial Hypercholesterolemia in Children and Adolescents: Effect of Lovastatin. *Pediatrics* **1996**, *97*, 619–628. [CrossRef]
60. Stein, E.A.; Illingworth, D.R.; Kwiterovich, J.P.O.; Liacouras, C.A.; Siimes, M.A.; Jacobson, M.S.; Brewster, T.G.; Hopkins, P.; Davidson, M.; Graham, K.; et al. Efficacy and Safety of Lovastatin in Adolescent Males with Heterozygous Familial Hypercholesterolemia. *JAMA* **1999**, *281*, 137–144. [CrossRef] [PubMed]
61. Clauss, S.B.; Holmes, K.W.; Hopkins, P.; Stein, E.; Cho, M.; Tate, A.; Johnson-Levonas, A.O.; Kwiterovich, P.O. Efficacy and Safety of Lovastatin Therapy in Adolescent Girls with Heterozygous Familial Hypercholesterolemia. *Pediatrics* **2005**, *116*, 682–688. [CrossRef] [PubMed]
62. Braamskamp, M.J.; Stefanutti, C.; Langslet, G.; Drogari, E.; Wiegman, A.; Hounslow, N.; Kastelein, J.J. Efficacy and Safety of Pitavastatin in Children and Adolescents at High Future Cardiovascular Risk. *J. Pediatr.* **2015**, *167*, 338–343.e5. [CrossRef]
63. Knipscheer, H.C.; Boelen, C.C.A.; Kastelein, J.J.P.; Van Diermen, D.E.; Groenemeijer, B.E.; Ende, A.V.D.; Büller, H.R.; Bakker, H.D. Short-Term Efficacy and Safety of Pravastatin in 72 Children with Familial Hypercholesterolemia. *Pediatr. Res.* **1996**, *39*, 867–871, Erratum in *Pediatr. Res.* **1996**, *40*, 866. [CrossRef]
64. Avis, H.J.; Hutten, B.A.; Gagné, C.; Langslet, G.; McCrindle, B.W.; Wiegman, A.; Hsia, J.; Kastelein, J.J.; Stein, E.A. Efficacy and Safety of Rosuvastatin Therapy for Children with Familial Hypercholesterolemia. *J. Am. Coll. Cardiol.* **2010**, *55*, 1121–1126. [CrossRef]
65. Couture, P.; Brun, L.D.; Szots, F.; Lelièvre, M.; Gaudet, D.; Després, J.-P.; Simard, J.; Lupien, P.J.; Gagné, C. Association of Specific LDL Receptor Gene Mutations with Differential Plasma Lipoprotein Response to Simvastatin in Young French Canadians with Heterozygous Familial Hypercholesterolemia. *Arter. Thromb. Vasc. Biol.* **1998**, *18*, 1007–1012. [CrossRef]
66. de Jongh, S.; Ose, L.; Szamosi, T.; Gagné, C.; Lambert, M.; Scott, R.; Perron, P.; Dobbelaere, D.; Saborio, M.; Tuohy, M.B.; et al. Efficacy and Safety of Statin Therapy in Children with Familial Hypercholesterolemia. *Circulation* **2002**, *106*, 2231–2237. [CrossRef] [PubMed]
67. Bays, H.E.; Jones, P.H.; Orringer, C.E.; Brown, W.V.; Jacobson, T.A. National Lipid Association Annual Summary of Clinical Lipidology 2016. *J. Clin. Lipidol.* **2016**, *10*, S1–S43. [CrossRef]
68. Miller, M.L.; Wright, C.C.; Rodriguez, B. Use of Lipid Lowering Medications in Youth. In *Endotext [Internet]*; Feingold, K.R., Anawalt, B., Boyce, A., Chrousos, G., de Herder, W.W., Dhatariya, K., Dungan, K., Hershman, J.M., Hofland, J., Kalra, S., et al., Eds.; MDText.com, Inc.: South Dartmouth, MA, USA, 2000.
69. De Ferranti, S.D.; Steinberger, J.; Ameduri, R.; Baker, A.; Gooding, H.; Kelly, A.S.; Mietus-Snyder, M.; Mitsnefes, M.M.; Peterson, A.L.; St-Pierre, J.; et al. Cardiovascular Risk Reduction in High-Risk Pediatric Patients: A Scientific Statement from the American Heart Association. *Circulation* **2019**, *139*, e603–e634. [CrossRef] [PubMed]
70. NCEP Expert Panel on Blood Cholesterol Levels in Children and Adolescents. National Cholesterol Education Program (NCEP): Highlights of the report of the Expert Panel on Blood Cholesterol Levels in Children and Adolescents. *Pediatrics* **1992**, *89*, 495–501. [CrossRef]
71. American Academy of Pediatrics; Committee on Nutrition. American Academy of Pediatrics. Committee on Nutrition. Cholesterol in childhood. *Pediatrics* **1998**, *101*, 141–147.
72. McCrindle, B.W.; Urbina, E.M.; Dennison, B.A.; Jacobson, M.S.; Steinberger, J.; Rocchini, A.P.; Hayman, L.L.; Daniels, S.R. Drug Therapy of High-Risk Lipid Abnormalities in Children and Adolescents. *Circulation* **2007**, *115*, 1948–1967. [CrossRef]

73. Ramaswami, U.; Humphries, S.E.; Priestley-Barnham, L.; Green, P.; Wald, D.S.; Capps, N.; Anderson, M.; Dale, P.; Morris, A.A. Current management of children and young people with heterozygous familial hypercholesterolaemia—HEART UK statement of care. *Atherosclerosis* **2019**, *290*, 1–8. [CrossRef]
74. Mach, F.; Baigent, C.; Catapano, A.L.; Koskinas, K.C.; Casula, M.; Badimon, L.; Chapman, M.J.; De Backer, G.G.; Delgado, V.; Ference, B.A.; et al. 2019 ESC/EAS Guidelines for the management of dyslipidaemias: Lipid modification to reduce cardiovascular risk: The Task Force for the management of dyslipidaemias of the European Society of Cardiology (ESC) and European Atherosclerosis Society (EAS): The Task Force for the management of dyslipidaemias of the European Society of Cardiology (ESC) and European Atherosclerosis Society (EAS). *Eur. Heart J.* **2020**, *41*, 111–188, Erratum in *Eur. Heart J.* **2020**, *41*, 4255. [CrossRef]
75. Cuchel, M.; Bruckert, E.; Ginsberg, H.N.; Raal, F.J.; Santos, R.D.; Hegele, R.A.; Kuivenhoven, J.A.; Nordestgaard, B.G.; Descamps, O.S.; Steinhagen-Thiessen, E.; et al. Homozigot ailevi hiperkolesterolemi: Klinisyenlerin taniyi ve klinik yönetimi geliştirmelerine yönelik yeni anlayişlar ve rehberlik. Avrupa Ateroskleroz Derneği'nin Ailevi Hiperkolesterolemi Üzerine Uzlaşi Paneli yazili görüşü [Homozygous familial hypercholesterolaemia: New insights and guidance for clinicians to improve detection and clinical management. A position paper from the Consensus Panel on Familial Hypercholesterolaemia of the European Atherosclerosis Society]. *Turk Kardiyol. Dern. Ars.* **2015**, *43* (Suppl. 1), 1–14. (In Turkish)
76. McGowan, M.P.; Dehkordi, S.H.H.; Moriarty, P.M.; Duell, P.B. Diagnosis and Treatment of Heterozygous Familial Hypercholesterolemia. *J. Am. Heart Assoc.* **2019**, *8*, e013225. [CrossRef] [PubMed]
77. Awad, K.; Serban, M.-C.; Penson, P.; Mikhailidis, D.P.; Toth, P.P.; Jones, S.R.; Rizzo, M.; Howard, G.; Lip, G.Y.; Banach, M. Effects of morning vs evening statin administration on lipid profile: A systematic review and meta-analysis. *J. Clin. Lipidol.* **2017**, *11*, 972–985.e9. [CrossRef] [PubMed]
78. Yoon, J.M. Dyslipidemia in Children and Adolescents: When and How to Diagnose and Treat? *Pediatr. Gastroenterol. Hepatol. Nutr.* **2014**, *17*, 85–92. [CrossRef] [PubMed]
79. Balla, S.; Ekpo, E.P.; Wilemon, K.A.; Knowles, J.W.; Rodriguez, F. Women Living with Familial Hypercholesterolemia: Challenges and Considerations Surrounding Their Care. *Curr. Atheroscler. Rep.* **2020**, *22*, 60. [CrossRef] [PubMed]
80. Hirota, T.; Ieiri, I. Drug–drug interactions that interfere with statin metabolism. *Expert Opin. Drug Metab. Toxicol.* **2015**, *11*, 1435–1447. [CrossRef]
81. Hu, P.; Dharmayat, K.I.; Stevens, C.A.; Sharabiani, M.T.; Jones, R.S.; Watts, G.F.; Genest, J.; Ray, K.; Vallejo-Vaz, A.J. Prevalence of Familial Hypercholesterolemia Among the General Population and Patients with Atherosclerotic Cardiovascular Disease. *Circulation* **2020**, *141*, 1742–1759. [CrossRef]
82. Bouhairie, V.E.; Goldberg, C. Familial hypercholesterolemia. *Clin. Cardiol.* **2015**, *33*, 169–179. [CrossRef]
83. Ramaswami, U.; Futema, M.; Bogsrud, M.P.; Holven, K.B.; van Lennep, J.R.; Wiegman, A.; Descamps, O.S.; Vrablik, M.; Freiberger, T.; Dieplinger, H.; et al. Comparison of the characteristics at diagnosis and treatment of children with heterozygous familial hypercholesterolaemia (FH) from eight European countries. *Atherosclerosis* **2019**, *292*, 178–187. [CrossRef]
84. O'Gorman, C.; Conwell, L.; O'Neill, M.B. Considering statins for cholesterol-reduction in children if lifestyle and diet changes do not improve their health: A review of the risks and benefits. *Vasc. Health Risk Manag.* **2010**, *7*, 1–14. [CrossRef]
85. Tada, H.; Takamura, M.; Kawashiri, M.-A. Familial Hypercholesterolemia: A Narrative Review on Diagnosis and Management Strategies for Children and Adolescents. *Vasc. Health Risk Manag.* **2021**, *17*, 59–67. [CrossRef]

International Journal of
Molecular Sciences

MDPI

Article

Nicotinamide Mononucleotide Protects against Retinal Dysfunction in a Murine Model of Carotid Artery Occlusion

Deokho Lee [1,2], Yohei Tomita [1,2], Yukihiro Miwa [1,2,3], Heonuk Jeong [1,2], Ari Shinojima [1,2], Norimitsu Ban [2], Shintaro Yamaguchi [4], Ken Nishioka [4], Kazuno Negishi [2], Jun Yoshino [4] and Toshihide Kurihara [1,2,*]

1 Laboratory of Photobiology, Keio University School of Medicine, Tokyo 160-8582, Japan
2 Department of Ophthalmology, Keio University School of Medicine, Tokyo 160-8582, Japan
3 Aichi Animal Eye Clinic, Nagoya 466-0827, Japan
4 Department of Internal Medicine, Keio University School of Medicine, Tokyo 160-8582, Japan
* Correspondence: kurihara@z8.keio.jp

Abstract: Cardiovascular abnormality-mediated retinal ischemia causes severe visual impairment. Retinal ischemia is involved in enormous pathological processes including oxidative stress, reactive gliosis, and retinal functional deficits. Thus, maintaining retinal function by modulating those pathological processes may prevent or protect against vision loss. Over the decades, nicotinamide mononucleotide (NMN), a crucial nicotinamide adenine dinucleotide (NAD^+) intermediate, has been nominated as a promising therapeutic target in retinal diseases. Nonetheless, a protective effect of NMN has not been examined in cardiovascular diseases-induced retinal ischemia. In our study, we aimed to investigate its promising effect of NMN in the ischemic retina of a murine model of carotid artery occlusion. After surgical unilateral common carotid artery occlusion (UCCAO) in adult male C57BL/6 mice, NMN (500 mg/kg/day) was intraperitoneally injected to mice every day until the end of experiments. Electroretinography and biomolecular assays were utilized to measure ocular functional and further molecular alterations in the retina. We found that UCCAO-induced retinal dysfunction was suppressed, pathological gliosis was reduced, retinal NAD^+ levels were preserved, and the expression of an antioxidant molecule (nuclear factor erythroid-2-related factor 2; *Nrf2*) was upregulated by consecutive administration of NMN. Our present outcomes first suggest a promising NMN therapy for the suppression of cardiovascular diseases-mediated retinal ischemic dysfunction.

Keywords: nicotinamide mononucleotide; oxidative stress; common carotid artery occlusion; retinal ischemia; neuroprotection

Citation: Lee, D.; Tomita, Y.; Miwa, Y.; Jeong, H.; Shinojima, A.; Ban, N.; Yamaguchi, S.; Nishioka, K.; Negishi, K.; Yoshino, J.; et al. Nicotinamide Mononucleotide Protects against Retinal Dysfunction in a Murine Model of Carotid Artery Occlusion. *Int. J. Mol. Sci.* **2022**, *23*, 14711. https://doi.org/10.3390/ijms232314711

Academic Editors: Simona Gabriela Bungau and Vesa Cosmin

Received: 2 November 2022
Accepted: 23 November 2022
Published: 25 November 2022

Publisher's Note: MDPI stays neutral with regard to jurisdictional claims in published maps and institutional affiliations.

1. Introduction

Cardiovascular dysfunction-induced retinal injury is a type of the severe vision-threatening diseases. As blood (including oxygen) to the retina is directly supplied through the ophthalmic artery stretched from the internal carotid artery (a branch of the common carotid artery), the retina is highly susceptible to cardiovascular dysfunction [1]. When cardiovascular abnormality (including carotid artery stenosis or occlusion) occurs, retinal ischemia can be evoked in human and murine [2–4], finally leading to transient or permanent visual impairment.

Retinal ischemia contains complicated pathological processes including oxidative stress, reactive gliosis, and retinal functional deficits [5,6]. As effective treatment has not yet been developed for retinal ischemia-induced retinal dysfunction, studies on searching for promising therapeutics to protect against retinal ischemia-induced retinal dysfunction have been attempted at the preclinical stage [6].

Nicotinamide mononucleotide (NMN) is an important precursor of nicotinamide adenine dinucleotide (NAD^+), an essential molecule required for various cellular functions [7]. Maintaining NAD^+ biosynthesis has been considered important for the prevention of a variety of age-related metabolic diseases and disorders, including obesity, insulin resistance, and diabetes [8–10]. Previously, NMN was also reported to prevent or protect against retinal damages (the inner and/or outer retina) in light-induced retinopathy, retinal detachment, or retinal ischemia/reperfusion injury [11–13]. It has been speculated that therapeutic effects of NMN might be associated with maintenance of levels of NAD^+. However, a protective effect of NMN has not yet been examined in cardiovascular disease-mediated retinal dysfunction.

Therefore, in the current study, we first investigated whether NMN treatment could show a promising protective effect in the ischemic retina of a murine model of carotid artery occlusion.

2. Results

2.1. Consecutive Treatment of NMN Protects against Retinal Dysfunction in a Mouse Model of Unilateral Common Carotid Artery Occlusion

To examine whether NMN treatment could show a protective effect against retinal dysfunction caused by unilateral common carotid artery occlusion (UCCAO) in adult mice, NMN (500 mg/kg) was intraperitoneally injected to mice after UCCAO (Figure 1A). NMN was continuously provided every day until the end of whole experiments. The concentration of NMN was determined based on our previous paper and other groups' reports [11–14], which is general for various experimental murine models. UCCAO was conducted using permanent surgical occlusion in the right common carotid artery (Figure 1B), according to our previous publications [15–17]. Retinal ischemia could be evoked as retinal blood is supplied through the ophthalmic artery (OpA) connected with the internal carotid artery (ICA), one of the branches of the common carotid artery (CCA). As eyelid drooping was generally detected as a sign for a successful UCCAO surgery [15–17], we grossly examined eyelids of all mice after UCCAO, and found 100% rate of eyelid drooping in UCCAO groups, similar to our previous observations.

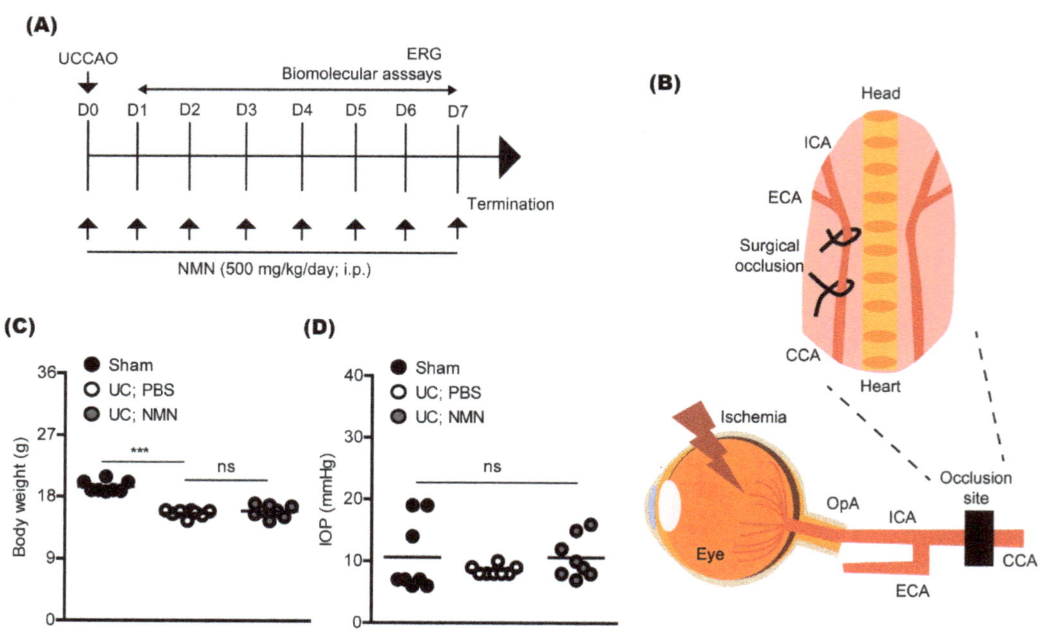

Figure 1. *Cont.*

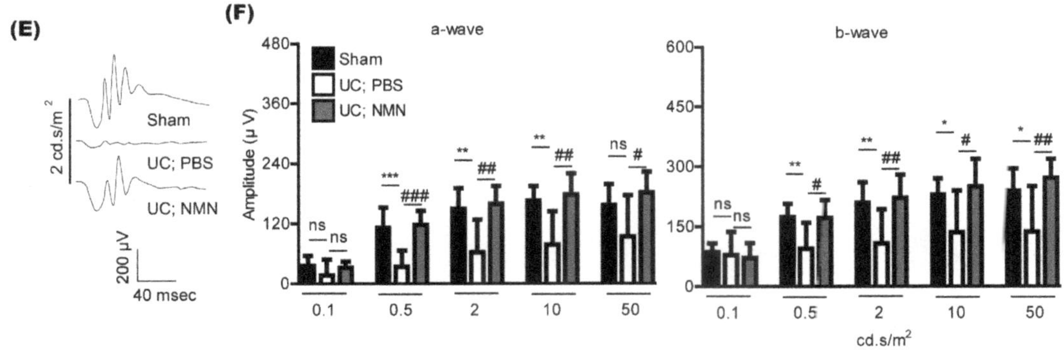

Figure 1. Retinal functional protection by continuous NMN treatment. (**A**) A general experimental scheme. UCCAO, unilateral common carotid artery occlusion; D, day; ERG, electroretinography; i.p., intraperitoneal injection; NMN, nicotinamide mononucleotide. (**B**) An anatomical view on the induction of retinal ischemia by surgical UCCAO. Surgical permanent occlusion of the common carotid artery (CCA) could induce occlusion of the internal carotid artery (ICA), finally leading to occlusion of the ophthalmic artery (OpA) which supplies the retina. (**C**) Quantitative analyses (n = 8 per group) demonstrated that the body weight was significantly reduced by UCCAO. There was no dramatic difference in the body weight between PBS-treated and NMN-treated UCCAO-operated mice. *** $p < 0.001$; ns, not significant. One-way ANOVA followed by a Bonferroni post hoc test. A graph is shown as mean with actual values (mean and standard deviation, sham: 19.41, 0.83; UC; PBS: 15.60, 0.50; UC; NMN: 15.94, 0.86). (**D**) Quantitative analyses (n = 8 per group) demonstrated that there was no difference in intraocular pressure (IOP) among all group. ns, not significant. One-way ANOVA followed by a Bonferroni post hoc test. A graph is shown as mean with actual values (mean and standard deviation, sham: 10.63, 5.78; UC; PBS: 8.50, 0.75; UC; NMN: 10.63, 3.37). (**E**,**F**) Representative waveforms (2 cd.s/m^2) of ERG and quantitative analyses (n = 7–12 per group) demonstrated that the ERG amplitudes decreased 7 days after UCCAO. NMN treatment suppressed reductions in the ERG amplitudes, flashed with various standardized intensities (0.1, 0.5, 2, 10, or 50 cd.s/m^2). * $p < 0.05$, ** $p < 0.01$, *** $p < 0.001$; # $p < 0.05$, ## $p < 0.01$, ### $p < 0.001$. One-way ANOVA followed by a Bonferroni post hoc test. Graphs are shown as mean ± standard deviation. UC, UCCAO.

As a general parameter, we screened body weight (Figure 1C) and intraocular pressure (Figure 1D) after UCCAO. After UCCAO, the body weight was reduced, which has been consistently seen in the murine UCCAO model [18,19]. There was no change in the body weight by NMN treatment. There was no dramatic difference in intraocular pressure among groups, similar with that observed in the murine UCCAO model.

Previously, we found that retinal dysfunction was seen 7 days after UCCAO [16]. We reproduced this finding in our current system (Figure 1E,F). Continuous NMN treatment significantly suppressed its reductions.

2.2. *Consecutive Treatment of NMN Reduces Pathological Gliosis in a Mouse Model of Unilateral Common Carotid Artery Occlusion*

We examined whether NMN treatment could reduce pathological gliosis caused by UCCAO in mice. Previously, there were increases in retinal glial activation (especially, an increased immunoreactivity in glial fibrillary acidic protein; GFAP) after UCCAO [16]. Consistent with the previous report, we could detect glial activation, analyzed with the GFAP immunoreactivity (Figure 2A). Furthermore, NMN treatment reduced pathological gliosis. Additionally, we screened two inflammatory glial chemokines (*Ccl2* and *Ccl12*) as their expressions have been reported to increase after UCCAO [17,18]. Induction in chemokine ligands *Ccl2* and *Ccl12* mRNA expressions was reduced by NMN treatment under the same condition (Figure 2B).

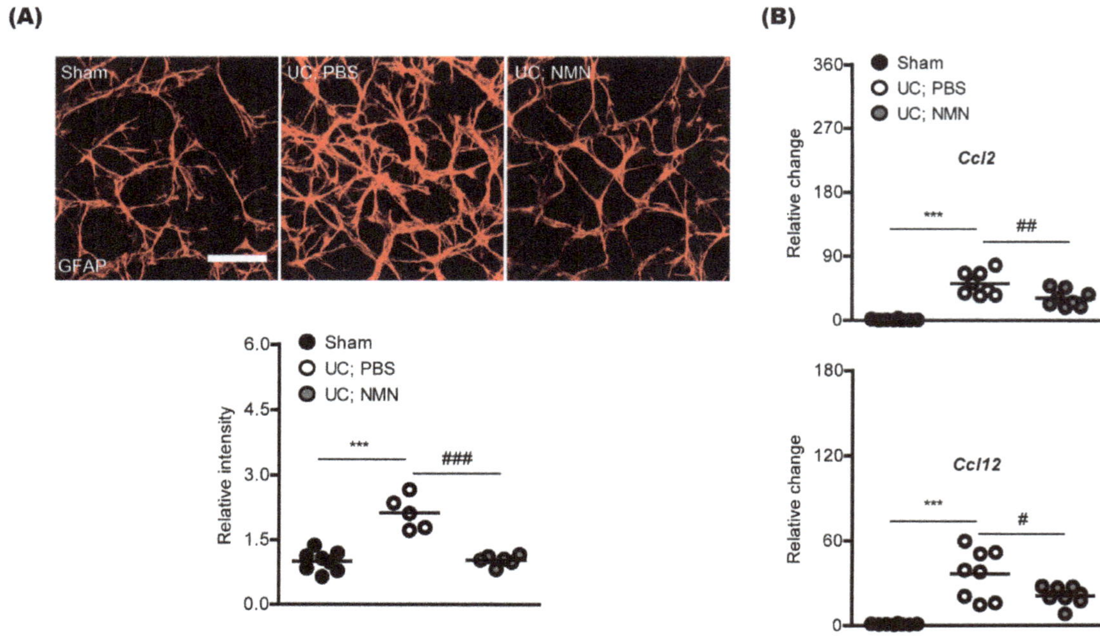

Figure 2. Modulation of pathological retinal gliosis and inflammation by consecutive NMN treatment. (**A**) Representative images and quantitative analyses (n = 5–8 per group) demonstrated that NMN treatment suppressed increases in GFAP immunoreactivities induced by UCCAO. Scale bar, 50 μm. *** $p < 0.001$; ### $p < 0.001$. One-way ANOVA followed by a Bonferroni post hoc test. A graph is shown as mean with actual values (mean and standard deviation, sham: 1.00, 0.23; UC; PBS: 2.12, 0.39; UC; NMN: 1.02, 0.12). (**B**) Quantitative analyses (n = 8 per group) demonstrated that NMN administration reduced upregulation in *Ccl2* and *Ccl12* mRNA expressions in the ischemic retina. *** $p < 0.001$; # $p < 0.05$, ## $p < 0.01$. One-way ANOVA followed by a Bonferroni post hoc test. Graphs are shown as mean with actual values (mean and standard deviation, sham: 1.00, 1.02; UC; PBS: 51.95, 16.42; UC; NMN: 31.46, 11.72; sham: 1.00, 0.38; UC; PBS: 36.54, 17.33; UC; NMN: 21.50, 6.54). UC, UCCAO.

2.3. Consecutive Treatment of NMN Preserves Redox Balance and Activates an Antioxidant Pathway in a Mouse Model of Unilateral Common Carotid Artery Occlusion

We examined whether NMN treatment could preserve redox balance and activate an antioxidant pathway. Numerous previous reports demonstrated that supplementation with NMN could increase NAD^+ biosynthesis in various cell types [20–22]. We applied this concept to our current system, and found the preservation of NAD^+ levels in the UCCAO-induced ischemic retina by continuous NMN treatment (Figure 3A).

Next, as NMN treatment has also been reported to increase antioxidant genes (especially, nuclear factor erythroid-2-related factor 2; *Nrf2*) in various cell types in vitro and in vivo [12,23–26]. Therefore, we examined whether NMN treatment could increase *Nrf2* mRNA expression under our current condition (Figure 3B). On day 2 after UCCAO, we found that *Nrf2* mRNA expression increased by NMN treatment, while there was no significant difference in its expression between sham and UCCAO groups (Figure 3B). On day 7 after UCCAO, we found decreases in its expression, while NMN treatment dramatically suppressed the reductions.

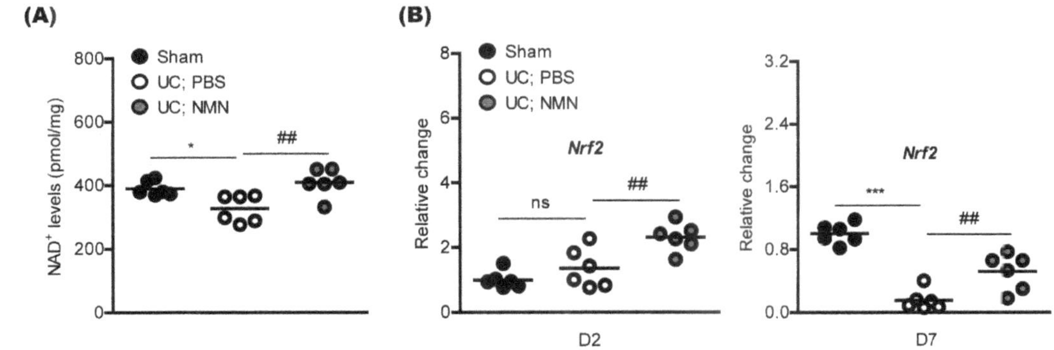

Figure 3. Modulation of redox balance and an antioxidant pathway by NMN treatment. (**A**) Quantitative analyses (n = 6 per group) demonstrated that retinal NAD$^+$ levels were significantly reduced by UCCAO. NMN treatment significantly suppressed its reduction. * $p < 0.05$; ## $p < 0.01$. One-way ANOVA followed by a Bonferroni post hoc test. A graph is shown as mean with actual values (mean and standard deviation, sham: 390.0, 23.38; UC; PBS: 326.2, 43.12; UC; NMN: 408.7, 44.13). (**B**) Quantitative analyses (n = 6 per group) demonstrated that NMN administration increased *Nrf2* mRNA expression in the ischemic retina. *** $p < 0.001$; ## $p < 0.01$; ns, not significant. One-way ANOVA followed by a Bonferroni post hoc test. A graph is shown as mean with actual values (mean and standard deviation, sham: 1.00, 0.26; UC; PBS: 1.35, 0.60; UC; NMN: 2.30, 0.43; sham: 1.00, 0.12; UC; PBS: 0.15, 0.12; UC; NMN: 0.51, 0.23). UC, UCCAO; D, day.

2.4. Consecutive Treatment of NMN Does Not Affect Retinal Thickness in a Mouse Model of Unilateral Common Carotid Artery Occlusion

According to our previous UCCAO papers, retinal thickness was not dramatically altered by UCCAO in mice. Nonetheless, we examined whether NMN treatment may affect retinal thickness in UCCAO-operated mice (Figure 4). There was no dramatic difference in retinal thickness (total, outer, and inner) among all groups. Based on our current functional and histological outcomes, NMN treatment may work on retinal functional protection without affecting retinal thickness in UCCAO-operated mice.

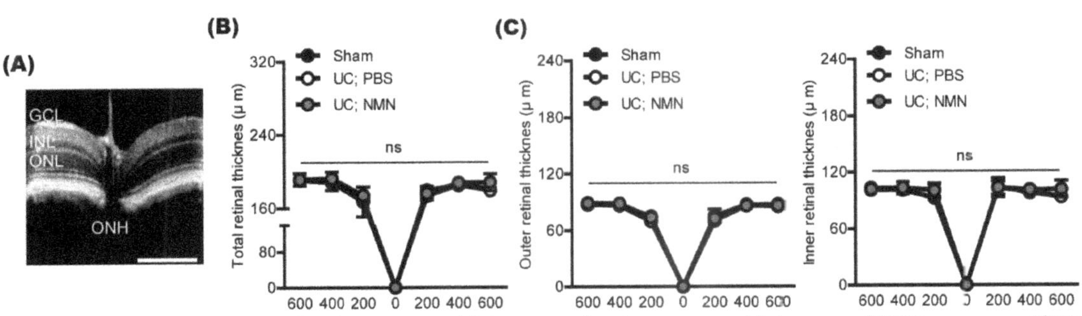

Figure 4. Screening of alterations in retinal thickness by NMN treatment. (**A**) A representative retinal picture of the optical coherence tomography (OCT). GCL, the ganglion cell layer; INL, the inner nuclear layer; ONL, the outer nuclear layer; ONH, the optic nerve head. Scale bar, 500 μm. (**B**,**C**) Quantitative analyses showed that NMN treatment did not change retinal thickness in UCCAO-operated mice, analyzed at 200, 400, and 600 μm from ONH (n = 5–6 per group). ns, not significant. Two-way ANOVA followed by a Bonferroni post hoc test. Graphs are shown as mean ± standard deviation. UC, UCCAO.

3. Discussion

The present study demonstrated that retinal functional impairment and pathological retinal gliosis could be induced by cardiovascular dysfunction, and continuous NMN treatment could show therapeutic effects against such injuries. Furthermore, redox imbalance under the retinal ischemic condition was modulated by continuous NMN treatment. Although promising NMN therapy in not only age-dependent or systemic metabolism-mediated diseases and disorders but also several retinopathies has been recently suggested [8–10], as far as we know, our current report is the first to expand the protective role of NMN in a murine model of permanent surgical occlusion of the unilateral common carotid artery.

We found that UCCAO-induced retinal dysfunction and pathological gliosis were reduced by consecutive NMN treatment. Similar aspects were described in previous studies from ours and others. We recently found that retinal ischemia/reperfusion injury-induced retinal dysfunction was lessened by consecutive NMN treatment [12]. Chen et al. showed that NMN treatment reduced cell death in the outer layer at the acute stage of retinal detachment [11]. Furthermore, NMN treatment suppressed pathological gliosis at the late stage of retinal detachment. Lin et al. showed that NMN treatment reduced retinal dysfunction in mice lacking *Nampt* and in mice of light-induced retinal damages [13]. In this aspect, NMN therapy might be promising for prevention or protection of various retinal diseases.

NMN is one of the natural compounds to boost NAD^+ biosynthesis. In mammals, NMN can be produced from nicotinamide (NAM, a form of vitamin B3 found in various foods) by nicotinamide phosphoribosyltransferase (NAMPT), the rate-limiting enzyme [21,27,28]. NMN can also be synthesized from nicotinamide riboside (NR) through NR kinase-mediated phosphorylation reaction. NMN can be converted into NAD^+ by NMN adenylyltransferase (NMNAT). In this regard, NMN supplementation could increase NAD^+ biosynthesis. In fact, numerous experimental studies showed its relationship under their experimental conditions in not only the central nervous system but also other types of body systems [21,27,28]. Chen et al. showed that retinal NAD^+ levels increased by NMN treatment in the retinal detachment-induced damaged retina [11]. In our current study, we found that retinal NAD^+ levels also increased by NMN treatment in the UCCAO-operated ischemic retina. Taken together, its event might be highly conserved. Nonetheless, more studies are needed on which strategies (administration methods, concentrations, or durations for NMN treatment) could effectively activate this biosynthesis.

Therapeutic roles of NMN are generally explained with upregulation of antioxidants to protect cells against redox imbalance. Based on our and other previous reports, the antioxidant function of NMN has been associated with the *Nrf2* antioxidant pathway [12,23–26]. We recently demonstrated that NMN treatment increased *Nrf2* expression and cellular protective properties in retinal 661W cells under $CoCl_2$-induced pseudohypoxic oxidative stress conditions [12]. Pu et al. demonstrated that NMN treatment showed protective effects against high glucose-evoked cellular damages in human corneal epithelial cells via the *Nrf2* signaling pathway [25]. Wei et al. demonstrated that NMN treatment attenuated intracerebral hemorrhage-induced cellular damages in the central nervous system through the *Nrf2* signaling pathway [26]. In our current data, NMN treatment constantly increased *Nrf2* expression in the ischemic retina under the UCCAO-induced redox imbalanced condition. Taken together, our present outcome also supports the notion that the therapeutic role of NMN might be associated with the *Nrf2* antioxidant signaling pathway in various eukaryotic cells and tissues.

In the murine model of UCCAO itself, there was no dramatic change in retinal thickness in our current experimental system. Its finding has been consistently detected in our previous publications [15–17]. Nonetheless, retinal thickness could be affected by retinal ischemic insults in humans [29–32]. Although there exist gaps between mice and humans, it can be speculated that retinal ischemia in humans is closely combined with systemic metabolic stress [33–35], while the experimental UCCAO murine model has no such stress.

To solve this issue, the UCCAO model can be combined with systemic metabolic dysregulation such as streptozotocin injection (to induce systemic diabetic conditions) or genetic manipulation (to cause metabolic diseases or conditions of interest) [36–39]. With this novel model of UCCAO, more clinically relevant findings could be obtained with NMN therapy, which will be further studied.

In summary, we firstly attempted to apply the currently promising NMN therapy to a murine model of unilateral common carotid artery occlusion. As the outcomes, retinal functional damages and pathological gliosis were suppressed by consecutive NMN treatment. Furthermore, retinal redox balance was preserved by consecutive NMN treatment along with upregulating the antioxidant pathway. Although further understandings on the mechanism of action of NMN therapy in retinal ischemia are desired, we suggest that NMN supplements can be one of the promising therapeutic strategies for protecting against ischemic retinopathy based on our brief observations.

4. Materials and Methods

4.1. Animal, Unilateral Common Carotid Artery Occlusion, and NMN Treatment

Whole animal experimental procedures were followed by the Ethics Committee on Animal Research of Keio University School of Medicine (#16017), the International Standards of Animal Care and Use, Animal Research: Reporting in Vivo Experiments, and the ARVO Statement for the Use of Animals in Ophthalmic and Vision Research.

Adult black male mice (C57BL/6, 5 weeks old) were obtained from CLEA Japan (Tokyo, Japan). After randomizing mice, we performed the UCCAO surgery, as previously described [15–17]. Briefly, the neck areas of mice were open to observe the right carotid arteries. The right CCA was tightly occluded using 6–0 silk sutures. After wound suturing, mice were recovered with warm pads to maintain body temperature. When mice were awake, they were back to cages for the further experiments. After mice were fully recovered (around 4 h after the surgery), intraperitoneal NMN (500 mg/kg) injection was consistently performed every day for 24 h interval.

4.2. Electroretinography (ERG) and Optical Coherence Tomography (OCT)

We performed ERG, as previously indicated in our papers [15,16]. Briefly, after dark adaptation for more than 12 h, mice were anesthetized using a mixture of medetomidine (7.5 μg/100 μL; Orion, Espoo, Finland), midazolam (40 μg/100 μL; Sandoz, Tokyo, Japan), and butorphanol tartrate (50 μg/100 μL; Meiji Seika Pharma, Tokyo, Japan) [40]. Scotopic ERG responses were recorded using a PuREC ERG acquisition system (MAYO, Inazawa, Japan). Various standardized light stimuli were evoked to measure the amplitudes of a-wave and b-wave.

Envisu R4310 OCT (Leica, Wetzlar, Germany) was performed as previously indicated in our papers [15,16]. Briefly, after anesthesia, mice were immediately placed on an OCT platform. After adjustment for positioning the retina to the OCT camera, retinal images at 200, 400, and 600 μm from the optic nerve head were captured. Retinal thickness was measured using the software provided by Envisu R4310 OCT.

4.3. Immunohistochemistry (IHC)

We performed IHC, as previously described in our papers [15–17]. After making flat-mounted retinas from the eyes, a primary antibody (GFAP 1:400, Cat #13-0300, Thermo Fisher Scientific, Waltham, MA, USA) was added to the retinas. After overnight incubation, the retinas were washed with PBS including 0.1% Triton three times. Species-appropriate fluorescence-conjugated secondary antibodies (Thermo Fisher Scientific, Waltham, MA, USA) were added to the retinas for 2 h at room temperature. After generally washing with PBS including 0.1% Triton three times, the retinas were mounted with cover slips and examined using the LSM710 microscope (Carl Zeiss, Jena, Germany).

4.4. Quantitative PCR (qPCR)

Whole procedures for qPCR analysis were as same as described in our previous publications [15–17]. A series of RNA-cDNA-qPCR kits (Qiagen, Velno, Netherlands; TOYOBO, Osaka, Japan; Applied Biosystems, Waltham, MA, USA) were utilized to prepare RNA samples, synthesize cDNA, and conduct qPCR. The primer sequence used in the present study is entered in Table 1. The generally used ΔΔCT analysis protocol was utilized to determine relative fold changes.

Table 1. Primer list.

Name	Direction	Sequence (5′ → 3′)	Accession Number
Hprt	Forward Reverse	TCAGTCAACGGGGGACATAAA GGGGCTGTACTGCTTAACCAG	NM_013556.2
Nrf2	Forward Reverse	TAGATGACCATGAGTCGCTTGC GCCAAACTTGCTCCATGTCC	NM_010902.4
Ccl2	Forward Reverse	CCCAATGAGTAGGCTGGAGA TCTGGACCCATTCCTTCTTG	NM_011333.3
Ccl12	Forward Reverse	GCTACAGGAGAATCACAAGCAGC ACGTCTTATCCAAGTGGTTTATGG	NM_011331.3

4.5. Nicotinamide Adenine Dinucleotide (NAD^+) Assay

We performed NAD^+ assay using a commercially available kit by following manufacturer's instructions (Cat #ab65348, Abcam, San Francisco, CA, USA). Briefly, fresh retinal samples (within 1 h after the last NMN treatment) were homogenized in extraction buffer in the kit. Then, concentrations of total NAD and decomposed NAD^+ were measured using the extracted samples by a plate reader (Synergy HT Multi-Mode, Winooski, VT, USA). NAD^+ was calculated with the equation NAD^+ = total NAD − decomposed NAD^+.

4.6. Statistical Analysis

Statistical significance for all experimental values was determined using one or two-way ANOVA followed by a Bonferroni post hoc test depending on the data set. Statistical significance was regarded when $p < 0.05$.

Author Contributions: D.L. designed the current study, conducted the experiments, and wrote the manuscript. Y.T., Y.M., H.J., A.S., N.B., S.Y., K.N. (Ken Nishioka), and J.Y. provided extensive experimental supports. K.N. (Kazuno Negishi) reviewed and revised the manuscript. T.K. supervised the whole study. All authors have read and agreed to the published version of the manuscript.

Funding: The present study was supported by Grants-in-Aid for Scientific Research (KAKENHI, number 15K10881, and 18K09424) from the Ministry of Education, Culture, Sports, Science and Technology (MEXT) to T.K. and JST SPRING (number JPMJSP2123) to D.L.

Institutional Review Board Statement: Whole animal experimental procedures were followed by the Ethics Committee on Animal Research of Keio University School of Medicine (#16017), the ARVO Statement for the Use of Animals in Ophthalmic and Vision Research, and the International Standards of Animal Care and Use, Animal Research: Reporting in Vivo Experiments.

Informed Consent Statement: Not applicable.

Data Availability Statement: The current study's data are available on request from the corresponding author.

Conflicts of Interest: Y.M. is employed by Aichi Animal Eye Clinic. The other authors declare no conflict of interest.

References

1. Vilares-Morgado, R.; Nunes, H.M.M.; Dos Reis, R.S.; Barbosa-Breda, J. Management of ocular arterial ischemic diseases A review. *Graefe's Arch. Clin. Exp.* **2022**. [CrossRef] [PubMed]
2. Long, C.P.; Chan, A.X.; Bakhoum, C.Y.; Toomey, C.B.; Madala, S.; Garg, A.K.; Freeman, W.R.; Goldbaum, M.H.; DeMaria, A.N.; Bakhoum, M.F. Prevalence of subclinical retinal ischemia in patients with cardiovascular disease-a hypothesis driven study. *EClinicalMedicine* **2021**, *33*, 100775. [CrossRef]
3. Lee, D.; Tomita, Y.; Yang, L.; Negishi, K.; Kurihara, T. Ocular Ischemic Syndrome and Its Related Experimental Models. *Int. J. Mol. Sci.* **2022**, *23*, 5249. [CrossRef]
4. Vestergaard, N.; Cehofski, L.J.; Honoré, B.; Aasbjerg, K.; Vorum, H. Animal Models Used to Simulate Retinal Artery Occlusion: A Comprehensive Review. *Transl. Vis. Sci. Technol.* **2019**, *8*, 23. [CrossRef]
5. Osborne, N.N.; Casson, R.J.; Wood, J.P.; Chidlow, G.; Graham, M.; Melena, J. Retinal ischemia: Mechanisms of damage and potential therapeutic strategies. *Prog. Retin. Eye Res.* **2004**, *23*, 91–147. [CrossRef]
6. Minhas, G.; Morishita, R.; Anand, A. Preclinical models to investigate retinal ischemia: Advances and drawbacks. *Front. Neurol.* **2012**, *3*, 75. [CrossRef] [PubMed]
7. Nadeeshani, H.; Li, J.; Ying, T.; Zhang, B.; Lu, J. Nicotinamide mononucleotide (NMN) as an anti-aging health product-Promises and safety concerns. *J. Adv. Res.* **2022**, *37*, 267–278. [CrossRef]
8. Okabe, K.; Yaku, K.; Tobe, K.; Nakagawa, T. Implications of altered NAD metabolism in metabolic disorders. *J. Biomed. Sci.* **2019**, *26*, 34. [CrossRef]
9. Yoshino, J.; Mills, K.F.; Yoon, M.J.; Imai, S. Nicotinamide mononucleotide, a key NAD(+) intermediate, treats the pathophysiology of diet- and age-induced diabetes in mice. *Cell Metab.* **2011**, *14*, 528–536. [CrossRef] [PubMed]
10. Hong, W.; Mo, F.; Zhang, Z.; Huang, M.; Wei, X. Nicotinamide Mononucleotide: A Promising Molecule for Therapy of Diverse Diseases by Targeting NAD+ Metabolism. *Front. Cell Dev. Biol.* **2020**, *8*, 246. [CrossRef] [PubMed]
11. Chen, X.; Amorim, J.A.; Moustafa, G.A.; Lee, J.J.; Yu, Z.; Ishihara, K.; Iesato, Y.; Barbisan, P.; Ueta, T.; Togka, K.A.; et al. Neuroprotective effects and mechanisms of action of nicotinamide mononucleotide (NMN) in a photoreceptor degenerative model of retinal detachment. *Aging* **2020**, *12*, 24504–24521. [CrossRef]
12. Lee, D.; Tomita, Y.; Miwa, Y.; Shinojima, A.; Ban, N.; Yamaguchi, S.; Nishioka, K.; Negishi, K.; Yoshino, J.; Kurihara, T. Nicotinamide Mononucleotide Prevents Retinal Dysfunction in a Mouse Model of Retinal Ischemia/Reperfusion Injury. *Int. J. Mol. Sci.* **2022**, *23*, 11228. [CrossRef] [PubMed]
13. Lin, J.B.; Kubota, S.; Ban, N.; Yoshida, M.; Santeford, A.; Sene, A.; Nakamura, R.; Zapata, N.; Kubota, M.; Tsubota, K.; et al. NAMPT-Mediated NAD(+) Biosynthesis Is Essential for Vision In Mice. *Cell Rep.* **2016**, *17*, 69–85. [CrossRef]
14. Mills, K.F.; Yoshida, S.; Stein, L.R.; Grozio, A.; Kubota, S.; Sasaki, Y.; Redpath, P.; Migaud, M.E.; Apte, R.S.; Uchida, K.; et al. Long-Term Administration of Nicotinamide Mononucleotide Mitigates Age-Associated Physiological Decline in Mice. *Cell Metab.* **2016**, *24*, 795–806. [CrossRef]
15. Lee, D.; Nakai, A.; Miwa, Y.; Tomita, Y.; Serizawa, N.; Katada, Y.; Hatanaka, Y.; Tsubota, K.; Negishi, K.; Kurihara, T. Retinal Degeneration in a Murine Model of Retinal Ischemia by Unilateral Common Carotid Artery Occlusion. *BioMed Res. Int.* **2021**, *2021*, 7727648. [CrossRef]
16. Lee, D.; Jeong, H.; Miwa, Y.; Shinojima, A.; Katada, Y.; Tsubota, K.; Kurihara, T. Retinal dysfunction induced in a mouse model of unilateral common carotid artery occlusion. *PeerJ* **2021**, *9*, e11665. [CrossRef] [PubMed]
17. Lee, D.; Kang, H.; Yoon, K.Y.; Chang, Y.Y.; Song, H.B. A mouse model of retinal hypoperfusion injury induced by unilateral common carotid artery occlusion. *Exp. Eye Res.* **2020**, *201*, 108275. [CrossRef] [PubMed]
18. Lee, D.; Tomita, Y.; Miwa, Y.; Jeong, H.; Mori, K.; Tsubota, K.; Kurihara, T. Fenofibrate Protects against Retinal Dysfunction in a Murine Model of Common Carotid Artery Occlusion-Induced Ocular Ischemia. *Pharmaceuticals* **2021**, *14*, 223. [CrossRef]
19. Lee, D.; Tomita, Y.; Jeong, H.; Miwa, Y.; Tsubota, K.; Negishi, K.; Kurihara, T. Pemafibrate Prevents Retinal Dysfunction in a Mouse Model of Unilateral Common Carotid Artery Occlusion. *Int. J. Mol. Sci.* **2021**, *22*, 9408. [CrossRef] [PubMed]
20. Shade, C. The Science Behind NMN-A Stable, Reliable NAD+Activator and Anti-Aging Molecule. *Integr. Med.* **2020**, *19*, 12–14.
21. Yoshino, J.; Baur, J.A.; Imai, S.I. NAD(+) Intermediates: The Biology and Therapeutic Potential of NMN and NR. *Cell Metab.* **2018**, *27*, 513–528. [CrossRef] [PubMed]
22. Nagahisa, T.; Yamaguchi, S.; Kosugi, S.; Homma, K.; Miyashita, K.; Irie, J.; Yoshino, J.; Itoh, H. Intestinal Epithelial NAD+ Biosynthesis Regulates GLP-1 Production and Postprandial Glucose Metabolism in Mice. *Endocrinology* **2022**, *163*, bqac023. [CrossRef] [PubMed]
23. Liu, X.; Dilxat, T.; Shi, Q.; Qiu, T.; Lin, J. The combination of nicotinamide mononucleotide and lycopene prevents cognitive impairment and attenuates oxidative damage in D-galactose induced aging models via Keap1-Nrf2 signaling. *Gene* **2022**, *822*, 146348. [CrossRef]
24. Luo, C.; Ding, W.; Yang, C.; Zhang, W.; Liu, X.; Deng, H. Nicotinamide Mononucleotide Administration Restores Redox Homeostasis via the Sirt3-Nrf2 Axis and Protects Aged Mice from Oxidative Stress-Induced Liver Injury. *J. Proteome Res.* **2022**, *21*, 1759–1770. [CrossRef]
25. Pu, Q.; Guo, X.X.; Hu, J.J.; Li, A.L.; Li, G.G.; Li, X.Y. Nicotinamide mononucleotide increases cell viability and restores tight junctions in high-glucose-treated human corneal epithelial cells via the SIRT1/Nrf2/HO-1 pathway. *Biomed. Pharmacother.* **2022**, *147*, 112659. [CrossRef] [PubMed]

26. Wei, C.C.; Kong, Y.Y.; Li, G.Q.; Guan, Y.F.; Wang, P.; Miao, C.Y. Nicotinamide mononucleotide attenuates brain injury after intracerebral hemorrhage by activating Nrf2/HO-1 signaling pathway. *Sci. Rep.* **2017**, *7*, 717. [CrossRef] [PubMed]
27. She, J.; Sheng, R.; Qin, Z.H. Pharmacology and Potential Implications of Nicotinamide Adenine Dinucleotide Precursors. *Aging Dis.* **2021**, *12*, 1879–1897. [CrossRef]
28. Palmer, R.D.; Elnashar, M.M.; Vaccarezza, M. Precursor comparisons for the upregulation of nicotinamide adenine dinucleotide. Novel approaches for better aging. *Aging Med.* **2021**, *4*, 214–220. [CrossRef] [PubMed]
29. Shariati, M.A.; Park, J.H.; Liao, Y.J. Optical coherence tomography study of retinal changes in normal aging and after ischemia. *Investig. Ophthalmol. Vis. Sci.* **2015**, *56*, 2790–2797. [CrossRef] [PubMed]
30. Ho, J.K.; Stanford, M.P.; Shariati, M.A.; Dalal, R.; Liao, Y.J. Optical coherence tomography study of experimental anterior ischemic optic neuropathy and histologic confirmation. *Investig. Ophthalmol. Vis. Sci.* **2013**, *54*, 5981–5988. [CrossRef] [PubMed]
31. Ackermann, P.; Brachert, M.; Albrecht, P.; Ringelstein, M.; Finis, D.; Geerling, G.; Aktas, O.; Guthoff, R. Alterations of the outer retina in non-arteritic anterior ischaemic optic neuropathy detected using spectral-domain optical coherence tomography. *Clin. Exp. Ophthalmol.* **2017**, *45*, 496–508. [CrossRef] [PubMed]
32. Kwon, D.H.; Kim, Y.C.; Kang, K.T. Clinical Significance of Choroidal Thickness in Eyes with Ocular Ischemic Syndrome. *Korean J. Ophthalmol. KJO* **2022**, *36*, 66–73. [CrossRef] [PubMed]
33. Jeong, A.; Yao, X.; van Hemert, J.; Sagong, M. Clinical significance of metabolic quantification for retinal nonperfusion in diabetic retinopathy. *Sci. Rep.* **2022**, *12*, 9342. [CrossRef] [PubMed]
34. Singh, C. Metabolism and Vascular Retinopathies: Current Perspectives and Future Directions. *Diagnostics* **2022**, *12*, 903. [CrossRef] [PubMed]
35. Duh, E.J.; Sun, J.K.; Stitt, A.W. Diabetic retinopathy: Current understanding, mechanisms, and treatment strategies. *JCI Insight* **2017**, *2*, e93751. [CrossRef]
36. Tesch, G.H.; Allen, T.J. Rodent models of streptozotocin-induced diabetic nephropathy. *Nephrology* **2007**, *12*, 261–266. [CrossRef]
37. Rochlani, Y.; Pothineni, N.V.; Kovelamudi, S.; Mehta, J.L. Metabolic syndrome: Pathophysiology, management, and modulation by natural compounds. *Ther. Adv. Cardiovasc. Dis.* **2017**, *11*, 215–225. [CrossRef] [PubMed]
38. Barroso, I.; McCarthy, M.I. The Genetic Basis of Metabolic Disease. *Cell* **2019**, *177*, 146–161. [CrossRef]
39. Yilmaz, B.S.; Gurung, S.; Perocheau, D.; Counsell, J.; Baruteau, J. Gene therapy for inherited metabolic diseases. *J. Mother Child* **2020**, *24*, 53–64. [CrossRef]
40. Miwa, Y.; Tsubota, K.; Kurihara, T. Effect of midazolam, medetomidine, and butorphanol tartrate combination anesthetic on electroretinograms of mice. *Mol. Vis.* **2019**, *25*, 645–653.

International Journal of *Molecular Sciences*

MDPI

Review

Tirzepatide: A Systematic Update

Imma Forzano [1], Fahimeh Varzideh [2,3], Roberta Avvisato [1,2,3], Stanislovas S. Jankauskas [2,3], Pasquale Mone [2,3] and Gaetano Santulli [1,2,3,*]

1 Department of Advanced Biomedical Sciences, Division of Cardiology, "*Federico II*" University, 80131 Naples, Italy
2 Department of Medicine, Division of Cardiology, Wilf Family Cardiovascular Research Institute, Einstein Institute for Aging Research, Albert Einstein College of Medicine, New York, NY 10461, USA
3 Department of Molecular Pharmacology, Einstein Institute for Neuroimmunology and Inflammation (*INI*), Einstein-Mount Sinai Diabetes Research Center (*ES-DRC*), Fleischer Institute for Diabetes and Metabolism (*FIDAM*), Albert Einstein College of Medicine, New York, NY 10461, USA
* Correspondence: gaetano.santulli@einsteinmed.edu

Abstract: Tirzepatide is a new molecule capable of controlling glucose blood levels by combining the dual agonism of Glucose-Dependent Insulinotropic Polypeptide (GIP) and Glucagon-Like Peptide-1 (GLP-1) receptors. GIP and GLP1 are incretin hormones: they are released in the intestine in response to nutrient intake and stimulate pancreatic beta cell activity secreting insulin. GIP and GLP1 also have other metabolic functions. GLP1, in particular, reduces food intake and delays gastric emptying. Moreover, Tirzepatide has been shown to improve blood pressure and to reduce Low-Density Lipoprotein (LDL) cholesterol and triglycerides. Tirzepatide efficacy and safety were assessed in a phase III SURPASS 1–5 clinical trial program. Recently, the Food and Drug Administration approved Tirzepatide subcutaneous injections as monotherapy or combination therapy, with diet and physical exercise, to achieve better glycemic blood levels in patients with diabetes. Other clinical trials are currently underway to evaluate its use in other diseases. The scientific interest toward this novel, first-in-class medication is rapidly increasing. In this comprehensive and systematic review, we summarize the main results of the clinical trials investigating Tirzepatide and the currently available meta-analyses, emphasizing novel insights into its adoption in clinical practice for diabetes and its future potential applications in cardiovascular medicine.

Keywords: cardiovascular medicine; diabetes; hypertension; GIP; GLP-1; glucagon; incretins; LY3298176; meta-analysis; obesity; SUMMIT; SURMOUNT; SURPASS; SYNERGY; Tirzepatide

Citation: Forzano, I.; Varzideh, F.; Avvisato, R.; Jankauskas, S.S.; Mone, P.; Santulli, G. Tirzepatide: A Systematic Update. *Int. J. Mol. Sci.* **2022**, *23*, 14631. https://doi.org/10.3390/ijms232314631

Academic Editors: Simona Gabriela Bungau and Vesa Cosmin

Received: 31 October 2022
Accepted: 22 November 2022
Published: 23 November 2022

1. Introduction

Type 2 Diabetes Mellitus (T2DM) is an important independent risk factor for atherosclerotic cardiovascular disease [1], is frequently associated with obesity, and has been linked to high morbidity and mortality rates [2]. Glucose-Dependent Insulinotropic Polypeptide (GIP) and Glucagon-Like Peptide-1 (GLP-1), also known as incretins, are hormones released in the intestine in response to the intake of nutrients, and they are capable of stimulating pancreatic beta cells to release insulin, thereby participating in the regulation of glucose homeostasis [3–5].

GIP inhibits gastric secretion activity, stimulates insulin secretion, and has insulin-like action on adipose tissue inhibiting lipolysis and promoting lipogenesis [6–8]. GLP1 is able to stimulate insulin secretion and inhibit glucagon release; it also slows gastric emptying and induces a sense of satiety [9]. Incretins are rapidly degraded by dipeptidyl-peptidase 4 (DPP4) [10,11]. Thus, DPP4 inhibitors and GLP1 receptor agonists have been used to date as drugs for the treatment of T2DM acting on the incretin system [12–14]. Mounting evidence indicates that the contemporary administration of GIP and GLP1 has a synergistic effect by significantly increasing insulin response and glucagonostatic action [15–25].

A new molecule with a combined agonist action on both GPI and GLP1 receptors ("twincretin") has been developed to take advantage of this synergistic effect: Tirzepatide (LY3298176) [26–28] is the first "twincretin", a synthetic peptide composed of 39 amino acids based on the GIP native sequence [29,30], combining the dual agonism of GIP and GLP-1 receptors controlling glycemic blood level and reducing body weight (Figure 1). Tirzepatide has an affinity for the GIP receptor equal to that of native GIP and an affinity for the GLP-1 receptor ~5-fold weaker than that of native GLP-1 [31,32]. Tirzepatide can also improve parameters related to cardiovascular risk, including blood pressure (BP) [33], waist circumference, LDL, and circulating triglycerides [34–37]. Both its effectiveness and safety have been demonstrated in several clinical trials.

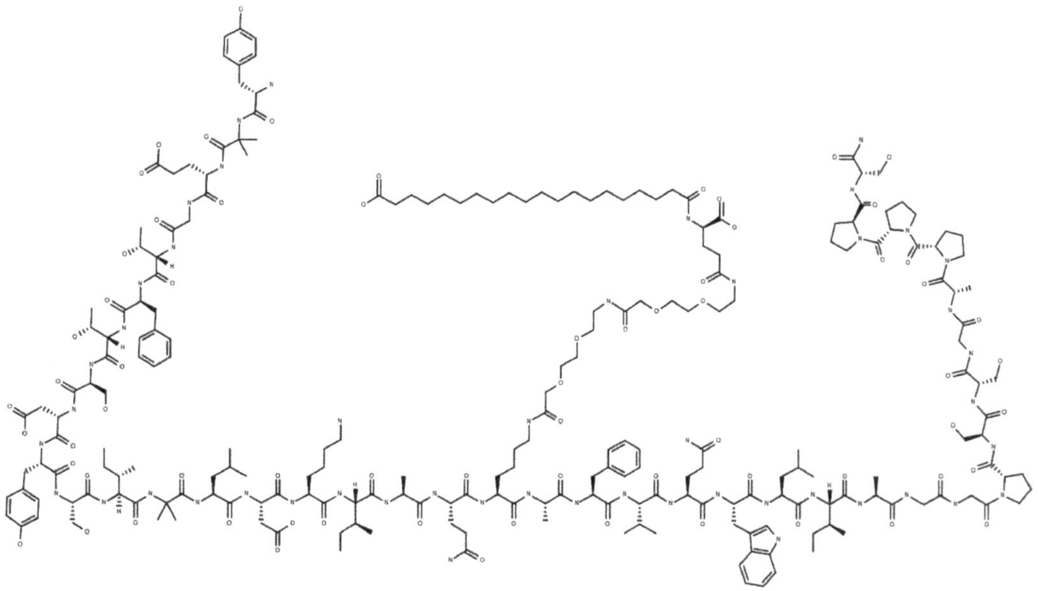

Figure 1. Structure of Tirzepatide (39 amino acids), a dual GIP/GLP-1 receptor agonist (twincretin) that has two non-coded amino acid residues (α-aminoisobutyric acid) at positions 2 and 13 and is acylated on K20 with a γGlu-2×OEG linker and C18 fatty diacid moiety.

In this comprehensive review, we analyze the main results of the trials investigating Tirzepatide and discuss its future use as a drug to tackle T2DM and cardiovascular disorders.

2. Tirzepatide in the SURPASS Trials

The *Study of Tirzepatide in Participants with T2DM Not Controlled with Diet and Exercise Alone* (SURPASS) clinical trials were designed to demonstrate the effectiveness and safety of Tirzepatide as a hypoglycemic drug in patients affected by T2DM. The primary outcome was the mean change in glycated hemoglobin (HbA1c) from baseline.

SURPASS-1 was a randomized, double blinded, clinical trial that evaluated the effectiveness of Tirzepatide (subcutaneous injections once weekly) compared to placebo in patients with T2DM inadequately controlled by diet and exercise [38]. This trial demonstrated that Tirzepatide is superior to placebo in improving glycemic control and reducing body weight [38]. Indeed, Tirzepatide, at all doses tested, was better than placebo in inducing changes from baseline in HbA1c, fasting serum glucose, body weight, and HbA1c targets of less than 7.0% (<53 mmol/mol) and less than 5.7% (<39 mmol/mol) at 40 weeks. Participants were randomly assigned to Tirzepatide 5 mg (n = 121), Tirzepatide 10 mg (n = 121), Tirzepatide 15 mg (n = 120), or placebo (n = 113) (Table 1).

Table 1. Results of the SURPASS-1 trial at 40 weeks [26]. Efficacy analyses were performed in the modified intention-to-treat (mITT) population, which includes all randomized participants receiving at least one treatment dose; * $p < 0.0001$; ** $p = 0.72$; *** $p = 0.22$.

Outcomes		Tirzepatide 5 mg (n = 121)	Tirzepatide 10 mg (n = 121)	Tirzepatide 15 mg (n = 120)	Placebo (n = 113)
HbA1c (%)	Baseline	7.97	7.88	7.88	8.08
	From baseline	−1.87 *	−1.89 *	−2.07 *	0.04 **
	Versus placebo	−1.91 *	−1.93 *	−2.11 *	-
Weight (kg)	From baseline	−7.0 *	−7.8 *	−9.5 *	−0.7 ***
	Versus placebo	−6.3 *	−7.1 *	−8.3 *	-

In summary, SURPASS-1 established the effectiveness of treatment with Tirzepatide once a week at doses of 5, 10, and 15 mg as monotherapy for T2DM compared with placebo in improving glycemic control. Tirzepatide exhibited robust effectiveness compared with placebo in glycemic control, with 31–52% of participants reaching normoglycemia (HbA1c < 5.7%; <39 mmol/mol) and meaningful reductions in body weight without an increased risk of clinically significant or severe hypoglycemia (<54 mg/dL; <3 mmol/L). The range of weight reduction was from 7 to 9.5 kg. The safety profile was consistent with GLP-1 receptor agonists [39,40].

SURPASS-2 was an open-label, phase III trial that compared Tirzepatide with the GLP-1R agonist Semaglutide once weekly in patients with T2DM [41]. Tirzepatide at all doses was noninferior and superior to Semaglutide with respect to the mean changes in HbA1c levels from baseline at 40 weeks [41]. Patients were randomly assigned in a 1:1:1:1 ratio to receive Tirzepatide at a dose of 5 mg (n = 470), 10 mg (n = 469), or 15 mg (n = 470) or Semaglutide (n = 469) at a dose of 1 mg. The mean changes from baseline in HbA1c levels were −2.01% with Tirzepatide 5 mg, −2.24% with Tirzepatide 10 mg, −2.30 % with Tirzepatide 15 mg, and −1.86% with Semaglutide; the differences between the 5 mg, 10 mg, and 15 mg Tirzepatide groups and the Semaglutide group were −0.15 %, −0.39 %, and −0.45%, respectively.

The HbA1c level target of less than 5.7% (normoglycemia) was met in 27 to 46% of the patients who received Tirzepatide and in 19% of the patients who received Semaglutide. Reductions in body weight were greater with Tirzepatide (from −7.6 to −11.2 kg) than with Semaglutide (−5.7 kg); the differences were −1.9 kg, −3.6 kg, and −5.5 kg with Tirzepatide 5 mg, 10 mg, and 15 mg, respectively (Table 2). These findings suggest that an increased prescription of Tirzepatide could have favorable effects on the control of diabetes and obesity [42].

Table 2. Results of the SURPASS-2 trial at 40 weeks [41]. Efficacy analyses were performed in the mITT population; * $p < 0.001$; ** $p = 0.02$.

Outcomes		Tirzepatide 5 mg (n = 470)	Tirzepatide 10 mg (n = 469)	Tirzepatide 15 mg (n = 470)	Semaglutide (n = 469)
HbA1c (%)	Baseline	8.32	8.30	8.26	8.25
	From baseline	−2.01	−2.24	−2.30	−1.86
	Versus Semaglutide	−0.15 **	−0.39 *	−0.45 *	-
Weight (kg)	From baseline	−7.6	−9.3	−11.2	−5.7
	Versus Semaglutide	−1.9 *	−3.6 *	−5.5 *	-

An adjusted indirect treatment comparison, using aggregate data from the SURPASS-2 study that met the HbA1c inclusion criterion of the *Research Study to Compare Two Doses of Semaglutide Taken Once Weekly in People with Type 2 Diabetes* (SUSTAIN FORTE trial) and from SUSTAIN FORTE metformin-only-treated patients, confirmed that HbA1c and weight reductions were significantly greater for Tirzepatide 10 and 15 mg versus Semaglutide 2 mg [43].

In the SURPASS-3 trial, once-weekly Tirzepatide was compared with once-daily insulin degludec as an add-on to metformin with or without inhibitors of Sodium/Glucose Transporter 2 (SGLT2) in patients with T2DM not adequately controlled [44]. In this open-label, phase 3 study, participants were randomly assigned (1:1:1:1) to once-weekly subcutaneous injections of Tirzepatide 5 (n = 358), 10 (n = 360), or 15 (n = 358) mg or once-daily subcutaneous injections of titrated insulin degludec (n = 359). The mean changes from baseline in the HbA1c levels were −1.93% with Tirzepatide 5 mg, −2.20% with Tirzepatide 10 mg, −2.37% with Tirzepatide 15 mg, and −1.34% with insulin degludec. The differences between the 5 mg, 10 mg, and 15 mg Tirzepatide groups and the degludec group were −0.59%, −0.86%, and −1.04%, respectively (Table 3). In this study, up to 93% of participants receiving Tirzepatide achieved the HbA1c target of less than 7.0%, and 26–48% of participants treated with Tirzepatide achieved normoglycemia (HbA1c < 5.7%; <39 mmol/mol) [45].

Table 3. Results of the SURPASS-3 trial at 52 weeks [44]. Efficacy analyses were performed in the mITT population; * $p < 0.0001$.

Outcomes		Tirzepatide 5 mg (n = 358)	Tirzepatide 10 mg (n = 360)	Tirzepatide 15 mg (n = 358)	Insulin Degludec (n = 359)
HbA1c (%)	Baseline	8.17	8.19	8.21	8.13
	From baseline	−1.93	−2.20	−2.37	−1.34
	Versus insulin degludec	−0.59 *	−0.86 *	−1.04 *	-
Weight (kg)	From baseline	−7.5	−10.7	−12.9	2.3
	Versus insulin degludec	−9.8 *	−13 *	−15.2 *	-

Reductions in body weight were greater with Tirzepatide than with insulin degludec. The differences were −9.8 kg with Tirzepatide 5 mg, −13 kg with Tirzepatide 10 mg, and −15.2 kg with Tirzepatide 15 mg. Once-weekly treatment with Tirzepatide also led to superior glycemic control, measured using continuous glucose monitoring (CGM), compared to insulin degludec in participants with T2DM on metformin, with or without an SGLT2 inhibitor [46]. A sub-study of the SURPASS-3 trial [47], using magnetic resonance imaging (MRI), demonstrated that Tirzepatide was also able to significantly reduce liver fat content (LFC) [48], the volume of visceral adipose tissue (VAT), and abdominal subcutaneous adipose tissue (ASAT) [47].

In order to provide an integrated measure of the physiological effects of Tirzepatide on glucose homoeostasis, a double-blind, randomized, parallel-arm, phase 1 study that compared the hormonal and metabolic effects of Tirzepatide (titrated to once-weekly 15 mg) with those of Semaglutide (titrated to once-weekly 1 mg) or placebo in patients with T2DM was designed, using gold-standard dynamic stimulatory tests to assess insulin secretion and insulin sensitivity (i.e., hyperglycemic and hyperinsulinemic euglycemic clamps) and using mixed-meal tolerance tests [49]. The trial revealed that the increase in the clamp disposition index, which adjusts insulin secretion for concurrent insulin sensitivity, was significantly larger in the Tirzepatide arm compared to the Semaglutide arm ($p < 0.0001$), with improvements in both the total insulin secretion rate and insulin sensitivity [49], the latter possibly being related to a greater weight loss (–11.2 kg vs. –6.9 kg) [50]. The responses to the mixed-meal tolerance test indicated reduced glucose excursions with

Tirzepatide compared to Semaglutide, associated with lower insulin and glucagon plasma concentrations, strongly suggesting a diminished insulin resistance and a reduced workload for pancreatic β cells [49–51].

In the SURPASS-4 trial, Tirzepatide was compared to insulin glargine in patients with a high cardiovascular risk, suffering from T2DM inadequately controlled by oral glucose-lowering medications [52]. Patients were randomly assigned to receive at least one dose of Tirzepatide 5 mg (n = 326), 10 mg (n = 321), 15 mg (n = 334), or insulin glargine (n = 978). The study demonstrated that all three doses of Tirzepatide markedly improved glucose control, reduced body weight, and improved the cardiovascular risk profile in these patients. Indeed, at 52 weeks, the mean HbA1c changes with Tirzepatide were −2.24% with 5 mg, -2.43% with 10 mg, and −2.58% with 15 mg versus −1.44% with insulin glargine [53]. The treatment difference versus insulin glargine was −0.80% for Tirzepatide 5 mg, −0.99% for Tirzepatide 10 mg, and −1.14% for Tirzepatide 15 mg [54]. The percentage of patients achieving normoglycemia using Tirzepatide was 23–43%. The weight loss from baseline was −7.1 kg, −9.5 kg, and −11.7 kg for patients treated with Tirzepatide 5 mg, 10 mg, and 15 mg, respectively, and the weight gain from baseline was +1.9 kg in patients treated with insulin glargine. The treatment differences versus insulin glargine were −9 kg for Tirzepatide 5 mg, −11.4 kg for Tirzepatide 10 mg, and −13.5 kg for Tirzepatide 15 mg (Table 4).

Table 4. Results of the SURPASS-4 trial at 52 weeks [52]. Efficacy analyses were performed in the mITT population; * $p < 0.0001$.

Outcomes		**Tirzepatide 5 mg (n = 326)**	**Tirzepatide 10 mg (n = 321)**	**Tirzepatide 15 mg (n = 334)**	**Insulin Glargine (n = 978)**
HbA1c (%)	Baseline	8.52	8.60	8.52	8.51
	From baseline	−2.24	−2.43	−2.58	−1.44
	Versus insulin glargine	−0.80 *	−0.99 *	−1.14 *	
Weight (kg)	From baseline	−7.1	−9.5	−11.7	1.9
	Versus insulin glargine	−9.0 *	−11.4 *	−13.5 *	

Important limitations of the SURPASS-4 trial should be noted, including the fact that the interventions were not blinded because of differences in devices or dose escalation schemes and that not all participants were treated for 104 weeks; moreover, the study only included patients with a body mass index (BMI) $\geq$ 25 kg/m^2 and a stable weight ($\leq$5% fluctuation in either direction) during the previous 3 months; thus, the effects in subjects with a lower BMI remain to be determined. Intriguingly, a post hoc analysis of the SURPASS-4 trial demonstrated that, in people with T2DM and a high cardiovascular risk, Tirzepatide slowed the rate of the decline of the estimated glomerular filtration rate (eGFR) and reduced the urine albumin–creatinine ratio in clinically meaningful ways compared with insulin glargine [55].

In the SURPASS-5 phase III trial, among patients with T2DM and inadequate glycemic control despite treatment with insulin glargine, the addition of subcutaneous Tirzepatide, compared with placebo, to titrated insulin glargine resulted in statistically significant improvements in glycemic control after 40 weeks [56]. Patients were randomized in a 1:1:1:1 ratio to receive once-weekly subcutaneous injections of 5 mg (n = 116), 10 mg (n = 119), or 15 mg (n = 120) Tirzepatide or volume-matched placebo (n = 120) over 40 weeks. At the end of the study, the mean HbA1c changes from baseline were −2.11% with 5 mg Tirzepatide, −2.40% with 10 mg Tirzepatide, and −2.34% with 15 mg Tirzepatide versus −0.86% with placebo. The differences between the 5 mg, 10 mg, and 15 mg Tirzepatide groups and the placebo group were −1.24%, −1.53%, and −1.47%, respectively. The proportion of patients treated with Tirzepatide, at all doses, reaching normoglycemia was 24.4–49.6%. The mean body weight changes from baseline were −5.4 kg with 5 mg Tirzepatide, −7.5 kg with 10 mg Tirzepatide, −8.8 kg with 15 mg Tirzepatide, and +1.6 kg with placebo [57]. The differences between the 5 mg, 10 mg, and 15 mg Tirzepatide groups and the placebo group were −7.1 kg, −9.1 kg, and −10.5 kg, respectively (Table 5).

Table 5. Results of the SURPASS-5 at 40 weeks [56]. Efficacy analyses were performed in the mITT population; * $p < 0.001$.

Outcomes		Tirzepatide 5 mg (n= 116)	Tirzepatide 10 mg (n= 119)	Tirzepatide 15 mg (n= 120)	Placebo (n= 120)
HbA1c (%)	Baseline	8.30	8.36	8.22	8.38
	From baseline	−2.11	−2.40	−2.34	−0.86
	Versus placebo	−1.24 *	−1.53 *	−1.47 *	
Weight (kg)	From baseline	−5.4	−7.5	−8.8	1.6
	Versus placebo	−7.1 *	−9.1 *	−10.5 *	

In SURPASS J-mono, a randomized phase III clinical trial conducted on Japanese patients affected by T2DM, Tirzepatide at doses of 5 mg, 10 mg, and 15 mg was compared to the GLP-1 agonist Dulaglutide (0.75 mg). Tirzepatide was superior to Dulaglutide for both glycemic control and reductions in body weight. The safety profile of Tirzepatide was consistent with that of GLP-1 receptor agonists, indicating a potential therapeutic use in Japanese patients with T2DM [58]. Compared to Dulaglutide, Tirzepatide showed greater potential for normalizing metabolic factors after a standardized meal, significantly reducing body weight and body fat mass [59].

Similarly, the SURPASS J-combo assessed the safety and glycemic efficacy of Tirzepatide as an add-on treatment in Japanese patients with T2DM who had inadequate glycemic control with stable doses of various oral antihyperglycemic monotherapies. This trial confirmed that Tirzepatide is well-tolerated as an add-on to oral antihyperglycemic monotherapy in Japanese participants with T2DM and demonstrated a significant improvement in glycemic control and reductions in body weight, irrespective of the background oral antihyperglycemic medication [60].

3. Tirzepatide and Obesity: The SURMOUNT-1 Trial

Obesity and, consequently, excess adiposity are related to numerous complications, including hypertension, dyslipidemia, and T2DM. In particular, body mass index (BMI) and waist circumference are known to be associated with T2DM and cardiovascular disease [61–64]. Obesity contributes to the development of cardiovascular disease and to cardiovascular mortality independently of other cardiovascular risk factors [65]. In the SURPASS 1–5 trials, weight reduction after Tirzepatide treatment was significant; BMI and waist circumference were analyzed showing a significant reduction in these parameters, suggesting a better metabolic profile in response to Tirzepatide. These results are very encouraging for the use of Tirzepatide to treat patients affected by obesity with or without T2DM.

The *Study of Tirzepatide in Participants with Obesity or Overweight* (SURMOUNT-1), a phase III, double-blind, randomized, controlled clinical trial, revealed that, in participants with a confirmed diagnosis of obesity, Tirzepatide 5 mg, 10 mg, or 15 mg once a week for 72 weeks provided substantial and sustained reductions in body weight [66]. Tirzepatide improved cardiometabolic measures, such as waist circumference. The exclusion criteria were diabetes, a change in body weight of more than 5 kg within 90 days before screening, previous or planned surgical treatment for obesity, and treatment with a medication that promotes weight loss within 90 days before screening. The co-primary endpoints were the percentage change in weight from baseline and a weight reduction of 5% or more.

Participants were randomly assigned to receive Tirzepatide administered subcutaneously in a 1:1:1:1 ratio at a dose of 5 mg (n = 630), 10 mg (n = 636), or 15 mg (n = 636) or placebo (n = 643) once a week for 72 weeks as an adjunct to lifestyle intervention. To obtain a meaningful effect on the basis of improvement in metabolic health, there should be a body weight reduction of 5% or more [67]. In the SURMOUNT-1 trial, the mean percentage changes in weight at the end of the study were 15 0%, −19.5%, and −20.9% with Tirzepatide 5 mg, 10 mg, and 15 mg, respectively, and −3.1% with placebo. The percentages of participants who had a weight reduction of 5% or more were 85%, 89%, and 91% with Tirzepatide 5 mg, 10 mg, and 15 mg, respectively, and 35% with placebo; 50% of participants in the 10 mg group and 57% of participants in the 15 mg group had a reduction in body weight of 20% or more, as compared with 3% in the placebo group [68–70]. Reductions in waist circumferences were also observed: in the Tirzepatide group, there was a mean reduction in waist circumference of −14 cm with 5 mg, −17.7 with 10 mg, and 18.5 cm with 15 mg versus a reduction of 4 cm with placebo. The estimated differences between the 5 mg, 10 mg, and 15 mg Tirzepatide groups and the placebo group were −10.1 cm, −13.8 cm, and −14.5 cm, respectively (Table 6).

Table 6. Results of the SURMOUNT-1 trial at 72 weeks [66]. Efficacy analyses were done in the mITT population; * $p < 0.001$.

Outcomes		**Tirzepatide 5 mg (n = 630)**	**Tirzepatide 10 mg (n = 636)**	**Tirzepatide 15 mg (n = 630)**	**Placebo (n = 643)**
Body Weight (kg)	Baseline	102.9	105.8	105.6	104.8
	From baseline (%)	−15	−19.5	−20.9	−3.1
	Versus placebo (%)	−11.9 *	−16.4 *	−17.8 *	
Waist circumference (cm)	From baseline	−14	−17.7	−18.5	−4
	Versus placebo	−10.1 *	−13.8 *	−14.5 *	

Of course, further studies are necessary to fully establish the effectiveness of Tirzepatide in improving lipid profile and metabolic syndrome and in reducing cardiovascular risk independently of glycemic status. Ideally, future studies should have a longer follow-up period. Indeed, Benfluorex, a fenfluramine derivative designed for weight loss, was associated with mitral regurgitation after two years of use in a case–control retrospective study [71].

4. Tirzepatide and Hypertension

Although hypertension is known to be associated with an increased risk of cardiovascular disease and death, BP control remains suboptimal [72,73]. Tirzepatide has been shown to favorably affect BP [33], and this result has been consistently confirmed in all trials. Indeed, in the SURPASS-1 trial, the mean systolic BP (SBP) decrease ranged from −4.7 to −5.2 mm Hg versus −2.00 mm Hg with placebo, whereas diastolic BP (DBP) did not significantly differ with placebo; in the SURPASS-2 trial, SBP and DBP decreased to −4.8 mm Hg and −1.9 mm Hg, respectively, with Tirzepatide at a dose of 5 mg; −5.3 mm Hg and −2.5 mm Hg, respectively, at a dose of 10 mg; and −6.5 mm Hg and −2.9 mm Hg, respectively, at a dose of 15 mg; versus −3.6 mm Hg and −1.0 mm Hg, respectively, with Semaglutide. In the SURPASS-3 trial, significant decreases in mean SBP of −4.9 to −6.6 mm Hg and in mean DBP of −1.9 to −2.5 mm Hg were observed for Tirzepatide, while no significant changes were detected for insulin degludec. In the SURPASS-4 trial, mean SBP and DBP decreased with Tirzepatide from −2.8 to −4.8 mm Hg and from −0.80 to −1.0 mm Hg, respectively, and increased with insulin glargine: a +1.3 mm Hg increase in SBP and a +0.7 mm Hg increase in DBP. In the SURPASS-5 trial, the mean changes in SBP and DBP were −6.1 to −12.6 mm Hg and −2.0 to −4.5 mm Hg, respectively, for the Tirzepatide groups and −1.7 mm Hg and −2.1 mm Hg for the placebo group. Similarly, in the SURMOUNT-1 trial, participants treated with Tirzepatide reported a mean reduction in SBP of −7.2 mm Hg and a mean reduction in DBP of −4.8 mm Hg. In the placebo group, a mean reduction in SBP of −1 mmHg and a reduction in DBP of −0.8 mm Hg were observed (Figure 2).

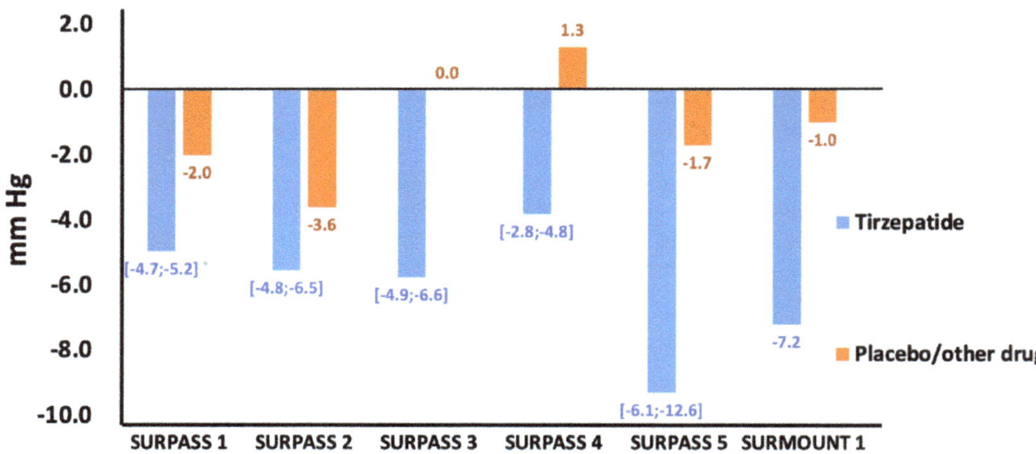

Figure 2. Mean variations in systolic blood pressure (SBP) with Tirzepatide vs. placebo/other drugs in the SURPASS-1 [38], SURPASS-2 [41], SURPASS-3 [44], SURPASS-4 [52], SURPASS-5 [56], and SURMOUNT-1 [66] clinical trials.

The beneficial effects of Tirzepatide on BP could be attributable to an amelioration of endothelial function and/or to reduced inflammation [33], although dedicated studies are needed to define the exact mechanisms underlying these findings. Nevertheless, a post hoc analysis from the phase IIb trial showed that Tirzepatide was associated with a dose-dependent decrease from baseline in the levels of high-sensitivity C-reactive protein, intercellular adhesion molecule 1 (ICAM-1), and YKL-40 at 26 weeks [26,74,75].

5. Data from Meta-Analyses

Different meta-analyses have been conducted to analyze the effectiveness and safety of Tirzepatide. Sattar and collaborators, scrutinizing all studies of the SURPASS program, compared the time to first occurrence of the well-established four major adverse cardiovascular events (MACE-4; cardiovascular death, myocardial infarction, stroke, and hospitalized unstable angina) between pooled Tirzepatide groups and control groups [76]. The stratified Cox proportional hazards model was used (fixed effect: treatment; stratification factor: trial-level cardiovascular risk) for the estimation of hazard ratios (HRs) and confidence intervals (CIs) comparing Tirzepatide to controls. They underlined that, despite the favorable effects of Tirzepatide on a range of CV risk factors, only the results from the SURPASS-4 trial have hitherto reported data on the CV safety of the drug. After their analysis, they concluded that Tirzepatide does not increase the risk of MACE-4 in participants with T2DM (HRs comparing Tirzepatide versus controls were as follows: for MACE-4, 0.80, 95% CI, 0.57–1.11; for cardiovascular death, 0.90, 95% CI, 0.50–1.61; and for all-cause death, 0.80, 95% CI, 0.51–1.25), even if the exclusion of people with unstable cardiovascular disease (such as class IV heart failure) is a limitation [76].

Karagiannis and co-workers [77] conducted a meta-analysis to assess the efficacy and safety of Tirzepatide in T2DM. The results with Tirzepatide demonstrated a dose-dependent superiority on glycemic efficacy and body weight reduction compared to placebo, GLP-1 receptor agonists, and basal insulin. Tirzepatide was associated with an increased incidence of gastrointestinal adverse events: nausea was the most frequent event with all

Tirzepatide doses, especially 15 mg (OR 5.60, 95% CI 3.12 to 10.06); Tirzepatide 15 mg was associated with higher incidences of vomiting (OR 5.50, 95% CI 2.40 to 12.59) and diarrhea (OR 3.31, 95% CI 1.40 to 7.85). The risk of hypoglycemia did not increase with Tirzepatide. The presence of statistical heterogeneity in the meta-analyses for changes in HbA1c and body weight, the assessment of the risk of bias solely for the primary outcome, and the generalization of findings mainly to individuals with overweight or obesity and already on metformin-based background therapy were acknowledged as study limitations [77]. Another pooled analysis [78] confirmed that Tirzepatide treatment resulted in a greater lowering of HbA1c (−1.94%, 95% CI: −2.02 to −1.87), fasting serum glucose (−54.72 mg/dL, 95% CI: −62.05 to −47.39), and body weight (−8.47, 95% CI: −9.66 to −7.27); as far as the safety profile is concerned, the results of this meta-analysis were essentially consistent with those of the above-mentioned analysis conducted by Karagiannis et al. [77]. Similarly, Dutta and colleagues [79] found that individuals receiving Tirzepatide for over 1 year had a significantly greater lowering of HbA1c (−0.75%, 95% CI: −1.05 to −0.45; $p < 0.01$), fasting glucose (−0.75 mmol/L, 95% CI: −1.05 to −0.45; $p < 0.01$), 2 h post-prandial-glucose (−0.87 mmol/L, 95% CI: −1.12 to −0.61; $p < 0.01$), weight (−8.63 kg, 95% CI: −12.89 to −4.36; $p < 0.01$), body mass index (−1.80 kg/m^2, 95% CI: −2.39 to −1.21; $p < 0.01$), and waist circumference (−4.43 cm, 95% CI: −5.31 to −3.55; $p < 0.01$) than individuals receiving Dulaglutide, Semaglutide, insulin degludec, or glargine. According to Guan and collaborators [80] Tirzepatide 10 and 15 mg had better antidiabetic and weight-loss effects (especially the 15 mg dose) compared to insulin (glargine or degludec) and selective GLP1 receptor agonists (Dulaglutide or Semaglutide once a week); Tirzepatide 15 mg greatly reduced glycated hemoglobin (surface under the cumulative ranking curve value, SCURA probability: 93.5%), body weight (99.7%), and fasting serum glucose (86.6%). Insulin caused less gastrointestinal events (93.5%), and there was no statistical difference between GLP1-RA and Tirzepatide. They concluded that additional well-designed RCTs are needed to evaluate its clinical performance with higher doses of GLP1 receptor agonists and to definitively determine the potential cardiovascular benefits [80]. In an evaluation of the optimal dose of Tirzepatide for the treatment of T2DM using a meta-analysis and a trial sequential analysis (TSA), Tirzepatide 15 mg was superior to 10 mg and 5 mg for lowering glycemia and reducing weight; Tirzepatide 5 mg was superior to 10 mg and 15 mg (which appear to have the same effect of the 10 mg) in terms of safety [81].

In their meta-analysis of randomized clinical trials on the efficacy and safety of Tirzepatide as a novel treatment for T2DM, Permana and colleagues concluded that this drug has shown superiority in glycemic control and body weight reduction with a good safety profile in patients with T2DM [82]. Lisco et al. considered eleven clinical trials and concluded that Tirzepatide provides a weight loss that exceeds that obtained with GLP-1 receptor agonists; hence, Tirzepatide is presented as a potent tool to improve glucose control without increasing hypoglycemic risk in poorly controlled T2DM treated with basal insulin with or without other hypoglycemic oral agents with effects on body weight loss, despite the background therapy [83].

Exploring the impact of Tirzepatide on cardiovascular disorders, Patoulias and coworkers conducted a meta-analysis assessing the effect of Tirzepatide on the risk of atrial fibrillation in patients with T2DM. Pooling data from SURPASS-2 to -5, they demonstrated that Tirzepatide compared with placebo or an active comparator did not have a significantly different effect on the risk of atrial fibrillation (risk ratio = 1.59; 95% CI: 0.46 to 5.47; $p = 0.47$) [84]. These findings remain to be confirmed in the forthcoming SURPASS-CVOT, a large phase 3, randomized, double-blind, cardiovascular outcome trial, assessing both the noninferiority and superiority of Tirzepatide against Dulaglutide. Nevertheless, in a subsequent study analyzing eight trials, the same group evidenced that Tirzepatide resulted in a significantly reduced risk of major averse cardiovascular events by 48% compared to a control (RR 0.52, 95% CI 0.38 to 0.72); moreover, Tirzepatide displayed a significantly attenuated risk of cardiovascular death (by 49%; RR 0.51, 95% CI 0.29 to 0.89), as well as all-cause death (by 49%; RR 0.51, 95% CI 0.34 to 0.77) [85].

In conclusion, according to the data currently available, it is reasonable to speculate that, despite the warranted need of other randomized clinical trials, Tirzepatide may be considered a new powerful tool to treat patients with diabetes and/or obesity.

6. New Perspectives

The effectiveness and safety profile of Tirzepatide have opened a wide range of perspectives for its use not limited to patients with T2DM and patients with obesity [86,87].

For instance, in *post hoc* analyses, the effects of Tirzepatide on biomarkers of non-alcoholic fatty liver disease (NAFLD) and fibrosis in patients with T2DM compared to Dulaglutide and placebo for 26 weeks have also been investigated, showing that a higher Tirzepatide dose significantly decreases NAFLD-related biomarkers and increases adiponectin in patients with T2DM [88].

Metabolomics and lipidomics analyses conducted in >250 patients revealed that a 26-week treatment with Tirzepatide significantly modulated a cluster of metabolites and lipids associated with insulin resistance and obesity: 3-hydroxyisobutyrate, branched-chain amino acids, branched-chain ketoacids, and direct catabolic products decreased compared to baseline and placebo [89]; of note, these changes were significantly larger with Tirzepatide than with Dulaglutide, and they were directly proportional to reductions in HbA1c, to indices of insulin resistance, and to proinsulin levels [90].

The beneficial effects of Tirzepatide are associated with a good safety profile. In the SURPASS 1–5 and SURMOUNT-1 clinical trials, the main adverse events occurring with Tirzepatide were mild-to-moderate gastrointestinal disorders during the initial dose-escalation phase without clinically significance evidence or severe hypoglycemia (<54 mg/dL; <3 mmol/L). Specifically, nausea, diarrhea, and vomiting occurred. The safety profile was overall consistent with that of GLP-1 receptor agonists (Table 7).

Table 7. Most commonly reported adverse events of Tirzepatide in clinical trials.

	Nausea (%)	Diarrhea (%)	Vomiting (%)
SURPASS-1	12–18	12–14	2–6
SURPASS-2	17–22	13–16	6–10
SURPASS-3	12–24	15–17	6–10
SURPASS-4	12–23	13–22	5–9
SURPASS-5	13–18	12–21	7–13
SURMOUNT-1	25–33	19–23	8–12

The beneficial influence of Tirzepatide on cardiometabolic parameters suggests the availability, in the near future, of a new weapon effective against cardiovascular and metabolic disorders [88,91–93]. Different phase I, II, and III trials are now ongoing to test the effectiveness and safety of Tirzepatide for its potentials uses (Table 8).

Table 8. Ongoing clinical trials testing Tirzepatide.

SYNERGY-NASH (phase II)	A Randomized, Double-Blind, Placebo-Controlled Phase 2 Study Comparing the Efficacy and Safety of Tirzepatide Versus Placebo in Patients with Nonalcoholic Steatohepatitis (NASH).
SURPASS-CVOT (phase III)	The Effect of Tirzepatide Versus Dulaglutide on Major Adverse Cardiovascular Events in Patients with Type 2 Diabetes.
SURPASS-PEDS (phase III)	Efficacy, Safety, and Pharmacokinetics/Pharmacodynamics of Tirzepatide in Pediatric and Adolescent Participants with Type 2 Diabetes Mellitus Inadequately Controlled with Metformin, or Basal Insulin, or Both.

Table 8. *Cont.*

SURPASS-6 (phase III)	The Effect of the Addition of Tirzepatide Once Weekly Versus Insulin Lispro Three Times Daily in Participants with Type 2 Diabetes Inadequately Controlled on Insulin Glargine (U100) with or without Metformin.
SURMOUNT-2 (phase III)	Efficacy and Safety of Tirzepatide Once Weekly in Participants with Type 2 Diabetes Who Have Obesity or Are Overweight.
SURMOUNT-3 (phase III)	Efficacy and Safety of Tirzepatide Once Weekly Versus Placebo After an Intensive Lifestyle Program in Participants without Type 2 Diabetes Who Have obesity or Are Overweight with Weight-Related Comorbidities.
SURMOUNT-4 (phase III)	Efficacy and Safety of Tirzepatide Once Weekly Versus Placebo for Maintenance of Weight Loss in Participants Without Type 2 Diabetes Who Have Obesity or Are Overweight with Weight-Related Comorbidities: A Randomized, Double-Blind, Placebo-Controlled Trial.
SURMOUNT-CN (phase III)	Efficacy and Safety of Tirzepatide Once Weekly in Chinese Participants Without Type 2 Diabetes Who Have Obesity or Are Overweight with Weight-Related Comorbidities: A Randomized, Double-Blind, Placebo-Controlled Trial.
SURMOUNT-J (phase III)	Efficacy and Safety of Once-Weekly Tirzepatide in Participants with Obesity Disease.
SUMMIT (phase III)	Study Comparing the Efficacy and Safety of Tirzepatide Versus Placebo in Patients with Heart Failure with Preserved Ejection Fraction and Obesity.
SURPASS-EARLY (phase IV)	Study to Evaluate the Long-Term Efficacy and Safety of Tirzepatide Compared with Intensified Conventional Care in Adults When Initiating Treatment Early in the Course of Type 2 Diabetes.

7. Conclusions

Tirzepatide is the first-in-class twincretin, a peptide with agonist action on both GLP-1 and GIP receptors. The randomized phase III trials SURPASS 1–5 and SURPASS J-mono demonstrated its effectiveness and safety in once-weekly administration in patients with T2DM compared with placebo and other hypoglycemic drugs. In particular, in SURPASS-4, where it was compared to insulin glargine in patients with T2DM, Tirzepatide reduced HbA1c up to −2.58% from baseline values at the 15 mg dose. In the SURPASS-1 trial, up to 52% of participants affected by T2DM achieved normoglycemic status (HbA1c < 5.7%; <39 mmol/mol). Its safety profile was consistent with that of GLP-1 receptor agonists without significant episodes of hypoglycemia (<54 mg/dL; <3 mmol/L).

Tirzepatide was demonstrated to reduce body weight in a dose-dependent way. In SURMOUNT-1, a randomized phase III trial where Tirzepatide effectiveness was tested in patients with overweight and obesity without T2DM, the percentages of participants who had a significant weight reduction of 5% or more were 85%, 89%, and 91% with Tirzepatide 5 mg, 10 mg, and 15 mg, respectively, and a robust weight loss of −20.9% was achieved with Tirzepatide 15 mg [94]. In all studies, reductions in BMI and waist circumference were demonstrated, with improvements in BP values, LDL, and triglycerides. In particular, the SURMOUNT-1 trial revealed a mean reduction in waist circumference of 18.5 cm with Tirzepatide 15 mg; in the SURPASS-5 trial, the mean change in SBP was −6.1 to −12.6 mm Hg in patients treated with Tirzepatide.

On 13 May 2022, the Food and Drug Administration (FDA) approved Tirzepatide once-a-week subcutaneous injections (with the dose adjusted based on tolerability to achieve blood glucose targets) as monotherapy or combination therapy, with diet and physical exercise, to achieve better glycemic blood levels in patients affected by T2DM [95,96]. In trials comparing Tirzepatide to other diabetes drugs, patients who received Tirzepatide 15 mg had a 0.5% greater reduction in HbA1c than those who received Semaglutide, a 0.9% greater reduction than those who received insulin degludec, and a 1.0% greater reduction than those who received insulin glargine. It is not fully clear whether the other doses are useful for physicians in clinical practice. The adverse events associated with Tirzepatide include nausea, vomiting, diarrhea, constipation, upper abdominal discomfort, decreased appetite, and abdominal pain; these effects are often observed in irritable bowel syndrome and could suggest a potential action of Tirzepatide on the gut microbiota.

The encouraging results of the SURMOUNT-1 trial and the ongoing trials on Tirzepatide are promising for its future application as a medication for obesity, heart failure, and NAFLD. In addition to drugs that act as dual agonists, such as Tirzepatide, preclinical studies have demonstrated that triagonists, achieving a concurrent activation of GLP-1, GIP, and glucagon receptors, normalize body weight in obese mice and enhance energy expenditure in a manner superior to that of monoagonists of the GLP-1 receptor and dual agonists acting on GLP-1 and GIP receptors [97]. Therefore, the unimolecular triple agonism [98–100] could soon represent a new standard for pharmaceutical interventions.

Author Contributions: Conceptualization, I.F. and G.S.; methodology and formal analysis, I.F., F.V., R.A., S.S.J. and P.M., writing—original draft preparation, I.F.; writing—review and editing, G.S. All authors have read and agreed to the published version of the manuscript.

Funding: The Santulli's Lab is supported in part by the National Institutes of Health (NIH): National Institute of Diabetes and Digestive and Kidney Diseases (**NIDDK**: R01-DK123259, R01-DK033823), National Heart, Lung, and Blood Institute (**NHLBI**: R01-HL159062, R01HL164772, R01-HL146691, T32-HL144456), National Center for Advancing Translational Sciences (**NCATS**: UL1TR002556-06) (to G.S.), the Diabetes Action Research and Education Foundation (to G.S.), and the Monique Weill-Caulier and Irma T. Hirschl Trusts (to G.S). F.V. is supported by a postdoctoral fellowship of the American Heart Association (AHA-22POST995561); S.S.J. is supported by a postdoctoral fellowship of the American Heart Association (AHA-21POST836407).

Institutional Review Board Statement: Not applicable.

Informed Consent Statement: Not applicable.

Data Availability Statement: Not applicable.

Acknowledgments: We thank X. Wang for helpful discussions.

Conflicts of Interest: The authors declare no conflict of interest.

References

1. Visseren, F.L.J.; Mach, F.; Smulders, Y.M.; Carballo, D.; Koskinas, K.C.; Back, M.; Benetos, A.; Biffi, A.; Boavida, J.M.; Capodanno, D.; et al. 2021 ESC Guidelines on cardiovascular disease prevention in clinical practice: Developed by the Task Force for cardiovascular disease prevention in clinical practice with representatives of the European Society of Cardiology and 12 medical societies with the special contribution of the European Association of Preventive Cardiology (EAPC). *Rev. Esp. Cardiol.* **2022**, *75*, 429. [CrossRef] [PubMed]
2. Joseph, J.J.; Deedwania, P.; Acharya, T.; Aguilar, D.; Bhatt, D.L.; Chyun, D.A.; Di Palo, K.E.; Golden, S.H.; Sperling, L.S.; American Heart Association Diabetes Committee of the Council on Lifestyle and Cardiometabolic Health; et al. Comprehensive Management of Cardiovascular Risk Factors for Adults with Type 2 Diabetes: A Scientific Statement from the American Heart Association. *Circulation* **2022**, *145*, e722–e759. [CrossRef] [PubMed]
3. Bulum, T. Nephroprotective Properties of the Glucose-Dependent Insulinotropic Polypeptide (GIP) and Glucagon-like Peptide-1 (GLP-1) Receptor Agonists. *Biomedicines* **2022**, *10*, 2586. [CrossRef] [PubMed]
4. Rehfeld, J.F. The Origin and Understanding of the Incretin Concept. *Front. Endocrinol.* **2018**, *9*, 387. [CrossRef] [PubMed]
5. Holst, J.J. From the Incretin Concept and the Discovery of GLP-1 to Today's Diabetes Therapy. *Front. Endocrinol.* **2019**, *10*, 260. [CrossRef] [PubMed]
6. Nauck, M.A.; Quast, D.R.; Wefers, J.; Pfeiffer, A.F.H. The evolving story of incretins (GIP and GLP-1) in metabolic and cardiovascular disease: A pathophysiological update. *Diabetes Obes. Metab.* **2021**, *23* (Suppl. 3), 5–29. [CrossRef] [PubMed]
7. Kim, S.J.; Nian, C.; McIntosh, C.H. GIP increases human adipocyte LPL expression through CREB and TORC2-mediated trans-activation of the LPL gene. *J. Lipid Res.* **2010**, *51*, 3145–3157. [CrossRef]
8. Baggio, L.L.; Drucker, D.J. Biology of incretins: GLP-1 and GIP. *Gastroenterology* **2007**, *132*, 2131–2157. [CrossRef] [PubMed]
9. Kim, K.S.; Seeley, R.J.; Sandoval, D.A. Signalling from the periphery to the brain that regulates energy homeostasis. *Nat. Rev. Neurosci.* **2018**, *19*, 185–196. [CrossRef] [PubMed]
10. Lu, J.; Wang, L.; Zhou, S.; Zhou, C.; Xie, L.; Chen, J.; Tang, D.; Tian, X.; Xie, D.; Ding, J.; et al. A double-blind, randomized, placebo and positive-controlled study in healthy volunteers to evaluate pharmacokinetic and pharmacodynamic properties of multiple oral doses of cetagliptin. *Br. J. Clin. Pharmacol* **2022**, *88*, 2946–2958. [CrossRef] [PubMed]

11. Chai, S.; Zhang, R.; Zhang, Y.; Carr, R.D.; Zheng, Y.; Rajpathak, S.; Ji, L. Effect of dipeptidyl peptidase-4 inhibitors on postprandial glucagon level in patients with type 2 diabetes mellitus: A systemic review and meta-analysis. *Front. Endocrinol.* **2022**, *13*, 994944. [CrossRef] [PubMed]
12. Gilbert, M.P.; Pratley, R.E. GLP-1 Analogs and DPP-4 Inhibitors in Type 2 Diabetes Therapy: Review of Head-to-Head Clinical Trials. *Front. Endocrinol.* **2020**, *11*, 178. [CrossRef] [PubMed]
13. Sanada, J.; Kimura, T.; Shimoda, M.; Tomita, A.; Fushimi, Y.; Kinoshita, T.; Obata, A.; Okauchi, S.; Hirukawa, H.; Kohara, K.; et al. Switching from Daily DPP-4 Inhibitor to Once-Weekly GLP-1 Receptor Activator Dulaglutide Significantly Ameliorates Glycemic Control in Subjects with Poorly Controlled Type 2 Diabetes Mellitus: A Retrospective Observational Study. *Front. Endocrinol.* **2021**, *12*, 714447. [CrossRef] [PubMed]
14. Tan, Q.; Akindehin, S.E.; Orsso, C.E.; Waldner, R.C.; DiMarchi, R.D.; Muller, T.D.; Haqq, A.M. Recent Advances in Incretin-Based Pharmacotherapies for the Treatment of Obesity and Diabetes. *Front. Endocrinol.* **2022**, *13*, 838410. [CrossRef] [PubMed]
15. Samms, R.J.; Coghlan, M.P.; Sloop, K.W. How May GIP Enhance the Therapeutic Efficacy of GLP-1? *Trends Endocrinol. Metab.* **2020**, *31*, 410–421. [CrossRef] [PubMed]
16. Novikoff, A.; O'Brien, S.L.; Bernecker, M.; Grandl, G.; Kleinert, M.; Knerr, P.J.; Stemmer, K.; Klingenspor, M.; Zeigerer, A.; DiMarchi, R.; et al. Spatiotemporal GLP-1 and GIP receptor signaling and trafficking/recycling dynamics induced by selected receptor mono- and dual-agonists. *Mol. Metab.* **2021**, *49*, 101181. [CrossRef] [PubMed]
17. Thomas, M.K.; Nikooienejad, A.; Bray, R.; Cui, X.; Wilson, J.; Duffin, K.; Milicevic, Z.; Haupt, A.; Robins, D.A. Dual GIP and GLP-1 Receptor Agonist Tirzepatide Improves Beta-cell Function and Insulin Sensitivity in Type 2 Diabetes. *J. Clin. Endocrinol. Metab.* **2021**, *106*, 388–396. [CrossRef] [PubMed]
18. Baggio, L.L.; Drucker, D.J. Glucagon-like peptide-1 receptor co-agonists for treating metabolic disease. *Mol. Metab.* **2021**, *46*, 101090. [CrossRef] [PubMed]
19. Portron, A.; Jadidi, S.; Sarkar, N.; DiMarchi, R.; Schmitt, C. Pharmacodynamics, pharmacokinetics, safety and tolerability of the novel dual glucose-dependent insulinotropic polypeptide/glucagon-like peptide-1 agonist RG7697 after single subcutaneous administration in healthy subjects. *Diabetes Obes. Metab.* **2017**, *19*, 1446–1453. [CrossRef] [PubMed]
20. Schmitt, C.; Portron, A.; Jadidi, S.; Sarkar, N.; DiMarchi, R. Pharmacodynamics, pharmacokinetics and safety of multiple ascending doses of the novel dual glucose-dependent insulinotropic polypeptide/glucagon-like peptide-1 agonist RG7697 in people with type 2 diabetes mellitus. *Diabetes Obes. Metab.* **2017**, *19*, 1436–1445. [CrossRef] [PubMed]
21. Norregaard, P.K.; Deryabina, M.A.; Tofteng Shelton, P.; Fog, J.U.; Daugaard, J.R.; Eriksson, P.O.; Larsen, L.F.; Jessen, L. A novel GIP analogue, ZP4165, enhances glucagon-like peptide-1-induced body weight loss and improves glycaemic control in rodents. *Diabetes Obes. Metab.* **2018**, *20*, 60–68. [CrossRef] [PubMed]
22. Finan, B.; Ma, T.; Ottaway, N.; Muller, T.D.; Habegger, K.M.; Heppner, K.M.; Kirchner, H.; Holland, J.; Hembree, J.; Raver, C.; et al. Unimolecular dual incretins maximize metabolic benefits in rodents, monkeys, and humans. *Sci. Transl. Med.* **2013**, *5*, 209ra151. [CrossRef] [PubMed]
23. Fisman, E.Z.; Tenenbaum, A. The dual glucose-dependent insulinotropic polypeptide (GIP) and glucagon-like peptide-1 (GLP-1) receptor agonist tirzepatide: A novel cardiometabolic therapeutic prospect. *Cardiovasc. Diabetol.* **2021**, *20*, 225. [CrossRef] [PubMed]
24. Samms, R.J.; Christe, M.E.; Collins, K.A.; Pirro, V.; Droz, B.A.; Holland, A.K.; Friedrich, J.L.; Wojnicki, S.; Konkol, D.L.; Cosgrove, R.; et al. GIPR agonism mediates weight-independent insulin sensitization by tirzepatide in obese mice. *J. Clin. Investig.* **2021**, *131*, e146353. [CrossRef] [PubMed]
25. Frias, J.P.; Bastyr, E.J., 3rd; Vignati, L.; Tschop, M.H.; Schmitt, C.; Owen, K.; Christensen, R.H.; DiMarchi, R.D. The Sustained Effects of a Dual GIP/GLP-1 Receptor Agonist, NNC0090-2746, in Patients with Type 2 Diabetes. *Cell Metab.* **2017**, *26*, 343–352.e342. [CrossRef]
26. Frias, J.P.; Nauck, M.A.; Van, J.; Kutner, M.E.; Cui, X.; Benson, C.; Urva, S.; Gimeno, R.E.; Milicevic, Z.; Robins, D.; et al. Efficacy and safety of LY3298176, a novel dual GIP and GLP-1 receptor agonist, in patients with type 2 diabetes: A randomised, placebo-controlled and active comparator-controlled phase 2 trial. *Lancet* **2018**, *392*, 2180–2193. [CrossRef] [PubMed]
27. Frias, J.P.; Nauck, M.A.; Van, J.; Benson, C.; Bray, R.; Cui, X.; Milicevic, Z.; Urva, S.; Haupt, A.; Robins, D.A. Efficacy and tolerability of tirzepatide, a dual glucose-dependent insulinotropic peptide and glucagon-like peptide-1 receptor agonist in patients with type 2 diabetes: A 12-week, randomized, double-blind, placebo-controlled study to evaluate different dose-escalation regimens. *Diabetes Obes. Metab.* **2020**, *22*, 938–946. [CrossRef] [PubMed]
28. Frias, J.P. Tirzepatide: A glucose-dependent insulinotropic polypeptide (GIP) and glucagon-like peptide-1 (GLP-1) dual agonist in development for the treatment of type 2 diabetes. *Expert Rev. Endocrinol. Metab.* **2020**, *15*, 379–394. [CrossRef] [PubMed]
29. Sun, B.; Willard, F.S.; Feng, D.; Alsina-Fernandez, J.; Chen, Q.; Vieth, M.; Ho, J.D.; Showalter, A.D.; Stutsman, C.; Ding, L.; et al. Structural determinants of dual incretin receptor agonism by tirzepatide. *Proc. Natl. Acad. Sci. USA* **2022**, *119*, e2116506119. [CrossRef] [PubMed]
30. Zhao, F.; Zhou, Q.; Cong, Z.; Hang, K.; Zou, X.; Zhang, C.; Chen, Y.; Dai, A.; Liang, A.; Ming, Q.; et al. Structural insights into multiplexed pharmacological actions of tirzepatide and peptide 20 at the GIP, GLP-1 or glucagon receptors. *Nat. Commun.* **2022**, *13*, 1057. [CrossRef] [PubMed]

31. Coskun, T.; Sloop, K.W.; Loghin, C.; Alsina-Fernandez, J.; Urva, S.; Bokvist, K.B.; Cui, X.; Briere, D.A.; Cabrera, O.; Roell, W.C.; et al. LY3298176, a novel dual GIP and GLP-1 receptor agonist for the treatment of type 2 diabetes mellitus: From discovery to clinical proof of concept. *Mol. Metab.* **2018**, *18*, 3–14. [CrossRef] [PubMed]
32. Willard, F.S.; Douros, J.D.; Gabe, M.B.; Showalter, A.D.; Wainscott, D.B.; Suter, T.M.; Capozzi, M.E.; van der Velden, W.J.; Stutsman, C.; Cardona, G.R.; et al. Tirzepatide is an imbalanced and biased dual GIP and GLP-1 receptor agonist. *JCI Insight* **2020**, *5*, e140532. [CrossRef] [PubMed]
33. Santulli, G. Tirzepatide versus Semaglutide Once Weekly in Type 2 Diabetes. *N. Engl. J. Med.* **2022**, *386*, e17. [CrossRef] [PubMed]
34. De Block, C.; Bailey, C.; Wysham, C.; Hemmingway, A.; Allen, S.E.; Peleshok, J. Tirzepatide for the treatment of adults with type 2 diabetes: An endocrine perspective. *Diabetes Obes. Metab.* **2022**. [CrossRef] [PubMed]
35. Min, T.; Bain, S.C. The Role of Tirzepatide, Dual GIP and GLP-1 Receptor Agonist, in the Management of Type 2 Diabetes: The SURPASS Clinical Trials. *Diabetes Ther.* **2021**, *12*, 143–157. [CrossRef] [PubMed]
36. Greenhill, C. Testing a novel dual receptor agonist for treatment of type 2 diabetes mellitus. *Nat. Rev. Endocrinol.* **2018**, *14*, 687. [CrossRef] [PubMed]
37. Stumvoll, M.; Tschop, M. Twice the benefits with twincretins? *Lancet* **2018**, *392*, 2142–2144. [CrossRef] [PubMed]
38. Rosenstock, J.; Wysham, C.; Frias, J.P.; Kaneko, S.; Lee, C.J.; Fernandez Lando, L.; Mao, H.; Cui, X.; Karanikas, C.A.; Thieu, V.T. Efficacy and safety of a novel dual GIP and GLP-1 receptor agonist tirzepatide in patients with type 2 diabetes (SURPASS-1): A double-blind, randomised, phase 3 trial. *Lancet* **2021**, *398*, 143–155. [CrossRef] [PubMed]
39. Paneni, F.; Patrono, C. Is tirzepatide in the surpass lane over GLP-1 receptor agonists for the treatment of diabetes? *Eur. Heart J.* **2021**, *42*, 4211–4212. [CrossRef] [PubMed]
40. Tan, T.M.; Khoo, B. Tirzepatide and the new era of twincretins for diabetes. *Lancet* **2021**, *398*, 95–97. [CrossRef] [PubMed]
41. Frias, J.P.; Davies, M.J.; Rosenstock, J.; Perez Manghi, F.C.; Fernandez Lando, L.; Bergman, B.K.; Liu, B.; Cui, X.; Brown, K.; SURPASS-2 Investigators. Tirzepatide versus Semaglutide Once Weekly in Patients with Type 2 Diabetes. *N. Engl. J. Med.* **2021**, *385*, 503–515. [CrossRef] [PubMed]
42. Chiu, N.; Aggarwal, R.; Bhatt, D.L. Generalizability of the SURPASS-2 Trial and Effect of Tirzepatide on US Diabetes and Obesity Control. *J. Am. Heart Assoc.* **2022**, *11*, e026297. [CrossRef] [PubMed]
43. Vadher, K.; Patel, H.; Mody, R.; Levine, J.A.; Hoog, M.; Cheng, A.Y.; Pantalone, K.M.; Sapin, H. Efficacy of tirzepatide 5, 10 and 15 mg versus semaglutide 2 mg in patients with type 2 diabetes: An adjusted indirect treatment comparison. *Diabetes Obes. Metab.* **2022**, *24*, 1861–1868. [CrossRef] [PubMed]
44. Ludvik, B.; Giorgino, F.; Jodar, E.; Frias, J.P.; Fernandez Lando, L.; Brown, K.; Bray, R.; Rodriguez, A. Once-weekly tirzepatide versus once-daily insulin degludec as add-on to metformin with or without SGLT2 inhibitors in patients with type 2 diabetes (SURPASS-3): A randomised, open-label, parallel-group, phase 3 trial. *Lancet* **2021**, *398*, 583–598. [CrossRef]
45. Bailey, C.J. Tirzepatide: A new low for bodyweight and blood glucose. *Lancet Diabetes Endocrinol.* **2021**, *9*, 646–648. [CrossRef]
46. Battelino, T.; Bergenstal, R.M.; Rodriguez, A.; Fernandez Lando, L.; Bray, R.; Tong, Z.; Brown, K. Efficacy of once-weekly tirzepatide versus once-daily insulin degludec on glycaemic control measured by continuous glucose monitoring in adults with type 2 diabetes (SURPASS-3 CGM): A substudy of the randomised, open-label, parallel-group, phase 3 SURPASS-3 trial. *Lancet Diabetes Endocrinol.* **2022**, *10*, 407–417. [CrossRef] [PubMed]
47. Gastaldelli, A.; Cusi, K.; Fernandez Lando, L.; Bray, R.; Brouwers, B.; Rodriguez, A. Effect of tirzepatide versus insulin degludec on liver fat content and abdominal adipose tissue in people with type 2 diabetes (SURPASS-3 MRI): A substudy of the randomised, open-label, parallel-group, phase 3 SURPASS-3 trial. *Lancet Diabetes Endocrinol.* **2022**, *10*, 393–406. [CrossRef] [PubMed]
48. Targher, G. Tirzepatide adds hepatoprotection to its armoury. *Lancet Diabetes Endocrinol.* **2022**, *10*, 374–375. [CrossRef]
49. Heise, T.; Mari, A.; DeVries, J.H.; Urva, S.; Li, J.; Pratt, E.J.; Coskun, T.; Thomas, M.K.; Mather, K.J.; Haupt, A.; et al. Effects of subcutaneous tirzepatide versus placebo or semaglutide on pancreatic islet function and insulin sensitivity in adults with type 2 diabetes: A multicentre, randomised, double-blind, parallel-arm, phase 1 clinical trial. *Lancet Diabetes Endocrinol.* **2022**, *10*, 418–429. [CrossRef]
50. Scheen, A.J. Add-on value of tirzepatide versus semaglutide. *Lancet Diabetes Endocrinol.* **2022**, *10*, 377–378. [CrossRef]
51. Starling, S. GIP-GLP1 receptor agonist shows promise. *Nat. Rev. Endocrinol.* **2022**, *18*, 391. [CrossRef] [PubMed]
52. Del Prato, S.; Kahn, S.E.; Pavo, I.; Weerakkody, G.J.; Yang, Z.; Doupis, J.; Aizenberg, D.; Wynne, A.G.; Riesmeyer, J.S.; Heine, R.J.; et al. Tirzepatide versus insulin glargine in type 2 diabetes and increased cardiovascular risk (SURPASS-4): A randomised, open-label, parallel-group, multicentre, phase 3 trial. *Lancet* **2021**, *398*, 1811–1824. [CrossRef] [PubMed]
53. Singh, M. In type 2 diabetes with increased CV risk, tirzepatide reduced HbA1c vs. glargine at 52 wk. *Ann. Intern. Med.* **2022**, *175*, JC34. [CrossRef] [PubMed]
54. Khoo, B.; Tan, T.M. Surpassing insulin glargine in type 2 diabetes with tirzepatide. *Lancet* **2021**, *398*, 1779–1781. [CrossRef] [PubMed]
55. Heerspink, H.J.L.; Sattar, N.; Pavo, I.; Haupt, A.; Duffin, K.L.; Yang, Z.; Wiese, R.J.; Tuttle, K.R.; Cherney, D.Z.I. Effects of tirzepatide versus insulin glargine on kidney outcomes in type 2 diabetes in the SURPASS-4 trial: Post-hoc analysis of an open-label, randomised, phase 3 trial. *Lancet Diabetes Endocrinol.* **2022**, *10*, 774–785. [CrossRef] [PubMed]
56. Dahl, D.; Onishi, Y.; Norwood, P.; Huh, R.; Bray, R.; Patel, H.; Rodriguez, A. Effect of Subcutaneous Tirzepatide vs Placebo Added to Titrated Insulin Glargine on Glycemic Control in Patients with Type 2 Diabetes: The SURPASS-5 Randomized Clinical Trial. *JAMA* **2022**, *327*, 534–545. [CrossRef] [PubMed]

57. Chipkin, S.R. Tirzepatide for Patients with Type 2 Diabetes. *JAMA* **2022**, *327*, 529–530. [CrossRef]
58. Inagaki, N.; Takeuchi, M.; Oura, T.; Imaoka, T.; Seino, Y. Efficacy and safety of tirzepatide monotherapy compared with dulaglutide in Japanese patients with type 2 diabetes (SURPASS J-mono): A double-blind, multicentre, randomised, phase 3 trial. *Lancet Diabetes Endocrinol.* **2022**, *10*, 623–633. [CrossRef]
59. Yabe, D.; Kawamori, D.; Seino, Y.; Oura, T.; Takeuchi, M. Change of pharmacodynamics parameters following once-weekly tirzepatide treatment versus dulaglutide in Japanese patients with type 2 diabetes (SURPASS J-mono sub-study). *Diabetes Obes. Metab.* **2022**. [CrossRef]
60. Kadowaki, T.; Chin, R.; Ozeki, A.; Imaoka, T.; Ogawa, Y. Safety and efficacy of tirzepatide as an add-on to single oral antihyperglycaemic medication in patients with type 2 diabetes in Japan (SURPASS J-combo): A multicentre, randomised, open-label, parallel-group, phase 3 trial. *Lancet Diabetes Endocrinol.* **2022**, *10*, 634–644. [CrossRef] [PubMed]
61. Emerging Risk Factors, C.; Wormser, D.; Kaptoge, S.; Di Angelantonio, E.; Wood, A.M.; Pennells, L.; Thompson, A.; Sarwar, N.; Kizer, J.R.; Lawlor, D.A.; et al. Separate and combined associations of body-mass index and abdominal adiposity with cardiovascular disease: Collaborative analysis of 58 prospective studies. *Lancet* **2011**, *377*, 1085–1095. [CrossRef] [PubMed]
62. Siren, R.; Eriksson, J.G.; Vanhanen, H. Waist circumference a good indicator of future risk for type 2 diabetes and cardiovascular disease. *BMC Public Health* **2012**, *12*, 631. [CrossRef] [PubMed]
63. Csige, I.; Ujvarosy, D.; Szabo, Z.; Lorincz, I.; Paragh, G.; Harangi, M.; Somodi, S. The Impact of Obesity on the Cardiovascular System. *J. Diabetes Res.* **2018**, *2018*, 3407306. [CrossRef] [PubMed]
64. Lazzaroni, E.; Ben Nasr, M.; Loretelli, C.; Pastore, I.; Plebani, L.; Lunati, M.E.; Vallone, L.; Bolla, A.M.; Rossi, A.; Montefusco, L.; et al. Anti-diabetic drugs and weight loss in patients with type 2 diabetes. *Pharmacol. Res.* **2021**, *171*, 105782. [CrossRef] [PubMed]
65. Powell-Wiley, T.M.; Poirier, P.; Burke, L.E.; Despres, J.P.; Gordon-Larsen, P.; Lavie, C.J.; Lear, S.A.; Ndumele, C.E.; Neeland, I.J.; Sanders, P.; et al. Obesity and Cardiovascular Disease: A Scientific Statement from the American Heart Association. *Circulation* **2021**, *143*, e984–e1010. [CrossRef] [PubMed]
66. Jastreboff, A.M.; Aronne, L.J.; Ahmad, N.N.; Wharton, S.; Connery, L.; Alves, B.; Kiyosue, A.; Zhang, S.; Liu, B.; Bunck, M.C.; et al. Tirzepatide Once Weekly for the Treatment of Obesity. *N. Engl. J. Med.* **2022**, *387*, 205–216. [CrossRef] [PubMed]
67. Ryan, D.H.; Yockey, S.R. Weight Loss and Improvement in Comorbidity: Differences at 5%, 10%, 15%, and Over. *Curr. Obes. Rep.* **2017**, *6*, 187–194. [CrossRef] [PubMed]
68. Davidson, M.B. In adults with obesity without diabetes, adding tirzepatide to a lifestyle intervention increased weight loss at 72 wk. *Ann. Intern. Med.* **2022**, *175*, JC116. [CrossRef] [PubMed]
69. Hindson, J. Tirzepatide to treat obesity: Phase III results. *Nat. Rev. Gastroenterol. Hepatol.* **2022**, *19*, 488. [CrossRef] [PubMed]
70. Tysoe, O. Tirzepatide highly effective for weight loss. *Nat. Rev. Endocrinol.* **2022**, *18*, 520. [CrossRef]
71. Frachon, I.; Etienne, Y.; Jobic, Y.; Le Gal, G.; Humbert, M.; Leroyer, C. Benfluorex and unexplained valvular heart disease: A case-control study. *PLoS ONE* **2010**, *5*, e10128. [CrossRef] [PubMed]
72. Ruckert, I.M.; Baumert, J.; Schunk, M.; Holle, R.; Schipf, S.; Volzke, H.; Kluttig, A.; Greiser, K.H.; Tamayo, T.; Rathmann, W.; et al. Blood Pressure Control Has Improved in People with and without Type 2 Diabetes but Remains Suboptimal: A Longitudinal Study Based on the German DIAB-CORE Consortium. *PLoS ONE* **2015**, *10*, e0133493. [CrossRef] [PubMed]
73. Trimarco, V.; Izzo, R.; Mone, P.; Lembo, M.; Manzi, M.V.; Pacella, D.; Falco, A.; Gallo, P.; Esposito, G.; Morisco, C.; et al. Therapeutic concordance improves blood pressure control in patients with resistant hypertension. *Pharmacol. Res.* **2022**, 106557. [CrossRef] [PubMed]
74. Wilson, J.M.; Lin, Y.; Luo, M.J.; Considine, G.; Cox, A.L.; Bowsman, L.M.; Robins, D.A.; Haupt, A.; Duffin, K.L.; Ruotolo, G. The dual glucose-dependent insulinotropic polypeptide and glucagon-like peptide-1 receptor agonist tirzepatide improves cardiovascular risk biomarkers in patients with type 2 diabetes: A post hoc analysis. *Diabetes Obes. Metab.* **2022**, *24*, 148–153. [CrossRef] [PubMed]
75. Frias, J.P.; Fernandez Lando, L.; Brown, K. Tirzepatide versus Semaglutide Once Weekly in Type 2 Diabetes. Reply. *N. Engl. J. Med.* **2022**, *386*, e17. [CrossRef]
76. Sattar, N.; McGuire, D.K.; Pavo, I.; Weerakkody, G.J.; Nishiyama, H.; Wiese, R.J.; Zoungas, S. Tirzepatide cardiovascular event risk assessment: A pre-specified meta-analysis. *Nat. Med.* **2022**, *28*, 591–598. [CrossRef] [PubMed]
77. Karagiannis, T.; Avgerinos, I.; Liakos, A.; Del Prato, S.; Matthews, D.R.; Tsapas, A.; Bekiari, E. Management of type 2 diabetes with the dual GIP/GLP-1 receptor agonist tirzepatide: A systematic review and meta-analysis. *Diabetologia* **2022**, *65*, 1251–1261. [CrossRef] [PubMed]
78. Bhagavathula, A.S.; Vidyasagar, K.; Tesfaye, W. Efficacy and Safety of Tirzepatide in Patients with Type 2 Diabetes Mellitus: A Systematic Review and Meta-Analysis of Randomized Phase II/III Trials. *Pharmaceuticals* **2021**, *14*, 991. [CrossRef] [PubMed]
79. Dutta, D.; Surana, V.; Singla, R.; Aggarwal, S.; Sharma, M. Efficacy and safety of novel twincretin tirzepatide a dual GIP and GLP-1 receptor agonist in the management of type-2 diabetes: A Cochrane meta-analysis. *Indian J. Endocrinol. Metab.* **2021**, *25*, 475–489. [CrossRef]
80. Guan, R.; Yang, Q.; Yang, X.; Du, W.; Li, X.; Ma, G. Efficacy and safety of tirzepatide in patients with type 2 diabetes mellitus: A bayesian network meta-analysis. *Front. Pharmacol.* **2022**, *13*, 998816. [CrossRef] [PubMed]
81. Yu, Y.; Hu, G.; Yin, S.; Yang, X.; Zhou, M.; Jian, W. Optimal dose of tirzepatide for type 2 diabetes mellitus: A meta-analysis and trial sequential analysis. *Front. Cardiovasc. Med.* **2022**, *9*, 990182. [CrossRef] [PubMed]

82. Permana, H.; Yanto, T.A.; Hariyanto, T.I. Efficacy and safety of tirzepatide as novel treatment for type 2 diabetes: A systematic review and meta-analysis of randomized clinical trials. *Diabetes Metab. Syndr.* **2022**, *16*, 102640. [CrossRef] [PubMed]
83. Lisco, G.; De Tullio, A.; Disoteo, O.; De Geronimo, V.; Piazzolla, G.; De Pergola, G.; Giagulli, V.A.; Jirillo, E.; Guastamacchia, E.; Sabba, C.; et al. Basal insulin intensification with GLP-1RA and dual GIP and GLP-1RA in patients with uncontrolled type 2 diabetes mellitus: A rapid review of randomized controlled trials and meta-analysis. *Front. Endocrinol.* **2022**, *13*, 920541. [CrossRef] [PubMed]
84. Patoulias, D.; Doumas, M.; Papadopoulos, C. Meta-Analysis Assessing the Effect of Tirzepatide on the Risk for Atrial Fibrillation in Patients with Type 2 Diabetes Mellitus. *Am. J. Cardiol.* **2022**, *173*, 157–158. [CrossRef] [PubMed]
85. Patoulias, D.; Papadopoulos, C.; Fragakis, N.; Doumas, M. Updated Meta-Analysis Assessing the Cardiovascular Efficacy of Tirzepatide. *Am. J. Cardiol.* **2022**, *181*, 139–140. [CrossRef] [PubMed]
86. Valenzuela-Vallejo, L.; Guatibonza-Garcia, V.; Mantzoros, C.S. Recent guidelines for Non-Alcoholic Fatty Liver disease (NAFLD)/Fatty Liver Disease (FLD): Are they already outdated and in need of supplementation? *Metabolism* **2022**, *136*, 155248. [CrossRef] [PubMed]
87. Moura, F.A.; Scirica, B.M.; Ruff, C.T. Tirzepatide for diabetes: On track to SURPASS current therapy. *Nat. Med.* **2022**, *28*, 450–451. [CrossRef] [PubMed]
88. Hartman, M.L.; Sanyal, A.J.; Loomba, R.; Wilson, J.M.; Nikooienejad, A.; Bray, R.; Karanikas, C.A.; Duffin, K.L.; Robins, D.A.; Haupt, A. Effects of Novel Dual GIP and GLP-1 Receptor Agonist Tirzepatide on Biomarkers of Nonalcoholic Steatohepatitis in Patients with Type 2 Diabetes. *Diabetes Care* **2020**, *43*, 1352–1355. [CrossRef] [PubMed]
89. Pirro, V.; Roth, K.D.; Lin, Y.; Willency, J.A.; Milligan, P.L.; Wilson, J.M.; Ruotolo, G.; Haupt, A.; Newgard, C.B.; Duffin, K.L. Effects of Tirzepatide, a Dual GIP and GLP-1 RA, on Lipid and Metabolite Profiles in Subjects with Type 2 Diabetes. *J. Clin. Endocrinol. Metab.* **2022**, *107*, 363–378. [CrossRef] [PubMed]
90. Ferrannini, E. Tirzepatide as an Insulin Sensitizer. *J. Clin. Endocrinol. Metab.* **2022**, *107*, e1752–e1753. [CrossRef] [PubMed]
91. Koufakis, T.; Maltese, G.; Kotsa, K. Toward a new model for the management of type 2 diabetes: The mountain is there and there is no other option than to climb it. *Pharmacol. Res.* **2022**, *184*, 106443. [CrossRef] [PubMed]
92. Schnell, O.; Battelino, T.; Bergenstal, R.; Bluher, M.; Bohm, M.; Brosius, F.; Carr, R.D.; Ceriello, A.; Forst, T.; Giorgino, F.; et al. Report from the CVOT Summit 2021: New cardiovascular, renal, and glycemic outcomes. *Cardiovasc. Diabetol.* **2022**, *21*, 50. [CrossRef] [PubMed]
93. Gallwitz, B. Clinical perspectives on the use of the GIP/GLP-1 receptor agonist tirzepatide for the treatment of type-2 diabetes and obesity. *Front. Endocrinol.* **2022**, *13*, 1004044. [CrossRef] [PubMed]
94. Slomski, A. Tirzepatide Trial Demonstrates Substantial Weight Loss. *JAMA* **2022**, *328*, 322. [CrossRef] [PubMed]
95. Mullard, A. Lilly's tirzepatide secures first approval in diabetes, paving path for dual-acting hormone mimetics. *Nat. Rev. Drug Discov.* **2022**, *21*, 480. [CrossRef] [PubMed]
96. Syed, Y.Y. Tirzepatide: First Approval. *Drugs* **2022**, *82*, 1213–1220. [CrossRef]
97. Knerr, P.J.; Mowery, S.A.; Douros, J.D.; Premdjee, B.; Hjollund, K.R.; He, Y.; Kruse Hansen, A.M.; Olsen, A.K.; Perez-Tilve, D.; DiMarchi, R.D.; et al. Next generation GLP-1/GIP/glucagon triple agonists normalize body weight in obese mice. *Mol. Metab.* **2022**, *63*, 101533. [CrossRef] [PubMed]
98. Capozzi, M.E.; DiMarchi, R.D.; Tschop, M.H.; Finan, B.; Campbell, J.E. Targeting the Incretin/Glucagon System With Triagonists to Treat Diabetes. *Endocr. Rev.* **2018**, *39*, 719–738. [CrossRef] [PubMed]
99. Finan, B.; Yang, B.; Ottaway, N.; Smiley, D.L.; Ma, T.; Clemmensen, C.; Chabenne, J.; Zhang, L.; Habegger, K.M.; Fischer, K.; et al. A rationally designed monomeric peptide triagonist corrects obesity and diabetes in rodents. *Nat. Med.* **2015**, *21*, 27–36. [CrossRef] [PubMed]
100. Knerr, P.J.; Mowery, S.A.; Finan, B.; Perez-Tilve, D.; Tschop, M.H.; DiMarchi, R.D. Selection and progression of unimolecular agonists at the GIP, GLP-1, and glucagon receptors as drug candidates. *Peptides* **2020**, *125*, 170225. [CrossRef] [PubMed]

International Journal of *Molecular Sciences*

MDPI

Article

Synthesis of Glycolysis Inhibitor PFK15 and Its Synergistic Action with an Approved Multikinase Antiangiogenic Drug on Human Endothelial Cell Migration and Proliferation

Jana Zlacká [1], Miroslav Murár [2], Gabriela Addová [2], Roman Moravčík [1], Andrej Boháč [2,3,†] and Michal Zeman [1,*]

[1] Department of Animal Physiology and Ethology, Faculty of Natural Sciences, Comenius University in Bratislava, Ilkovičova 6, 842 15 Bratislava, Slovakia
[2] Department of Organic Chemistry, Faculty of Natural Sciences, Comenius University in Bratislava, Ilkovičova 6, 842 15 Bratislava, Slovakia
[3] Biomagi, Ltd., Mamateyova 26, 851 04 Bratislava, Slovakia
* Correspondence: michal.zeman@uniba.sk
† Author responsible for the chemistry part of this work (synthesis and description of PFK15).

Abstract: Activated endothelial, immune, and cancer cells prefer glycolysis to obtain energy for their proliferation and migration. Therefore, the blocking of glycolysis can be a promising strategy against cancer and autoimmune disease progression. Inactivation of the glycolytic enzyme PFKFB3 (6-phosphofructo-2-kinase/fructose-2,6-biphosphatase) suppresses glycolysis level and contributes to decreased proliferation and migration of cancer (tumorigenesis) and endothelial (angiogenesis) cells. Recently, several glycolysis inhibitors have been developed, among them (*E*)-1-(pyridin-4-yl)-3-(quinolin-2-yl)prop-2-en-1-one (PFK15) that is considered as one of the most promising. It is known that PFK15 decreases glucose uptake into the endothelial cells and efficiently blocks pathological angiogenesis. However, no study has described sufficiently PFK15 synthesis enabling its general availability. In this paper we provide all necessary details for PFK15 preparation and its advanced characterization. On the other hand, there are known tyrosine kinase inhibitors (e.g., sunitinib), that affect additional molecular targets and efficiently block angiogenesis. From a biological point of view, we have studied and proved the synergistic inhibitory effect by simultaneous administration of glycolysis inhibitor PFK15 and multikinase inhibitor sunitinib on the proliferation and migration of HUVEC. Our results suggest that suppressing the glycolytic activity of endothelial cells in combination with growth factor receptor blocking can be a promising antiangiogenic treatment.

Keywords: PFK15 synthesis; sunitinib *L*-malate; HUVEC; angiogenesis; synergy

Citation: Zlacká, J.; Murár, M.; Addová, G.; Moravčík, R.; Boháč, A.; Zeman, M. Synthesis of Glycolysis Inhibitor PFK15 and Its Synergistic Action with an Approved Multikinase Antiangiogenic Drug on Human Endothelial Cell Migration and Proliferation. *Int. J. Mol. Sci.* **2022**, 23, 14295. https://doi.org/10.3390/ijms232214295

Academic Editors: Simona Gabriela Bungau and Vesa Cosmin

Received: 28 September 2022
Accepted: 15 November 2022
Published: 18 November 2022

Publisher's Note: MDPI stays neutral with regard to jurisdictional claims in published maps and institutional affiliations.

1. Introduction

The alteration of metabolism and increased level of glucose metabolism in activated endothelial cells points to glycolysis as a possible target to inhibit pathological angiogenesis. Endothelial and tumor cells utilize glycolysis to metabolize glucose to lactate instead of oxidative phosphorylation, a phenomenon known as the Warburg effect, which provides an anabolic substrate to cover the demand for rapid cell proliferation and migration. The key regulator of glycolytic flux is the enzyme PFKFB3, which catalyzes the synthesis of F2,6P2 (fructose-2,6-bisphosphate), activating phosphofructokinase 1 (PFK-1), the rate-limiting key enzyme of glycolysis. In recent years, several glycolysis inhibitors have been developed, such as (*E*)-3-(pyridin-3-yl)-1-(pyridin-4-yl)prop-2-en-1-one (**3PO**) [1], (*E*)-1-(pyridin-4-yl)-3-(quinolin-2-yl)prop-2-en-1-one (**PFK15**) [2], and (*E*)-1-(pyridin-4-yl)-3-(7-(trifluoromethyl)quinolin-2-yl)prop-2-en-1-one (**PFK158**) [3]. Previously published data demonstrate that glycolysis inhibitors decrease glucose uptake and lactate production, simultaneously reduce the intracellular concentration of F2,6P2, and finally also reduce the level of glycolysis in the cells. Moreover, in our previous study, we confirmed that simultaneous administration of glycolysis inhibitor **3PO** and inhibition of the post-receptor signal cascade has a synergic inhibitory effect on the proliferation and migration of endothelial cells [4].

Generally, **PFK15** is considered to be a more effective glycolysis inhibitor than **3PO**, and its effectivity has been described in several studies [5,6]. In gastric cancer cell lines, **PFK15** induced cell cycle arrest by inhibiting the cyclin-CDKs/Rb/E2F-signaling pathway. Moreover, **PFK15** suppressed tumor growth by inducing the apoptosis of hepatocellular cancer cells and reducing the tumor vessel diameter [7]. Additionally, previously published data confirmed that chronotherapy represents an additional promising approach to glycolysis inhibition [5,8].

PFK15 is not limited to cancer treatment but can also be related to understanding the activation of immune cells in chronic infection and autoimmune diseases. **PFK15** treatment reduces metabolic reprogramming in T-cells in vitro, leading to a delay in the onset of type I diabetes [9]. Therefore, the regulation of glycolysis metabolism in immune cells can be a new possible therapeutic target in autoimmune disease prevention. Additionally, uncontrolled angiogenesis contributes to the microvascular changes in diabetic patients, including the development of diabetic retinopathy [10]. Thus, angiogenesis inhibition is one of the potential targets in the diabetic nephropathy treatment [11].

The effectivity of **PFK15** was confirmed in several studies on different cell lines [2,12]; however, its synthesis has not yet been sufficiently described in the literature. The need for **PFK15** to be available on a larger scale, e.g., for in vivo assays, motivated us to develop the missing synthesis of this compound. Therefore, we here provide all necessary details for the synthesis of **PFK15** and its advanced characterization (e.g., by 2D NMR spectra), which are missing in the literature. To test the predicted antiangiogenic efficacy of **PFK15**, multiple assays were used. None of them were performed with a combination of **PFK15** and **sunitinib** before. The tests were performed on human umbilical vein endothelial cells (HUVEC), which provide a useful model system to study prospective antiangiogenic inhibitors or their synergy in combination with differently operating compounds.

2. Results

The antiangiogenic potential (endothelial cells growth, proliferation, and migration) of glycolysis inhibitor **PFK15** and multikinase inhibitor **sunitinib** was tested on HUVEC isolated from human umbilical cords.

2.1. *Effect of PFK15 and Sunitinib on HUVEC Growth and Proliferation (IC_{50})*

The IC_{50} concentration of HUVEC proliferation was determined as 2.6 μM for **PFK15** [8] and 1.6 μM for **sunitinib** (Table 1).

Table 1. The IC_{50} values for **PFK15** and **sunitinib** determined on HUVEC.

	IC_{50} [μM]
PFK15	2.6
sunitinib	1.6

As demonstrated in Figure 1, the HUVEC showed a different shape of response curve for **PFK15** and **sunitinib**, reflecting a higher sensitivity of HUVEC on **PFK15** concentration on the one hand, and a higher potential of **sunitinib** to block cell growth and proliferation at lower concentrations on the other, underlining its multikinase properties.

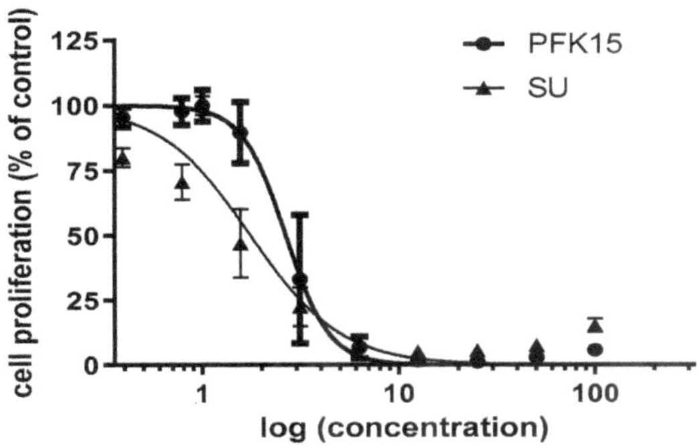

Figure 1. HUVEC response to **PFK15** ((*E*)-1-(pyridin-4-yl)-3-(quinolin-2-yl)prop-2-en-1-one) and **sunitinib** (SU) treatment after 72 h. Concentration values on the x-axis are on a logarithmic scale and are in μM.

2.2. *Effect of PFK15 and Sunitinib on HUVEC Migration (Wound Healing Assay)*

Cell migration was quantified after the cell treatment with **PFK15** and **sunitinib** alone or in combination. The combined action of **PFK15** (at 5 or 10 μM) with **sunitinib** (at 0.1 or 1 μM) more efficiently inhibited HUVEC migration compared to **PFK15** or **sunitinib** alone. Simultaneous administration of **PFK15** and **sunitinib** at specific concentrations clearly showed a synergistic effect on the inhibition of HUVEC migration (Figure 2; representative photos illustrating cell migration are available in Supplementary Material, Figure S8).

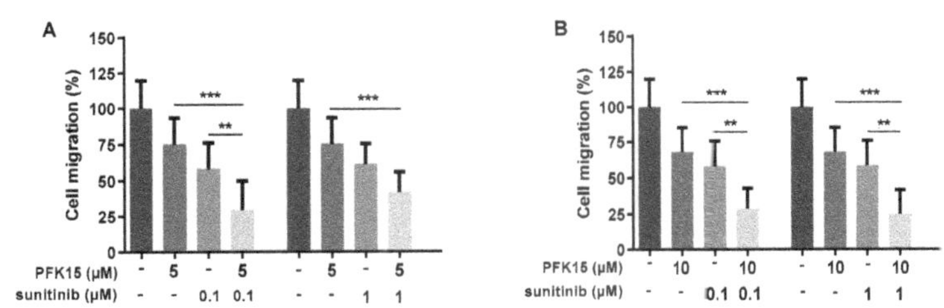

Figure 2. Evaluation of the administration of **PFK15** ((*E*)-1-(pyridin-4-yl)-3-(quinolin-2-yl)prop-2-en-1-one) and sunitinib, alone or combined, on HUVEC migration. HUVEC were treated with **PFK15** at 5 (**A**) and 10 μM (**B**) and sunitinib at 0.1 and 1 μM for 8 h and changes in cell migration were analyzed. Statistical differences among groups were determined by one-way ANOVA followed by Tukey's post hoc test. Data are presented as the mean ± SEM of three independent experiments performed in quadruplicate (** $p < 0.01$; *** $p < 0.001$).

2.3. *Effect of PFK15 and Sunitinib on HUVEC Proliferation*

In the next experiments, we studied changes in HUVEC proliferation in the presence of **PFK15** and **sunitinib** in order to confirm their inhibitory and synergic effect. In the preliminary experiment, we applied **PFK15** at concentrations of 5 and 10 μM and sunitinib at 0.1 and 1 μM (similarly to the previous wound healing assay), alone or in combination, but in this case, the effects of inhibitors applied alone were too strong to see any combined effects (see the pictures in Supplementary Materials, Figure S9).

Therefore, in the next experiments we selected lower concentrations of **PFK15** (2, 3, and 4 μM) and **sunitinib** (4, 5, and 6 μM) alone or in combinations. In these trials, both **PFK15** and **sunitinib** alone reduced cell proliferation compared to the control. Moreover, simultaneous administration of both inhibitors at specific concentrations showed a synergistic effect on cell proliferation compared to application of the inhibitors alone. **PFK15** at 3 and 4 μM in combination with **sunitinib** at 4, 5, and 6 μM most efficiently inhibited HUVEC proliferation (Figure 3B,C).

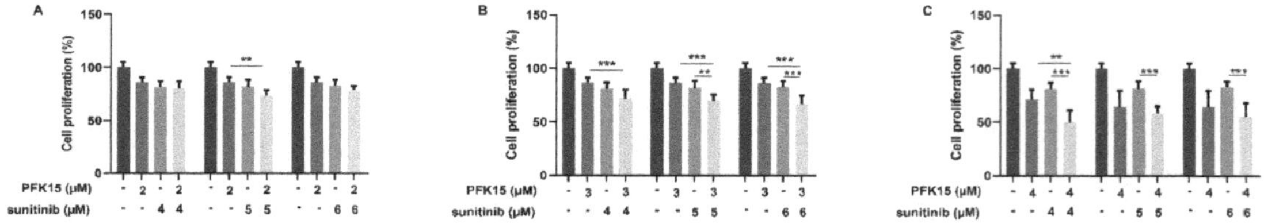

Figure 3. Effect of **PFK15** ((*E*)-1-(pyridin-4-yl)-3-(quinolin-2-yl)prop-2-en-1-one) and **sunitinib** administered alone or in combination on HUVEC proliferation. HUVEC were treated with **PFK15** at 2 (**A**), 3 (**B**), or 4 (**C**) μM and **sunitinib** at 4, 5, or 6 μM for 72 h. Changes in cell proliferation were recorded. Statistical differences among groups were determined by one-way ANOVA followed by Tukey's post hoc test. Data are presented as the mean ± SEM of three independent experiments (** $p < 0.01$; *** $p < 0.001$).

3. Discussion

The limited efficacy of the 1st generation of antiangiogenic therapy led to the development of selective PFKFB3 inhibitors that showed potential activity against the key glycolysis enzyme in different cell types, as well as in various animal models [2,5,6]. Recently, several glycolytic inhibitors were identified; among them, **PFK15** showed important antitumor efficiency. However, until now, no study has described the synthesis of this generally available glycolytic inhibitor in detail. Therefore, we performed an efficient synthesis of **PFK15** and confirmed its efficiency on activated HUVEC cells. Additionally, we observed more effective inhibition of HUVEC proliferation and migration using a combination of glycolysis inhibitor **PFK15** with multikinase inhibitor **sunitinib**, targeting different biological pathways and explaining the synergic effect of their inhibition.

3PO is one of the first synthesized blockers of glycolysis [1], and its effective synthesis has already been described by our research group [13]. Although its antiangiogenic potential was described in several studies, the concentration at which **3PO** is effective appears to be relatively high. Moreover, this high concentration is difficult to achieve in clinical studies because of the poor solubility of **3PO** in water [14]. Therefore, other inhibitors of PFKFB3 have been developed. Substitution of the pyridine-4-yl ring in the inhibitor **3PO** for the quinoline-2-yl group results in a more powerful inhibitor **PFK15**, which is also characterized by higher selectivity for the enzyme PFKFB3 [2]. The antitumor effect of **PFK15** was compared with the effect of chemotherapy drugs that are approved by the Food and Drug Administration (FDA) for the treatment of human cancer (e.g., irinotecan, temozolomide, and gemcitabine). The **PFK15** inhibitor suppressed pancreatic and colon adenocarcinoma growth, similarly to gemcitabine and irinotecan, while the effect of **PFK15** on glioblastoma growth was lower compared to the chemotherapeutic agent temozolomide [2]. The activity of the PFKFB3 enzyme can be regulated by several factors, including the AMPK signaling pathway, which can function as an upstream of the PFKFB3 regulator. It was confirmed that **PFK15** is able to inhibit AMPK and Akt-mTORC1 signaling in colorectal cancer cells [15], and thereby reduce tumor cell viability.

Limitations related to low inhibitor water solubility may be solved using nanocarriers or by chemical modification of the inhibitor structure [14]. Moreover, the low cytotoxic effect of glycolysis inhibitors on healthy cells [6], confirmed in several studies [16,17], suggests that the combination of a glycolysis inhibitor with other chemotherapy agents (e.g., multikinase inhibitor or cytostatic drug) may be an important strategy in advanced cancer treatment.

The additive inhibitory effect of the simultaneous administration of **3PO** and **sunitinib** on angiogenesis was demonstrated under in vivo conditions on a zebrafish embryo model, in which the formation of intersomitic vessels was evaluated [16]. Our previously published data [4] also confirmed that cells treated with the glycolysis inhibitor **3PO** at 10 μM and **sunitinib** at 1 μM, as well as 20 μM **3PO** in combination with **sunitinib** in a concentration range of 0.1–10 μM, significantly reduced cell proliferation compared to the inhibitors applied alone.

We determined the IC_{50} value for both **PFK15** and **sunitinib** (Figure 1) and have observed a significant decrease of HUVECs migration and proliferation after cell treatment with **PFK15** and **sunitinib** (Figures 2 and 3). To confirm the synergic effect of the combined administration of both inhibitors, **PFK15** and **sunitinib** were applied at different concentrations. **PFK15** applied at 3 and 4 μM and **sunitinib** at 4, 5, and 6 μM more efficiently lowered cell proliferation compared to the inhibitors alone (Figure 3).

In addition to the simultaneous inhibition of the dominant metabolic pathway (e.g., by **PFK15**) and the activity on human kinase receptors (e.g , by **sunitinib**), there are also other possibilities to reduce cell metabolism and other physiological processes in cells. The inhibition of glycolysis in tumor cells may lead to the cells showing the increased preference for another metabolic pathway for ATP production, e.g., oxidative phosphorylation. The simultaneous inhibition of glycolysis and oxidative phosphorylation using metformin and 2-DG (2-deoxyglucose) showed an additive effect compared to the application of these inhibitors alone [18]. Another possibility is the use of phenformin, which was clinically tested in the treatment of diabetes, and oxamate, which acts as an inhibitor of lactate dehydrogenase enzyme activity. A synergistic effect was observed in the combined effect of phenformin and oxamate, where a significant antitumor effect was confirmed by simultaneous inhibition of complex I in mitochondria and LDH activity in the cytosol of the cells [17].

We used the IC_{50} value, which expresses the concentration of tested substance causing inhibition of cell growth by 50%, to determine the toxicity of **sunitinib** and the glycolysis blocker **PFK15**. The measured values show that HUVEC are more sensitive to **sunitinib** compared to **PFK15**. Similar results were obtained with bovine aortic endothelial cells (BAEC) at the IC_{50} concentration of 2.2 μM of **sunitinib** and 1.6 μM for HUVEC, respectively (Table 1). This reflects the multi-targeted tyrosine kinase inhibitory action of **sunitinib** compared with **PFK15**. The multikinase inhibitor **sunitinib** is used in the clinical treatment of several types of cancer and its effect has been documented by many published data [19,20]. The mechanism of action of **sunitinib** includes several signaling pathways, possessing receptor as well as non-receptor tyrosine kinases (e.g., VEGF-R, PDGF-R, CSFR, c-KIT, etc.) [21].

Consistent with our results, the different effectivity of **3PO** and **PFK15** was shown in a model of an immortalized human T-lymphocyte line (Jurkat cells) and non-small cell lung carcinoma cells (H522) after 48 h of treatment, with **PFK15** having a more pronounced effect than **3PO** [2]. Previously published data suggest that the effect of these inhibitors varies significantly depending on the cell type and origin (human vs. animal cells), degree of differentiation, inhibitor concentrations, and time of incubation [1,2,12,22]. Our previous study showed that **PFK15** reduced HUVEC and colorectal adenocarcinoma DLD1 cell proliferation at the same level, indicating that the same molecular targets are affected by this treatment [8]. However, compared to an animal model of BAEC, the effectivity of **PFK15** was quite different, as the IC_{50} value for BAEC was determined to be 7.4 μM. **3PO** showed similar results to **PFK15**, with an IC_{50} value of 6.9 μM, indicating that the effectivity of glycolysis inhibition depends on several additional factors.

The simultaneous administration of **PFK15** and the chemotherapeutic drug paclitaxel (PTX) showed a synergistic effect and may be an effective strategy in cancer treatment. Both drugs were carried to the cells by solid lipid nanoparticles encapsulated in membranes formulated by fusing breast cancer cell and activated fibroblast membranes [23]. The cytotoxic effect of PTX/**PFK15** in solid lipid nanoparticles was confirmed in vitro, as well as in vivo. In tumor-bearing mice, the monotherapy of nanoparticles with PTX or **PFK15** did not induce a significant therapeutic effect; however, mice treated with PTX/**PFK15** nanoparticles showed enhanced tumor inhibition. These data suggest that the dual targeting of cancer cell and cancer-associated fibroblasts (CAF) is more effective, as the energy supply from CAF to cancer cells is blocked by glycolysis inhibition [23].

In summary, we developed the missing synthesis of **PFK15** with the aim of making this glycolysis inhibitor broadly available in large amounts for in vitro and in vivo biological studies. Additionally, **PFK15** was also thoroughly physicochemically characterized. From a biological point of view, we proved sufficient sensitivity of HUVEC proliferation and migration to both inhibitors alone. More importantly, we have proven the synergistic effect on HUVEC proliferation and migration with the combination of **PFK15** and **sunitinib** at their appropriate concentrations. In future research, the pharmacokinetic properties of **PFK15** should be improved by developing more biologically active form of inhibitor, which can be tested either alone and/or in combination with other clinically used drugs in different pathophysiological conditions.

4. Materials and Methods

4.1. Chemistry

4.1.1. General Conditions

The starting chemicals for syntheses were purchased from Sigma-Aldrich (St. Louis, MO, USA). Other chemicals and solvents were purchased from local commercial sources and were of analytical grade quality. The melting point was measured by a digital melting point apparatus Barnstead Electrothermal IA9200 and is uncorrected. ^{1}H- and ^{13}C-NMR spectra were recorded on a Varian Gemini (600 and 150 MHz, respectively); chemical shifts are given in parts per million (ppm); tetramethylsilane was used as an internal standard; and DMSO-d_6 was used as the solvent. Infrared (IR) spectra were acquired by Fourier transform infrared spectroscopy (FT-IR) attenuated total reflectance REACT IR 1000 (ASI Applied Systems) with a diamond probe and MTS detector. Mass spectra were performed on a liquid chromatography mass spectrometer LC-MS using an Agilent Technologies 1200 Series equipped with mass spectrometer Agilent Technologies 6100 Quadrupole LC-MS. The course of the reactions was followed by thin-layer chromatography (TLC) (Merck Silica gel 60 F254), and a UV lamp (254 nm), and iodine vapors were used for visualization of the TLC spots.

4.1.2. Synthetic Pathway to (*E*)-1-(Pyridin-4-yl)-3-(quinolin-2-yl)prop-2-en-1-one (**PFK15**)

Inhibitor **PFK15** (**6**) was prepared in four reaction steps, according to Scheme 1. After bromination of commercially available 4-acetylpyridine (**1**) with bromine in a mixture of conc. acids HBr/HOAc (1/3, respectively) at room temperature (rt), the hydrobromide of 2-bromo-1-(pyridin-4-yl)ethan-1-one (**2**) was obtained in 94% yield. Salt **2** was converted to compound **3** in 83% yield by PPh_3 and Et_3N in THF under reflux overnight. Compound **4** was prepared from salt **3** in 60% yield by NaOH in a mixture of $MeOH/H_2O$ at rt overnight. The desired **PFK15** inhibitor (**6**) was finally synthetized from commercially available 2-quinolinecarboxaldehyde (**5**) and the previously prepared compound **4** in refluxing toluene overnight, with 78% yield (or 36.5% overall yield via 4 steps). **PFK15** was purified by crystallization from EtOH or ethyl acetate (EA).

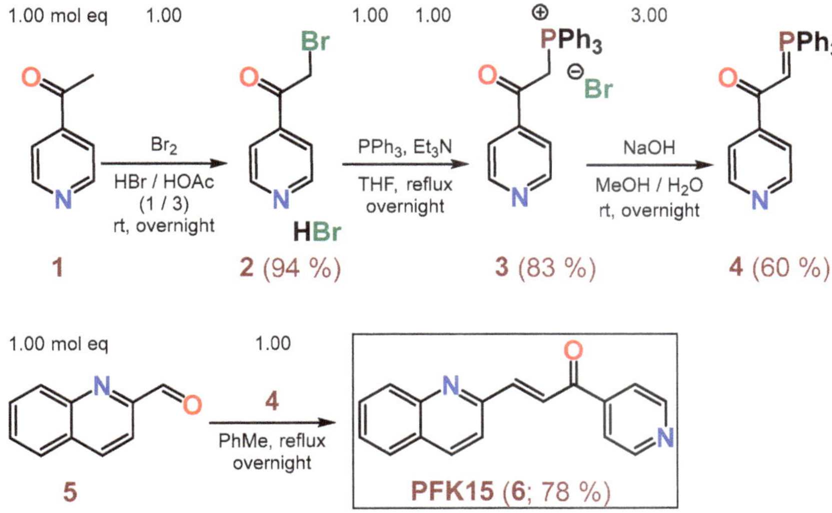

Scheme 1. Synthesis of inhibitor **PFK15** (**6**), starting from commercially available 1-(pyridin-4-yl)ethan-1-one (**1**) and 2-quinolinecarboxaldehyde (**5**).

4.1.3. Synthesis of 2-Bromo-1-(pyridin-4-yl)ethan-1-one hydrobromide (**2**)

A solution of 32.0 g (264 mmol, 1.00 mol eq) 1-(pyridin-4-yl)ethan-1-one (**1**) in 250 mL glacial acetic acid (100%) was cooled in an ice bath, and 150 mL conc. HBr (48 *w*/*w* %) was carefully added. Subsequently, a solution of 42.2 g (264 mmol, 1.00 mol eq) Br_2 in 20 mL glacial acetic acid was added dropwise to the reaction mixture. The reaction was stirred overnight at rt while a white precipitate formed. Afterwards, 250 mL Et_2O was added, and the mixture was stirred for 30 min at rt and filtered. The obtained solid material was washed with 150 mL Et_2O, dried to yield 70.0 g (249 mmol, 94%) of **2** in the form of a white solid powder. Salt **2** was used in the next step without further purification. Its m.p., ^{1}H-NMR [24] and MS [25] spectra are described in the literature.

4.1.4. Synthesis of (2-Oxo-2-(pyridin-4-yl)ethyl)triphenylphosphonium bromide (**3**)

First of all, 65.4 g (249 mmol, 1.00 mol eq) PPh_3 and 34.7 mL (25.2 g, 249 mmol, 1.00 mol eq) Et_3N were sequentially added to a solution of 70.0 g (249 mmol, 1.00 mol eq) **2** in 400 mL of THF. The reaction mixture was refluxed overnight while an orange precipitate formed. After cooling the mixture to rt, the solid material was filtered off, washed with 150 mL Et_2O, and dried under reduced pressure to yield 95.9 g (207 mmol, 83%) of an orange product **3**, which was used without further purification in the next reaction.

4.1.5. Synthesis of 1-(Pyridin-4-yl)-2-(triphenyl-λ^5-phosphanylidene)ethan-1-one (**4**)

Within 30 min, 10.6 g (265 mmol, 3.00 mol eq) NaOH in 303 mL H_2O was added to a solution of 40.9 g (88.5 mmol, 1.00 mol eq) **3** in a mixture of solvents (82 mL of MeOH and 151 mL of H_2O). The mixture was stirred overnight at rt. The brown precipitate was filtered off, and washed sequentially with 100 mL H_2O and 100 mL Et_2O and dried under reduced pressure. A yield of 20.11 g (52.7 mmol, 60%) of **4** was obtained and used without further purification in the next step. Its m.p., ^{1}H-NMR, and IR spectra are described in the literature [26].

4.1.6. Synthesis of (*E*)-1-(Pyridin-4-yl)-3-(quinolin-2-yl)prop-2-en-1-one (**6, PFK15**)

First of all, 8.86 g (23.2 mmol, 1.00 mol eq) of a previously prepared compound **4** was added to a solution of 3.65 g (23.2 mmol, 1.00 mol eq) 2-quinolinecarboxaldehyde (**5**) in 250 mL toluene. The reaction mixture was stirred and refluxed overnight while a precipitated material appeared in the reaction. The mixture was cooled down and volatile parts were evaporated by rotary vacuum evaporator (RVE). The obtained crude reaction product (4.71 g, 78%) was crystallized from EtOH, yielding 2.54 g (9.76 mmol, 42%) of the required inhibitor **6** (**PFK15**), possessing an HPLC-UV (254 nm) purity of 99.2%. The synthesis and other characteristics of compound **6** are not yet described in the literature.

M.p.: 159.4–160.6 °C [EtOH].

The spectral diagrams (^{1}H- and ^{13}C-NMR) of **6** (**PFK15**) are based on 1D- and 2D-NMR techniques and are shown in Figure 4 below (see also the Supplementary Material to this paper).

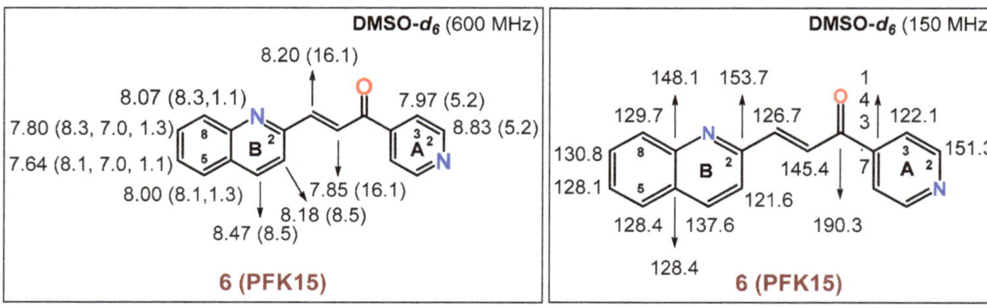

Figure 4. The spectral diagrams describing ^{1}H- and ^{13}C-NMR characteristics of **6** (**PFK15**).

^{1}H-NMR (600 MHz, DMSO-d_6): δ 8.83 (d, 2H, $J(A_2,A_3) = 5.2$ Hz, 2 × H-C_A(2)), 8.47 (d, 1H, $J(B_3,B_4) = 8.5$ Hz, H-C_B(4)), 8.20 (d, 1H, J(CH=CH) = 16.1 Hz, -C<u>H</u>=CH-C=O), 8.18 (d, 1H, $J(B_3,B_4) = 8.5$ Hz, H-C_B(3)), 8.07 (dd, 1H, $J(B_7,B_8) = 8.3$ Hz, $J(B_6,B_8) = 1.1$ Hz, H-C_B(8)), 8.00 (dd, 1H, $J(B_5,B_6) = 8.1$ Hz, $J(B_5,B_7) = 1.3$ Hz, H-C_B(5)), 7.97 (d, 2H, $J(A_2,A_3) = 5.2$ Hz, 2 × H-C_A(3)), 7.85 (d, 1H, J(CH=CH) = 16.1 Hz, -CH=C<u>H</u>-C=O), 7.80 (ddd, 1H, $J(B_7,B_8) = 8.3$ Hz, $J(B_6,B_7) = 7.00$ Hz, $J(B_5,B_7) = 1.3$ Hz, H-C_B(7)), 7.64 (ddd, 1H, $J(B_5,B_6) = 8.1$ Hz, $J(B_6,B_7) = 7.00$ Hz, $J(B_6,B_8) = 1.1$ Hz, H-C_B(6)).

^{13}C-NMR (150 MHz, DMSO-d_6): δ 190.3 (-<u>C</u>=O), 153.7 C_B(2), 151.3 (2 × C_A(2)), 148.1 C_B(9), 145.4 (-CH=<u>C</u>H-C=O), 143.7 C_A(4), 137.6 C_B(4), 130.8 C_B(7), 129.7 C_B(8), 128.4 C_B(10), 128.4 C_B(5), 128.1 C_B(6), 126.7 (-<u>C</u>H=CH-C=O), 122.1 (2 × C_A(3)), 121.6 C_B(3).

FT IR (solid sample, cm^{-1}): 3051 (m), 2111 (w), 2054 (w), 1930 (w), 1858 (w), 1661 (m), 1589 (s), 1551 (m), 1502 (m), 1433 (w), 1408 (m), 1377 (w), 1351 (m), 1306 (m), 1286 (m), 1255 (m), 1209 (s), 1149 (m), 1117 (m), 1060 (w), 1034 (s), 980 (s), 951 (m), 897 (m), 872 (m), 842 (m), 820 (s), 877 (s), 770 (s), 738 (s), 698 (s), 655 (s), 554 (m), 524 (m), 495 (m), 477 (m), 448 (m), 428 (m).

MS of **6** (**PFK15**, calculated $FW_{exact} = 260.09$; ESI/positive mode: found $[M+H]^+ = 261.1$).

Elemental Analysis: Calculated for $C_{17}H_{12}N_2O$ (260.30): C, 78.44; H, 4.65; N, 10.76. Found: C, 78.12; H, 4.38; N, 10.99.

The spectra and detailed spectral characteristics (1D-, 2D-NMR: NOESY, HSQC, HMBC, IR and MS) for **PFK15** (**6**), not yet published in the literature, are presented in the Supplementary Material to this paper (Figures S1–S7).

4.2. Biology

4.2.1. Cell Culture and Cultivation

Human umbilical vein endothelial cells (HUVEC) were isolated from fresh umbilical cords, as described previously [4], and cultured in endothelial cell growth medium (ECGM; PromoCell, Heidelberg, Germany) supplemented with 100 U/mL penicillin and 100 µg/mL streptomycin (Biosera, Nuaillé, France). Cells were incubated in a humidified incubator (Heal Force Bio-meditech, Qingdao, China) at standard conditions (in 5% CO_2 at 37 °C). Cells were used between passages 1 and 6.

4.2.2. Drug Preparation

Stock solutions of the multikinase inhibitor **sunitinib *L*-malate** (Pfizer, New York, NY, USA) (abbreviated to **sunitinib** or **SU** in the text for simplification) and the glucose metabolism inhibitor (*E*)-1-(pyridin-4-yl)-3-(quinolin-2-yl)prop-2-en-1-one (**PFK15**) were prepared in dimethyl sulfoxide (DMSO; Sigma-Aldrich, USA) at a concentration of 10 mM. The stock solutions were subsequently diluted in an appropriate concentration in ECGM (and finally in test solutions containing 1% DMSO) for the in vitro experiments. **Sunitinib** was obtained as a gift from a pharmaceutical company (Pfizer Inc., New York, NY, USA). The synthesis and exact structure assignments of **PFK15** were performed as described above.

4.2.3. IC_{50} Evaluation

Cells were seeded in a 96-well plate with serial dilutions of **PFK15** and **sunitinib** (starting at 100 μM solution containing 1% DMSO) at a density of 3×10^3 cells/well. The MTS assay was performed according to the manufacturer's instructions, as previously described [8]. The half maximal inhibitory concentration (IC_{50}) value was calculated using Graph Pad Prism 6 software (Graph Pad, San Diego, CA, USA).

4.2.4. Cell Proliferation Assay

The MTS assay is generally used for the quantification of cell proliferation, viability, or cytotoxicity. Cell proliferation was determined by colorimetric MTS assay, according to the manufacturer´s instructions. Briefly, cells were seeded in a 96-well plate at a density of 3×10^3 cells/well and incubated with serial dilutions of **PFK15** and **sunitinib** for 3 days. After 3 days, 10 μL of yellow MTS (5 mg/mL) (Promega, Madison, WI, USA, CAS: 138169-43-4) was added to each well and cells were incubated for 2–3 h at 37 °C. The absorbances were measured at a wavelength of 490 nm (green color). All determinations were performed in quadruplicate, with three independent experiments.

4.2.5. Migration Assay (Wound Healing Assay)

Cells were seeded in 24-well plates coated with 1.5% gelatin (Sigma-Aldrich, USA) at a density of 5×10^5 cells/well (HUVEC). After reaching a confluent monolayer the medium was replaced with a starvation medium (ECGM supplemented with 2% FBS) and cells were incubated for a further 17 h. The confluent cell monolayer was wounded using pipet tips and washed twice with PBS. Subsequently, cells were treated with different doses of inhibitors diluted in ECGM medium. The migration of HUVEC was observed with an Olympus IMT2 inverted optical microscope (Olympus, Tokyo, Japan) and recorded by a Moticam 1000 camera system (Motic Incorporation, Hong Kong) at 0 h and 8 h after treatment. Changes in cell migration were evaluated using the software Motic Images Plus 2.0 PL (Motic Incorporation, Hong Kong).

4.3. Statistical Analysis

The results are expressed as a mean ± SEM. Each value represents the average of at least three independent experiments. Statistical analysis was performed using STATISTICA 7.0 (StatSoft Inc., Tulsa, OK, USA) and GraphPad Prism 6 (GraphPad Software, Inc.). Statistical differences among groups were determined by one-way ANOVA followed by Tukey's post hoc test. The value $p < 0.05$ was considered as significant.

5. Conclusions

In conclusion, we described the synthesis of the glycolysis inhibitor PFK15 in detail. Our results confirmed the effectivity of PFK15 in combination with sunitinib. **Sunitinib** is an approved drug with therapeutic applications against tumor growth and tumor angiogenesis. The glycolysis inhibitor **PFK15** reduces glucose uptake and causes starvation of activated endothelial and tumor cells. Both tumor and activated endothelial cells are dependent on a high influx of glucose as an important source of energy and building blocks.

Therefore, the synergistic effect of the above inhibitors to block different biological targets and mechanisms of action on HUVEC-based angiogenesis is promising from a therapeutical point of view.

Supplementary Materials: The following supporting information can be downloaded at: https://www.mdpi.com/article/10.3390/ijms232214295/s1.

Author Contributions: Conceptualization, J.Z., R.M., M.Z., M.M., G.A. and A.B.; methodology, J.Z., R.M., M.M., G.A. and A.B.; formal analysis, J.Z., R.M. and A.B.; writing—original draft preparation, J.Z., A.B. and M.Z.; writing—review and editing, J.Z., A.B. and M.Z.; visualization, J.Z. and A.B.; supervision, M.Z. and A.B.; funding acquisition, M.Z. and A.B. All authors have read and agreed to the published version of the manuscript.

Funding: This study was supported by grants APVV-17-0178 and by the Operation Program of Integrated Infrastructure for the Project, Advancing University Capacity and Competence in Research, Development and Innovation, ITMS2014+: 313021X329, co-financed by the European Regional Development Fund, VEGA 1/0670/18 and Biomagi, Ltd.

Institutional Review Board Statement: Not applicable.

Informed Consent Statement: Not applicable.

Acknowledgments: We are grateful to the pharmaceutical company Pfizer Inc., USA, for supporting us with sunitinib *L*-malate that was obtained as a gift.

Conflicts of Interest: The authors declare no conflict of interest.

References

1. Clem, B.; Telang, S.; Clem, A.; Yalcin, A.; Meier, J.; Simmons, A.; Rasku, M.A.; Arumugam, S.; Dean, W.L.; Eaton, J.; et al. Small-Molecule Inhibition of 6-Phosphofructo-2-Kinase Activity Suppresses Glycolytic Flux and Tumor Growth. *Mol. Cancer Ther.* **2008**, *7*, 110–120. [CrossRef] [PubMed]
2. Clem, B.F.; O'neal, J.; Tapolsky, G.; Clem, A.L.; Imbert-Fernandez, Y.; Ii, D.A.K.; Klarer, A.C.; Redman, R.; Miller, D.M.; Trent, J.O.; et al. Targeting 6-Phosphofructo-2-Kinase (PFKFB3) as a Therapeutic Strategy against Cancer. *Mol. Cancer Ther.* **2013**, *12*, 1461–1470. [CrossRef] [PubMed]
3. Mondal, S.; Roy, D.; Sarkar Bhattacharya, S.; Jin, L.; Jung, D.; Zhang, S.; Kalogera, E.; Staub, J.; Wang, Y.; Xuyang, W.; et al. Therapeutic Targeting of PFKFB3 with a Novel Glycolytic Inhibitor PFK158 Promotes Lipophagy and Chemosensitivity in Gynecologic Cancers. *Int. J. Cancer* **2019**, *144*, 178–189. [CrossRef]
4. Horváthová, J.; Moravčík, R.; Boháč, A.; Zeman, M. Synergic Effects of Inhibition of Glycolysis and Multikinase Receptor Signalling on Proliferation and Migration of Endothelial Cells. *Gen. Physiol. Biophys.* **2019**, *38*, 157–163. [CrossRef] [PubMed]
5. Chen, L.; Zhao, J.; Tang, Q.; Li, H.; Zhang, C.; Yu, R.; Zhao, Y.; Huo, Y.; Wu, C. PFKFB3 Control of Cancer Growth by Responding to Circadian Clock Outputs. *Sci. Rep.* **2016**, *6*, 24324. [CrossRef] [PubMed]
6. Li, H.M.; Yang, J.G.; Liu, Z.J.; Wang, W.M.; Yu, Z.L.; Ren, J.G.; Chen, G.; Zhang, W.; Jia, J. Blockage of Glycolysis by Targeting PFKFB3 Suppresses Tumor Growth and Metastasis in Head and Neck Squamous Cell Carcinoma. *J. Exp. Clin. Cancer Res.* **2017**, *36*, 7. [CrossRef]
7. Matsumoto, K.; Noda, T.; Kobayashi, S.; Sakano, Y.; Yokota, Y.; Iwagami, Y.; Yamada, D.; Tomimaru, Y.; Akita, H.; Gotoh, K.; et al. Inhibition of Glycolytic Activator PFKFB3 Suppresses Tumor Growth and Induces Tumor Vessel Normalization in Hepatocellular Carcinoma. *Cancer Lett.* **2021**, *500*, 29–40. [CrossRef]
8. Horváthová, J.; Moravčík, R.; Matúšková, M.; Šišovský, V.; Boháč, A.; Zeman, M. Inhibition of Glycolysis Suppresses Cell Proliferation and Tumor Progression in Vivo: Perspectives for Chronotherapy. *Int. J. Mol. Sci.* **2021**, *22*, 4390. [CrossRef]
9. Martins, C.P.; New, L.A.; O'Connor, E.C.; Previte, D.M.; Cargill, K.R.; Tse, I.L.; Sims-Lucas, S.; Piganelli, J.D. Glycolysis Inhibition Induces Functional and Metabolic Exhaustion of CD4+ T Cells in Type 1 Diabetes. *Front. Immunol.* **2021**, *12*, 669456. [CrossRef]
10. Cheng, R.; Ma, J.-X. Angiogenesis in Diabetes and Obesity. *Rev. Endocr. Metab. Disord.* **2015**, *16*, 67–75. [CrossRef]
11. Min, J.; Zeng, T.; Roux, M.; Lazar, D.; Chen, L.; Tudzarova, S. The Role of HIF1α-PFKFB3 Pathway in Diabetic Retinopathy. *J. Clin. Endocrinol. Metab.* **2021**, *106*, 2505. [CrossRef] [PubMed]
12. Zhu, W.; Ye, L.; Zhang, J.; Yu, P.; Wang, H.; Ye, Z.; Tian, J. PFK15, a Small Molecule Inhibitor of PFKFB3, Induces Cell Cycle Arrest, Apoptosis and Inhibits Invasion in Gastric Cancer. *PLoS ONE* **2016**, *11*, e0163768. [CrossRef] [PubMed]
13. Murár, M.; Horvathová, J.; Moravčík, R.; Addová, G.; Zeman, M.; Boháč, A. Synthesis of Glycolysis Inhibitor (E)-3-(Pyridin-3-Yl)-1-(Pyridin-4-Yl)Prop-2-En-1-One (3PO) and Its Inhibition of HUVEC Proliferation Alone or in a Combination with the Multi-Kinase Inhibitor Sunitinib. *Chem. Pap.* **2018**, *72*, 2979–2985. [CrossRef]

14. Kotowski, K.; Supplitt, S.; Wiczew, D.; Przystupski, D.; Bartosik, W.; Saczko, J.; Rossowska, J.; Drag-Zalesińska, M.; Michel, O.; Kulbacka, J. 3PO as a Selective Inhibitor of 6-Phosphofructo- 2-Kinase/Fructose-2,6-Biphosphatase 3 in A375 Human Melanoma Cells. *Anticancer Res.* **2020**, *40*, 2613–2625. [CrossRef]
15. Yan, S.; Yuan, D.; Li, Q.; Li, S.; Zhang, F. AICAR Enhances the Cytotoxicity of PFKFB3 Inhibitor in an AMPK Signaling-Independent Manner in Colorectal Cancer Cells. *Med. Oncol.* **2021**, *39*, 10. [CrossRef]
16. Schoors, S.; de Bock, K.; Cantelmo, A.R.; Georgiadou, M.; Ghesquière, B.; Cauwenberghs, S.; Kuchnio, A.; Wong, B.W.; Quaegebeur, A.; Goveia, J.; et al. Partial and Transient Reduction of Glycolysis by PFKFB3 Blockade Reduces Pathological Angiogenesis. *Cell Metab.* **2014**, *19*, 37–48. [CrossRef]
17. Miskimins, W.K.; Ahn, H.J.; Kim, J.Y.; Ryu, S.; Jung, Y.S.; Choi, J.Y. Synergistic Anti-Cancer Effect of Phenformin and Oxamate. *PLoS ONE* **2014**, *9*, 85576. [CrossRef] [PubMed]
18. Sahra, I.B.; Laurent, K.; Giuliano, S.; Larbret, F.; Ponzio, G.; Gounon, P.; le Marchand-Brustel, Y.; Giorgetti-Peraldi, S.; Cormont, M.; Bertolotto, C.; et al. Targeting Cancer Cell Metabolism: The Combination of Metformin and 2-Deoxyglucose Induces P53-Dependent Apoptosis in Prostate Cancer Cells. *Cancer Res.* **2010**, *70*, 2465–2475. [CrossRef]
19. Hu, J.; Wang, W.; Liu, C.; Li, M.; Nice, E.; Xu, H. Retraction Note: Receptor Tyrosine Kinase Inhibitor Sunitinib and Integrin Antagonist Peptide HM-3 Show Similar Lipid Raft Dependent Biphasic Regulation of Tumor Angiogenesis and Metastasis. *J. Exp. Clin. Cancer Res.* **2020**, *39*, 381. [CrossRef]
20. Pla, A.F.; Brossa, A.; Bernardini, M.; Genova, T.; Grolez, G.; Villers, A.; Leroy, X.; Prevarskaya, N.; Gkika, D.; Bussolati, B. Differential Sensitivity of Prostate Tumor Derived Endothelial Cells to Sorafenib and Sunitinib. *BMC Cancer* **2014**, *14*, 939. [CrossRef]
21. le Tourneau, C.; Raymond, E.; Faivre, S. Sunitinib: A Novel Tyrosine Kinase Inhibitor. A Brief Review of Its Therapeutic Potential in the Treatment of Renal Carcinoma and Gastrointestinal Stromal Tumors (GIST). *Ther. Clin. Risk Manag.* **2007**, *3*, 341. [CrossRef] [PubMed]
22. de Bock, K.; Georgiadou, M.; Carmeliet, P. Role of Endothelial Cell Metabolism in Vessel Sprouting. *Cell Metab.* **2013**, *18*, 634–647. [CrossRef] [PubMed]
23. Zang, S.; Huang, K.; Li, J.; Ren, K.; Li, T.; He, X.; Tao, Y.; He, J.; Dong, Z.; Li, M.; et al. Metabolic Reprogramming by Dual-Targeting Biomimetic Nanoparticles for Enhanced Tumor Chemo-Immunotherapy. *Acta Biomater.* **2022**, *148*, 181–193. [CrossRef] [PubMed]
24. Turcotte, S.; Chan, D.A.; Sutphin, P.D.; Giaccia, A.J.; Hay, M.P.; Denny, W.A.; Bonnet, M.M. Heteroaryl Compounds, Compositions, and Methods of Use in Cancer Treatment. U.S. Patent WO2009114552A1, 19 September 2009.
25. Newcom, J.S.; Spear, K.L. P2x4 Receptor Modulating Compounds and Methods of Use Thereof. *Pain* **2015**, *161*, 1425.
26. Portevin, B.; Tordjman, C.; Pastoureau, P.; Bonnet, J.; de Nanteuil, G. 1,3-Diaryl-4,5,6,7-Tetrahydro-2H-Isoindole Derivatives: A New Series of Potent and Selective COX-2 Inhibitors in Which a Sulfonyl Group Is Not a Structural Requisite. *J. Med. Chem.* **2000**, *43*, 4582–4593. [CrossRef] [PubMed]

International Journal of *Molecular Sciences*

MDPI

Article

New Molecules of Diterpene Origin with Inhibitory Properties toward α-Glucosidase

Elena Tretyakova [1], Irina Smirnova [1], Oxana Kazakova [1,*], Ha Thi Thu Nguyen [2], Alina Shevchenko [3], Elena Sokolova [3], Denis Babkov [3] and Alexander Spasov [3]

1 Ufa Institute of Chemistry, Ufa Federal Research Centre, Russian Academy of Sciences, 71 Prospect Oktyabrya, 450054 Ufa, Russia
2 Institute of Chemistry, Vietnamese Academy of Science and Technology, 18 Hoang Quoc Viet Str., Cau Giay Dist., Hanoi 100000, Vietnam
3 Scientific Center for Innovative Drugs, Volgograd State Medical University, Novorossiyskaya Str. 39, 400087 Volgograd, Russia
* Correspondence: obf@anrb.ru; Tel.: +7-347-235-6066

Abstract: The incidence of diabetes mellitus (DM), one of the most common chronic metabolic disorders, has increased dramatically over the past decade and has resulted in higher rates of morbidity and mortality worldwide. The enzyme, α-Glucosidase (α-GLy), is considered a therapeutic target for the treatment of type 2 DM. Herein, we synthesized arylidene, heterocyclic, cyanoetoxy- and propargylated derivatives of quinopimaric acid (levopimaric acid diene adduct with *p*-benzoquinone) **1–50** and, first, evaluated their ability to inhibit α-GLy. Among the tested compounds, quinopimaric acid **1**, 2,3-dihydroquinopimaric acid **8** and its amide and heterocyclic derivatives **9**, **30**, **33**, **39**, **44**, with IC_{50} values of 35.57–65.98 μM, emerged as being good inhibitors of α-GLy. Arylidene 1β-hydroxy and 1β,13α-epoxy methyl dihydroquinopimarate derivatives **6**, **7**, **26–29**, thiadiazole **32**, 1a,4a-dehydroquinopimaric acid **40** and its indole, nitrile and propargyl hybrids **35–38**, **42**, **45**, **48**, and **50** showed excellent inhibitory activities. The most active compounds **38**, **45**, **48**, and **50** displayed IC_{50} values of 0.15 to 0.68 μM, being 1206 to 266 more active than acarbose (IC_{50} of 181.02 μM). Kinetic analysis revealed the most active diterpene indole with an alkyne substituent **45** as a competitive inhibitor with K_i of 50.45 μM. Molecular modeling supported this finding and suggested that the indole core plays a key role in the binding. Compound **45** also has favorable pharmacokinetic and safety properties, according to the computational ADMET profiling. The results suggested that quinopimaric acid derivatives should be considered as potential candidates for novel alternative therapies in the treatment of type 2 diabetes.

Keywords: diabetes mellitus; α-glucosidase; abietane diterpenoids; levopimaric acid; quinopimaric acid; molecular docking; ADMET

Citation: Tretyakova, E.; Smirnova, I.; Kazakova, O.; Nguyen, H.T.T.; Shevchenko, A.; Sokolova, E.; Babkov, D.; Spasov, A. New Molecules of Diterpene Origin with Inhibitory Properties toward α-Glucosidase. *Int. J. Mol. Sci.* **2022**, *23*, 13535. https://doi.org/10.3390/ijms232113535

Academic Editors: Simona Gabriela Bungau and Vesa Cosmin

Received: 12 October 2022
Accepted: 1 November 2022
Published: 4 November 2022

1. Introduction

Enzymes responsible for breaking down proteins, carbohydrates and lipids into smaller and more readily absorbable molecules are a key component of the digestive system. The main cause of many metabolic diseases is abnormal changes in the activity of these enzymes. Inhibition of metabolic enzymes, such as α-glucosidase (α-GLy), is one of the accepted approaches in the treatment of diabetes mellitus (DM), which is one of the most common chronic endocrine diseases, along with arterial hypertension and obesity [1]. The number of people with disorders of carbohydrate metabolism and the incidence of DM are constantly growing, which is primarily due to an increase in the number of patients with obesity, as well as in average life expectancy [2]. Type II diabetes mellitus (DM2), accounting for about 90% of all cases of diabetes, is caused by a decrease in insulin sensitivity in target organs, such as the liver, muscle, and adipose tissue, as well as a deficiency in insulin secretion [3–5]. Medicinal agents with the ability to stimulate glucose uptake in these tissues can be used to improve insulin resistance and, therefore, to treat DM2 [6]. Today, a vast number of synthetic antidiabetic agents, such as acarbose, miglitol, sulfonylurea, metformin, and thiozolidinedione, are readily available on the market [7–9]. However, their effectiveness is limited, due to low bioavailability and unwanted side effects [10–12]. Therefore, there is a great need to develop alternative and more active antidiabetic drugs from natural sources.

Abietane diterpenoids are classes of compounds which are mainly found in the conifer family and have long been used to treat a variety of ailments [13]. Their derivatives are characterized by a wide range of biological activities, like anticancer, antiviral, antimicrobial, antileishmanial, antiplasmodial, antifungal, antitumour, cytotoxicity, antiulcer, cardiovascular, antioxidant, anti-inflammatory and antidiabetic activities [14–21]. Abietic and dehydroabietic acids have been reported to decrease the activity of glucose-6-phosphatase and to stimulate glycogen synthase [22]. Carnosic acid derivatives are very effective in treating diabetic complications by improving insulin secretion [23] and glucose homeostasis or by stimulating glucose uptake by increasing peripheral glucose clearance in tissues [24]. Abietic and carnosic acids also significantly activate nuclear receptor peroxisome proliferator-activated receptor PPAR-γ by exerting its beneficial effect on lipid and glucose homeostasis through PPAR-γ-mediated pathways [25,26]. Carnosol stimulates glucose uptake [27], improves diabetes and its complications by the regulation of oxidative stress and inflammatory responses [28] and suppresses forskolin-induced luciferase expression, when monitored by the cAMP/response element, and glucose-6-phosphatase gene promoters [29–31]. Tanshinones exhibited potent protein tyrosine phosphatase 1B inhibitory activity [32] as well as increased the activity of insulin on the tyrosine phosphorylation of the insulin receptor in addition to the activation of the kinases Akt, ERK1/2, and GSK3beta and may be very useful for developing new anti-diabetic agents as specific insulin receptor activators [33].

Studies evaluating the antidiabetic properties of abietane diterpenoids in an animal model using rats and mice showed that dehydroabietic acid reduces plasma glucose and insulin levels, as well as plasma and hepatic triglyceride levels, by suppressing the production of monocyte chemoattractant protein-1 and tumor necrosis factor-alpha and increasing that of adiponectin, through decrease in macrophage infiltration into adipose tissues [34]. Carnosol and carnosic acid reduced plasma glucose, total cholesterol, and triglycerides in a diabetic group of rats and suppressed inflammation and lipogenesis in mice administered a high-fat diet, through C-kinase substrate regulation [23,28,35]. Tanshinone analogs also demonstrated a significant decrease in blood glucose level, total cholesterol and triglyceride, free fatty acids, and insulin receptor substrate 1 expression, body weight loss and higher insulin resistance when administered to type 2 diabetic rats, with oral administration, resulting in the activation of AMP-activated protein kinase in aortas from ob/ob or db/db mice [35,36]. Data on systematic studies of the enzymatic activity of abietane diterpenoid derivatives, obtained as a result of various modifications of the native core, in particular, on levopimaric acid derivatives as potential inhibitors of α-GLy, are practically absent in the literature. Therefore, herein, we describe the synthesis of abietane type derivatives with arylidene, heterocyclic, nitrile and acetylene fragments. These derivatives were, then, first evaluated for in-vitro α-GLy inhibition. The mechanism of inhibition and enzyme binding were investigated with kinetic and molecular modeling approaches.

2. Results and Discussion

2.1. Chemistry

Since the quinopimaric acid structure (the Diels-Alder reaction product of levopimaric acid and *p*-benzoquinone) contains major reaction centers at the C-1, C-3, C-4, and C-20 atoms, we planned to functionalize these positions for better understanding of the structure–activity relationship and to reveal new promising molecules with antidiabetic activity (Figure 1).

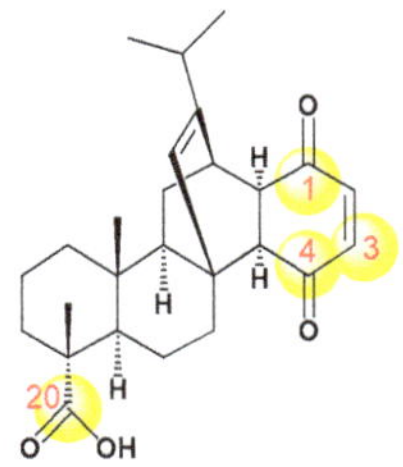

Figure 1. Quinopimaric acid reaction centers for the SAR studies of the current research.

Modifications of these sites involved the synthesis of arylidene and heterocyclic derivatives and quinopimaric acid cyanoetoxy- and propargylated analogs. Figure 2 shows the structures of quinopimaric acid **1** and its analogs **2–30**, modified at position C-1, C-3, C-4, and C-20.

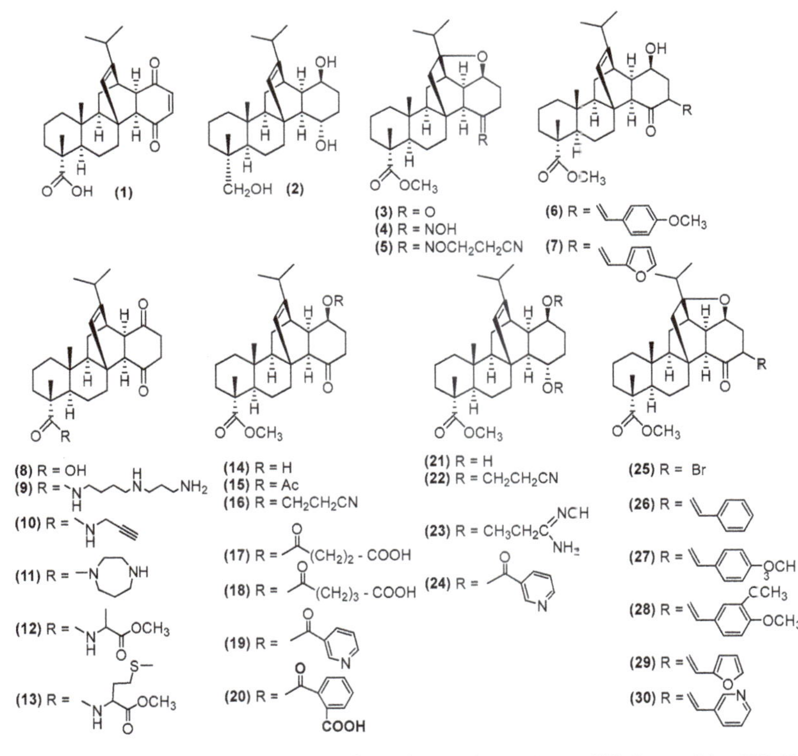

Figure 2. Structures of quinopimaric acid **1** and its analogs **2–30**, modified at position C-1, C-3, C-4, and C-20.

Figure 3 shows the structures of quinopimaric acid heterocyclic derivatives obtained as a result of interaction with hydrazine hydrate **31**, thiourea **32**, and using the Fischer reaction (indoles **33, 34**), Nenitzescu reaction (indoles **35–38**) and Beckmann rearrangement (lactam **39**).

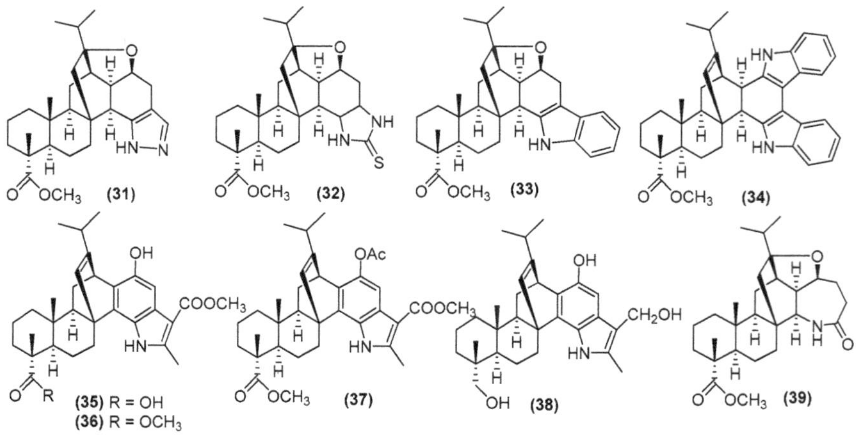

Figure 3. Structures of quinopimaric acid heterocyclic derivatives **31–39**.

We planned to use the Nenitzescu reaction [37] for the synthesis of new diterpene indoles. In this reaction, 1a,4a-dehydroquinopimaric acid **40**, easily formed in two steps from quinopimaric acid **1** [38], was used as the quinone component, as well as ethyl 3-aminocrotonate or 3-aminocrotononitrile being were used as the new enamine components. Under the conditions of the Nenitzescu indole synthesis by the reaction of 1a,4a-dehydroquinopimaric acid **40** with the corresponding enamine in glacial AcOH, at room temperature, diterpene indoles **42**, **44** were synthesized in 76 and 69% yields, respectively. Methyl ester of diterpene indole **43** was obtained in quantitative yield by treating compound **42** with methyl iodide during reflux in acetone for 2 h in the presence of potash (Scheme 1), or direct synthesis from 1a,4a-dehydroquinopimaric acid methyl ester **41**, similar to the preparation of compounds **42** and **44**.

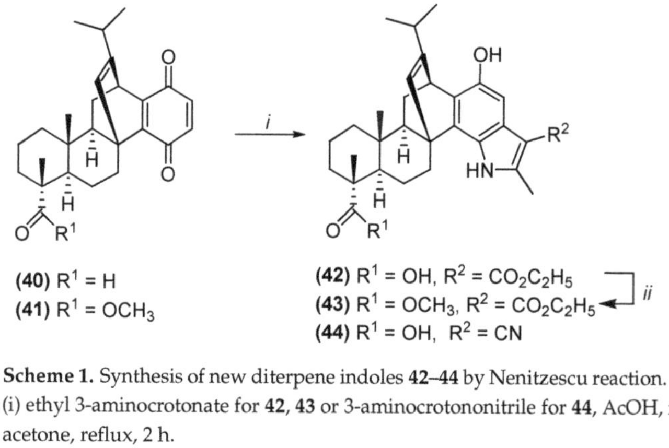

Scheme 1. Synthesis of new diterpene indoles **42–44** by Nenitzescu reaction. Reagents and conditions: (i) ethyl 3-aminocrotonate for **42**, **43** or 3-aminocrotononitrile for **44**, AcOH, rt, 20 h; (ii) CH_3I, K_2CO_3, acetone, reflux, 2 h.

Propargyl derivatives **45**, **46**, **48** were obtained in 79–83% yields by the reaction of diterpene indoles **36**, **43** and quinone **40** with propargyl bromide during reflux in dimethylformamide in the presence of K_2CO_3. Cyanoethyloxy derivatives **47**, **50** were prepared by adding acrylonitrile in 1,4-dioxane. at room temperature, to the diterpene indole **37** or aromatic derivative **49** in the occurrence of phase transfer catalyst triethylbenzylammonium chloride in combination with an alkali (30% KOH) (Scheme 2).

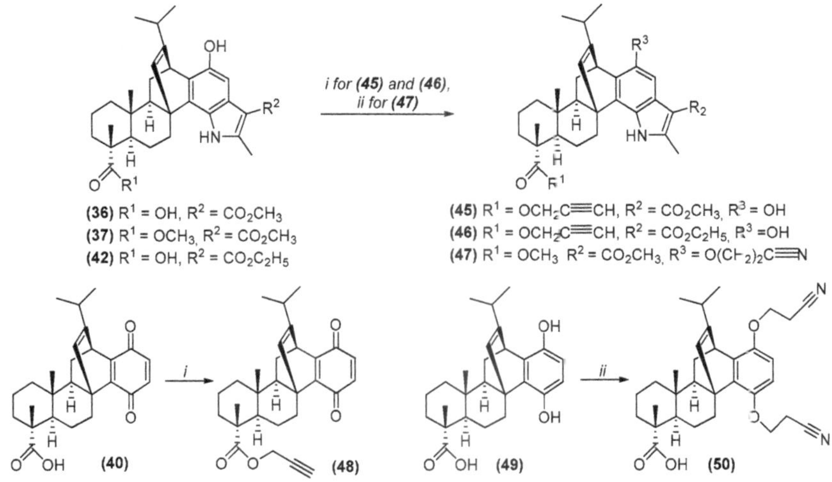

Scheme 2. Synthesis of propargyl **45**, **46**, **48** and cyanoethyloxy **47**, **50** quinopimaric acid derivatives. Reagents and conditions: (i) propargyl bromide, DMF, K_2CO_3, reflux, 2 h; (ii) acrylonitrile, 1,4-dioxane, KOH, BTAC, rt, 2 h.

The structures of the synthesized compounds were confirmed using mass spectrometry, and one- and two-dimensional (COSY, NOESY, ^{1}H–^{13}C HSQC, ^{1}H–^{13}C HMBC) NMR spectroscopy. Thus, the signal of the C-2 carbon atom of the aromatic ring in the ^{13}C NMR spectra of compound **42–44** appeared at δ 99.7-103.3 ppm, and correlated with the signal of the H-2 proton at δ 6.83–7.28 ppm in the ^{1}H–^{13}C HSQC spectra. The ^{1}H NMR spectra showed characteristic signals of methyl group protons at δ 2.51–2.71 (3′-CH_3), as well as broadened signals of the hydroxyl group and NH group at δ 9.12–9.35 and 12.13 ppm, respectively. The ^{1}H NMR spectra of compound **43** contained an additional signal of the protons of the methyl ester group at δ 3.76 ppm, which, in the ^{1}H–^{13}C HSQC spectrum, correlated with the signal of the C-21 atom at δ 15.5 ppm. In the ^{13}C NMR spectra of compound **44**, a carbon signal of the CN-group was observed at δ 117.8 ppm. The ^{1}H NMR spectra of propargyl derivatives **45**, **46**, **48** contained the methylene group proton signal in the region δ 4.67–4.82 ppm, while in the ^{13}C NMR the triple bond carbon signals appeared at δ 74.4–74.9 and 77.9–79.3 ppm, respectively. The signals of the cyanoethyl methylene groups in the ^{1}H NMR spectra of compounds **47**, **50** were observed in the region δ 2.80–2.91 and 4.05–4.30 ppm, and the characteristic carbon signal of the nitrile group in the ^{13}C NMR spectra was observed at δ 117.4–117.6 ppm (Figures S1–S18, Supplementary Materials).

2.2. Inhibition of Yeast α-Glucosidase

All the synthesized compounds **1–50** were tested for their inhibitory potential against yeast α-GLy. Acarbose served as a control drug in this experiment. The IC_{50} values of compounds are provided in Table 1.

Quinopimaric acid derivatives **2–5**, **10–25**, **27**, **31**, **34**, **41**, **43**, **46**, **49** showed moderate to poor α-GLy inhibition. Quinopimaric acid **1**, 2,3-dihydroquinopimaric acid **8** and its amide and heterocyclic derivatives **9**, **30**, **33**, **39**, **44**, with an IC_{50} value of 35.24 ± 0.71 Mm—65.98 ± 0.03 μM, emerged as good inhibitors of α-GLy. Arylidene 1β-hydroxy and 1β,13-epoxy methyl dihydroquinopimarate derivatives **6**, **7**, **26–29**, thiadiazole **32**, 1a,4a-dehydroquinopimaric acid **40** and its indole, nitrile and propargyl hybrids **35-38**, **42**, **45**, **48**, and **50** showed excellent inhibitory activities. The most active compounds, **38**, **45**, **48**, and **50**, displayed IC_{50} values of 0.15 ± 0.008 μM to 0.68 ± 0.045 μM, being 1206 to 266 more potent than acarbose (IC_{50} of 181.02 ± 3.1 μM).

Table 1. α-Glucosidase inhibitory potential of the synthesized compounds **1–50**.

Compound	$IC_{50} \pm SE$ (μM)
1	59.59 ± 0.18
2	>255
3	>255
4	>255
5	>255
6	1.63 ± 0.006
7	2.50 ± 0.011
8	35.57 ± 0.92
9	35.24 ± 0.71
10	>255
11	>255
12	>255
13	>255
14	>255
15	>255
16	>255
17	>255
18	>255
19	>255
20	>255
21	>255
22	>255
23	>255
24	>255
25	>255
26	13.08 ± 0.01
27	>255
28	12.73 ± 0.21
29	1.63 ± 0.041
30	38.80 ± 0.33
31	>255
32	9.66 ± 0.77
33	65.98 ± 0.03
34	>255
35	7.95 ± 0.20
36	8.94 ± 0.96
37	7.28 ± 0.40
38	0.39 ± 0.03
39	44.77 ± 0.96
40	7.088 ± 0.12
41	>255
42	2.52 ± 0.34
43	>255
44	68.22 ± 0.03
45	0.15 ± 0.008
46	>255
47	4.95 ± 0.25
48	0.68 ± 0.045
49	>255
50	0.23 ± 0.01
Acarbose (reference drug)	181.02 ± 3.1

As shown in Table 1, quinopimaric acid **1** had an activity against α-GLy three times higher than that of acarbose. Its simplest modifications, namely, the reduction of the C2-C3 bond (compound **8**), led to an increase in activity, and the introduction of a double bond into the C1a-C4a position (compound **40**) further enhanced this. Modifications at positions C-1, C-4, C-20 of dihydroquinopimaric acid (compounds **2–5, 9–14, 15–24**) were unsuccessful and led to a complete loss of activity. However, the introduction of arylidene substituents into the C-3 position of 1β,13α-epoxy methyldihydroquinopimarate **3** (compounds **26**, **28–30**) resulted mainly in active compounds, and the most successful, for this series of compounds, was the presence of a furfural fragment in the molecule. Heterocyclization of dihydroquinopimaric acid (compounds **31–34**, **39**) led to compounds with high activity against α-GLy only in the case of thiadiazole **32** and indole **34**.

The use of acid **40** as a starting compound for heterocyclization provided more active compounds. Indole and its derivatives **35–38** showed excellent activity, especially alcohol **38**. Inspired by these results, we carried out the synthesis of new indoles using other enamines. The obtained two new indoles **42** and **44** also had very good activity, and it was better for the indole with an ethyl substituent. Modification of the C-20 position by introducing a triple bond into the molecule (compounds **45**, **46**), and the C-1 position with a cyanoethyl fragment (compound **47**) in the case of indole **36**, further enhanced this activity.

Thus, from the studied series of 50 compounds synthesis of indoles based on 1a,4a-dehydroquinopimaric acid **40** proved to be the most successful approach to obtain highly active α-GLy inhibitors. Modification, according to the Nenitzescu reaction, and reduction of carboxyl and methoxycarboxyl groups, with the formation of trihydroxy derivative **38**, propargylation of C-20 positions in acid **40** and indole **36**, as well as cyanoethylation of the aromatic derivative **49**, realized compounds with IC_{50} values < 1 μM.

For compounds **38**, **45**, **48** and **50**, which showed the highest activity against α-GLy, studies of their anti-antioxidant, antimicrobial and cytotoxic activity were carried out (Tables S1–S3, see Supplementary Materials).

2.3. *The Mechanism of α-Glucosidase Inhibition by Compound* **45**

The mechanism of action for the most active diterpene indole with an alkyne substituent **45** was determined in a kinetic experiment using different 4-nitrophenyl β-D-glucopyranoside (pNPG) substrate concentrations. Nonlinear regression of kinetic curves, using the Michaelis–Menten equation (Figure 4), revealed that higher inhibitor concentrations increased K_m, while V_{max} remained constant, which rendered compound **45** as a competitive inhibitor. The inhibition constant K_i was estimated as 50.45 μM.

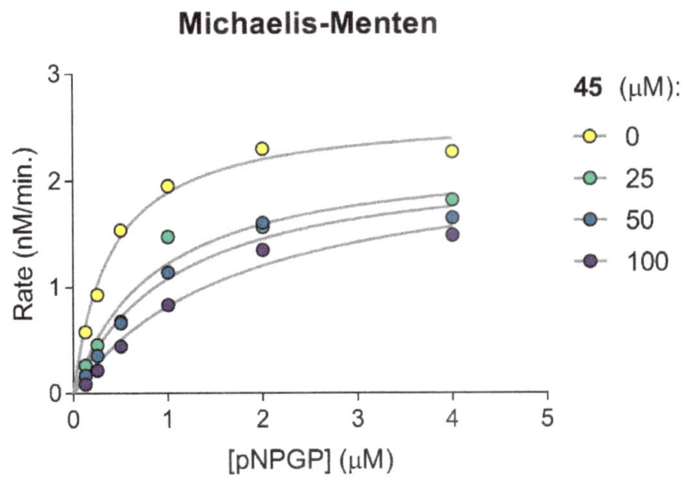

Compound **45** (μM)	0	25	50	100
V_{max} (nM/min.)	2.636	2.277	2.221	2.249
K_m (μM)	0.393	0.855	1.045	1.726

Figure 4. Michaelis—Menten kinetics for compound **45** indicates a competitive inhibition. The experiment was performed in two independent series.

The literature data regarding the mechanism of α-glucosidase inhibition for diterpenes is scarce. The majority of reported compounds are non-competitive inhibitors, e.g., ent-atisane-3-oxo-16β,17-acetonide [39], (E)-labda-8(17),12-diene-15,16-dial [40], ent-kaurane derivative of chepraecoxin A [41], and bis-labdanic diterpene were reported as mixed-type inhibitors [42]. Notably, these inhibitors share an alicyclic core. In contrast, diterpene carnosol, comprising aromatic catechol moiety, is a competitive inhibitor [43]. Aromatic rings of carnosol and compound **45** favor π-stacking interaction with phenylalanine, tyrosine, or tryptophan side chains [44], which might result in distinct binding patterns and inhibition kinetics.

2.4. Docking Studies for Compound **45**

We performed a molecular modeling study to gain insight into the structural basis of interactions between the lead compound **45** and α-GLy enzyme. Since "structure cannot be predicted from kinetics" [45], we avoided preconceived competitive mechanism assumptions and subjected the whole protein surface to a docking procedure (Figure 5).

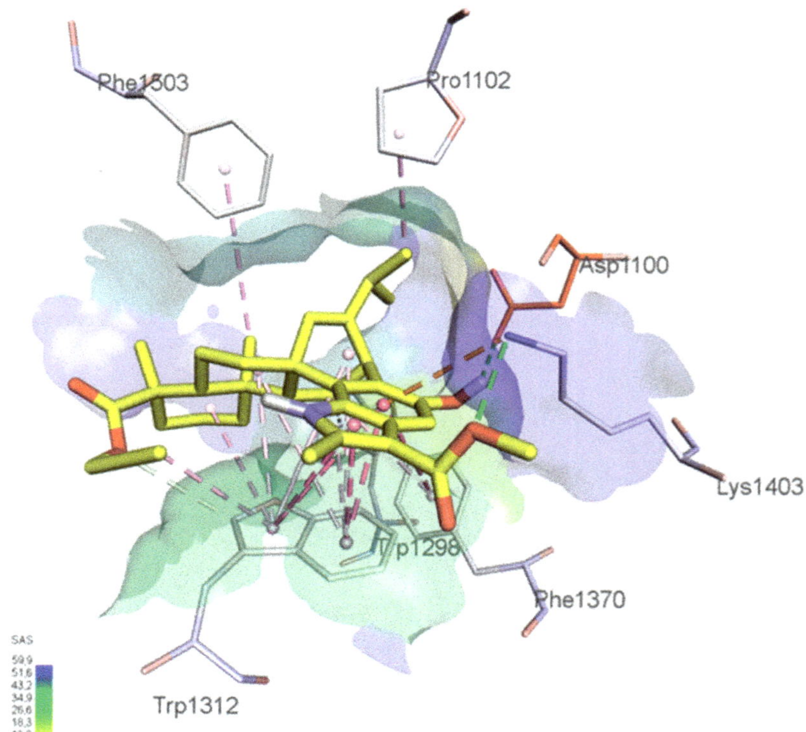

Figure 5. Proposed binding mode of compound **45** to yeast α-glucosidase. The inhibitor is shown in yellow carbons, catalytic Asp1100 is shown in orange carbons. Surface visualized solvent-accessible area. Dashed lines indicate key interactions with enzyme amino acids.

Nevertheless, docking proposed that diterpene derivative 45 shared a favorable binding site with acarbose. Moreover, the indole hydroxyl group appeared to form a conventional H-bond with carboxyl of the catalytic Asp1100 residue. The indole core itself contributed to the binding the most. It was anchored by strong π-π parallel stacking with a Trp1312 side chain and T-shaped π-stacking with a Phe1370 side chain. The amino group of Lys1403 formed an H-bond with the ester substituent and charged π-cation interaction with the indole aromatic system. The dodecahydrophenanthrene part of the molecule was also stabilized by Van der Waals forces with multiple lipophilic residues (Pro1102, Trp1298, Trp1312, Phe1503). Both ester moieties of compound 45 pointed towards the solvent-accessible area of the pocket, providing an opportunity for the introduction of polar fragments. This modification might improve water solubility without hampering enzyme binding. To sum up, molecular modeling confirmed the competitive mechanism of action revealed in the kinetic experiment and provided guidance for future structural optimization.

2.5. ADMET Profiling of Compound **45**

We assessed drug-like, pharmacokinetic and toxicological properties of the lead compound **45** using a consensus of predictive services that took into account different computational strategies (Table 2).

Table 2. ADMET properties predicted for compound **45**.

Property	ADMETlab [46]	ADMETlab 2.0 [47]	SwissADME [48]	ProTox-II [49]	Consensus Value
Physicochemical					
Water solubility (µg/mL)	1.94	2.57	0.02		1.51
LogP	6.21	4.93	6.30	6.69	5.81
Absorption					
Human intestinal absorption	Yes	No	Low		No
Human oral bioavailability	No	No			No
Caco-2 permeability	Yes	No			–
P-glycoprotein substrate	No	No	No		No
P-glycoprotein inhibitor	Yes	Yes			Yes
Distribution					
Plasma protein binding (%)	87.09	99.83			93.46
BBB permeability	No	No	No		No
Metabolism					
CYP1A2 inhibitor	No	No	No		No
CYP2C19 inhibitor	Yes	Yes	No		Yes
CYP2C9 inhibitor	Yes	Yes	Yes		Yes
CYP2D6 inhibitor	No	Yes	No		No
CYP2D6 substrate	No	No			No
CYP3A4 inhibitor	Yes	Yes	No		Yes
CYP3A4 substrate	Yes	Yes			Yes
Excretion					
Total Clearance (mL/min/kg)	1.77	4.03			2.9
$T_{1/2}$ (h)	2.13	0.15			1.14
Toxicity					
AMES toxicity	No	No		No	No
hERG inhibitor	Yes	No			–
Rat acute oral LD_{50} (mg/kg)	121.4			520	320.7
Hepatotoxicity	Yes	No		No	No
Skin Sensitisation	No	No			No
Carcinogenicity		No		No	No

Compound **45** fulfilled Lipinski's rule of 5 (with the exception of molecular weight <500) and Pfizer's rules. At the same time, GSK and "Golden triangle" rules were violated. Mean water solubility was acceptable. Importantly, there was a good consensus on low intestinal absorption and oral bioavailability, which could avoid systemic exposure to the substance. In the case of entering systemic circulation, indole **45** was anticipated to be bound to plasma proteins, likely due to high lipophilicity. Blood–brain barrier penetration was unlikely. Liver metabolism was to be mediated by cytochrome P450 3A4. Acute oral toxicity was predicted to be sufficiently low to achieve a wide therapeutic window. There were no alerts for toxicity to the liver and heart, nor mutagenicity and carcinogenicity. Hence, compound **45**'s calculated ADMET profile was favorable for the proposed mechanism of action.

3. Materials and Methods

3.1. General

The spectra were recorded at the Center for the Collective Use "Chemistry" of the UIC UFRC RAS and RCCU "Agidel" of the UFRC RAS. ^{1}H and ^{13}C-NMR spectra were recorded on a "Bruker AM-500" (Bruker, Billerica, MA, USA, 500 and 125.5 MHz respectively, δ, ppm, Hz) in $CDCl_3$, internal standard tetramethylsilane. Melting points were detected on a micro table "Rapido PHMK05" (Nagema, Dresden, Germany). Optical rotations were measured on a polarimeter "Perkin-Elmer 241 MC" (Perkin Elmer, Waltham, MA, USA) in a tube length of 1 dm. Elemental analysis was performed on a Euro EA-3000 CHNS analyzer (Eurovector, Milan, Italy); the main standard was acetanilide. Thin-layer chromatography analyses were performed on Sorbfil plates (Sorbpolimer, Krasnodar, Russian Federation), using the solvent system chloroform–ethyl acetate, 40:1. Substances were detected by 10% H_2SO_4 with subsequent heating to 100–120 °C for 2–3 min. All the reagents and solvents were purchased from standard commercial vendors and were used without any further purification. For the synthesis of quinopimaric acid **1** [38] pine resin Pinus silvestris (containing about 25% levopimaric acid) was used. Compounds **2, 14, 15, 17–21, 24** [50], **3** [51], **4** [52], **5** [53], **6, 9, 11, 26–28, 30** [16], **7** [54], **8** [50], **10** [55], **12, 13** [56], **16** [15], **22, 23** [57], **25** [58], **29** [59], **31, 32, 34** [60], **33** [14], **35** [61], **36–39** [38], **40, 49** [62] were obtained according to the methods previously described.

3.2. *Synthesis of Compounds* **42** *and* **44**

A threefold excess of ethyl 3-aminocrotonate (0.387 g, 3 mmol) or 3-aminocrotononitrile (0.246 g, 3 mmol) was added with stirring to a solution of compound **40** (0.408 g, 1 mmol) in glacial AcOH (20 mL). The reaction mixture was stirred at room temperature for 20 h, and then poured into H_2O. The precipitate was filtered off, washed until neutral, and the residue was air-dried. The reaction product was chromatographed on a silica gel column, eluent $CHCl_3$–MeOH, 40: 1.

1-Hydroxy-13-isopropyl-1′-(ethoxycarbonyl)-7,10a,2′-trimethyl-5,6,6b,7,8,9,10,10a, 10b, 11,12,13-dodecahydro-12,4b-ethenophenanthro-[2,1-g]indole-7-carboxylic acid (**42**). Yield 0.395 g (76%), mp 131–133°C, $[\alpha]_{D19}$ + 12.2° (c 1.5, $CHCl_3$). ^{1}H NMR spectrum, δ, ppm (J, Hz): 0.79 (3H, s, 18-CH_3); 0.86–0.99 (1H, m, 10-CH_2); 1.18 (3H, d, J = 7.0, 17-CH_3); 1.22 (3H, d, J = 7.0, 16-CH_3); 1.29 (3H, s, 19-CH_3); 1.22–1.69 (10H, m, 6,8,9,10,11-CH_2, 10b-CH); 1.77–1.80 (1H, m, 6b-CH); 1.98–2.18 (1H, m, 15-CH); 2.45–2.58 (1H, m, 5-CH_2); 2.67 (3H, s, 3′-CH_3); 2.82–2.91 (1H, m, 5-CH_2); 3.45 (3H, s, 6′-CH_3); 4.27 (1H, s, 12-CH); 4.31-4.35 (2H, m, 5′-CH_2); 5.68 (1H, s, 14-CH); 7.13 (1H, s, 2-CH); 9.65 (2H, br. s, OH); 9.80 (1H, br. s, NH). ^{13}C NMR spectrum, δ, ppm: 14.9 (C-3′); 16.4 (C-19); 16.7 (C-9), 17.2 (C-18); 20.5 (C-17); 20.8 (C-16); 22.0 (C-6); 28.3 (C-11); 32.0 (C-15); 33.5 (C-5); 36.1 (C-12); 36.8 (C-8); 38.2 (C-4b); 38.9 (C-10); 46.5 (C-7); 46.6 (C-10a); 49.8 (C-6b); 50.3 (C-6′); 54.9 (C-10b); 60.2 (C-5′), 102.9 (C-2); 108.0 (C-1′); 124.5 (C-3); 127.5 (C-13); 131.1 (C-1a); 133.6 (C-4a); 143.3 (C-4); 146.8 (C-1); 153.2 (C-14); 162.8 (C-2′); 165.3 (C-4′); 183.1 (C-20). Mass spectrum, m/z (Irel, %): 520 $[M+H]^+$ (100). Found, %: C 73.95; H 7.92; N 2.68. $C_{32}H_{41}NO_5$. Calculated, %: C 73.96; H 7.95; N 2.70.

1′-Cyano-1-hydroxy-13-isopropyl-7,10a,2′-trimethyl-5,6,6b,7,8,9,10,10a,10b,11,12,13-dodecahydro-12,4b-ethenophenanthro [2,1-g]indole-7-carboxylic acid (**44**). Yield 0.32 g (69%), mp 127–129°C, $[\alpha]_{D19}$ +25.5° (c 1.0, $CHCl_3$). ^{1}H NMR spectrum, δ, ppm (J, Hz): 0.78 (3H, s, 18-CH_3); 0.82–0.95 (1H, m, 10-CH_2); 1.18 (3H, d, J = 7.0, 17-CH_3); 1.20 (3H, d, J = 7.0, 16-CH_3); 1.25 (3H, s, 19-CH_3); 1.22–1.69 (10H, m, 6,8,9,10,11-CH_2, 10b-CH); 1.77–1.80 (1H, m, 6b-CH); 1.98–2.18 (1H, m, 15-CH); 2.45–2.58 (1H, m, 5-CH_2); 2.51 (3H, s, 3′-CH_3); 2.55–2.85 (1H, m, 5-CH_2); 4.31 (1H, s, 12-CH); 5.65 (1H, s, 14-CH); 6.55-6.83 (1H, m, 2-CH); 9.35 (3H, br. s, OH, NH). ^{13}C NMR spectrum, δ, ppm: 12.8 (C-3′); 16.4 (C-19); 16.7 (C-9), 17.1 (C-18); 20.1 (C-17); 20.3 (C-16); 22.0 (C-6); 28.1 (C-11); 32.0 (C-15); 34.6 (C-5); 36.8 (C-12); 37.6 (C-8); 38.2 (C-4b); 38.8 (C-10); 46.3 (C-7); 46.8 (C-10a); 49.9 (C-6b); 55.2 (C-10b); 83.3 (C-1′); 99.7 (C-2); 117.8 (C-4′); 124.4 (C-3); 126.8 (C-1a); 127.1 (C-14); 130.9 (C-4); 134.4 (C-4a); 144.3 (C-1); 146.4 (C-13); 153.7 (C-2′); 182.5 (C-20). Mass spectrum, m/z (Irel, %): 473 $[M+H]^+$ (100). Found, %: C 76.22; H 7.65; N 5.90. $C_{30}H_{36}N_2O_3$. Calculated, %: C 76.24; H 7.68; N 5.93.

3.3. *Synthesis of Methyl 1-Hydroxy-13-Isopropyl-1′-(Ethoxycarbonyl)-7,10a,2′-Trimethyl-5,6,6b,7,8,9,10,10a,10b,11,12,13-Dodecahydro-12,4b-Ethenophenanthro-[2,1-g]indole-7-Carboxylate (**43**)*

Procedure A. Methyl iodide (2 mL) and potassium carbonate (0.21 g) were added to a solution of **40** (0.408 g, 1 mmol) in acetone (15 mL), and the mixture was heated to reflux for 2 h. The reaction mixture was filtered. The filtrate was evaporated under reduced pressure, and the residue was purified on a silica gel column, eluent hexane–ethyl acetate, 5:1.

Procedure B. A threefold excess of 3-aminocrotononitrile (0.246 g, 3 mmol) was added with stirring to a solution of compound **41** (0.422 g, 1 mmol) in glacial AcOH (20 mL). The reaction mixture was stirred at room temperature for 20 h, and then poured into H_2O. The precipitate was filtered off, washed until neutral, and the residue was air-dried. The reaction product was chromatographed on a silica gel column, eluent $CHCl_3$–MeOH, 40: 1.

Yield 0.45 g (85%), mp 120–122°C, $[\alpha]_{D19}$ +12.9° (c 0.75, $CHCl_3$). ^{1}H NMR spectrum, δ, ppm (J, Hz): 0.79 (3H, s, 18-CH_3); 0.86–0.99 (1H, m, 10-CH_2); 1.18 (3H, d, J = 7.0, 17-CH_3); 1.22 (3H, d, J = 7.0, 16-CH_3); 1.29 (3H, s, 19-CH_3); 1.22–1.69 (10H, m, 6,8,9,10,11-CH_2, 10b-CH); 1.40–1.43 (3H, s, 6′-CH_3); 1.77–1.80 (1H, m, 6b-CH); 1.98–2.18 (1H, m, 15-CH); 2.45–2.58 (1H, m, 5-CH_2); 2.73 (3H, s, 3′-CH_3); 2.85–3.15 (1H, m, 5-CH_2); 3.76 (3H, s, 21-CH_3), 4.27 (1H, s, 12-CH); 4.31–4.39 (2H, m, 5′-CH_2); 5.71 (1H, s, 14-CH); 7.28 (1H, s, 2-CH); 5.95 (1H, br. s, NH); 12.13 (1H, br. s, OH). ^{13}C NMR spectrum, δ, ppm: 14.9 (C-3′); 15.5 (C-21), 16.4 (C-19); 16.8 (C-9), 17.2 (C-18); 20.2 (C-17); 20.5 (C-16); 22.1 (C-6); 28.3 (C-11); 32.2 (C-15); 33.2 (C-5); 36.3 (C-12); 36.9 (C-8); 38.4 (C-4b); 38.9 (C-10); 46.6 (C-7); 47.4 (C-10a); 49.9 (C-6b); 52.0 (C-6′); 54.8 (C-10b); 60.3 (C-5′), 103.3 (C-2); 108.2 (C-1′); 124.7 (C-3); 127.6 (C-13); 130.9 (C-1a); 133.6 (C-4a); 143.6 (C-4); 146.4 (C-1); 153.1 (C-14); 152.9 (C-2′); 165.2 (C-4′); 179.7 (C-20). Mass spectrum, m/z (I_{rel}, %): 534 [M] + (100). Found, %: C 74.25; H 8.10; N 2.60. $C_{33}H_{43}NO_5$. Calculated, %: C 74.27; H 8.12; N 2.62.

3.4. *Synthesis of Compounds **45**, **46** and **49***

To a solution containing 1 mmol of compound **36**, **37** or **40** in 5 mL of dimethylformamide, 1.2 mmol (0.09 mL) of propargyl bromide and 2.2 mmol (0.30 g) of K_2CO_3 were added. The reaction mixture was stirred for 18 h and evaporated at room temperature. The residue was diluted with $CHCl_3$, washed with 5% HCl and water, dried over $CaCl_2$, and evaporated in a vacuum. The residue was purified by column chromatography, eluent hexane: ethyl acetate, 5:1

Methyl 7-propynyl 1-hydroxy-13-isopropyl-7,10a,2′-trimethyl- 5,6,6b,7,8,9,10,10a,10b,11,12,13-dodecahydro-12,4b-ethenophenanthro [2,1-g]indole-7,1′-dicarboxylate (**45**). Yield 0.39 g (72%), mp 110–112°C, $[\alpha]_{D20}$ + 33.4° (c 0.10, $CHCl_3$). ^{1}H NMR spectrum, δ, ppm (J, Hz): 0.81 (3H, s, 18-CH_3); 0.86–0.96 (1H, m, 10-CH_2); 1.02 (3H, d, J = 7.0, 17-CH_3); 1.05 (3H, d, J = 7.0, 16-CH_3); 1.21 (3H, s, 19-CH_3); 1.31–1.69 (10H, m, 6,8,9,10,11-CH_2, 10b-CH); 1.77–1.80 (1H, m, 6b-CH); 1.98–2.18 (1H, m, 15-CH); 2.41–2.48 (1H, m, 5-CH_2); 2.50 (1H, br s., 8′-CH); 2.71 (3H, s, 3′-CH_3); 2.99–3.00 (1H, m, 5-CH_2); 3.91 (3H, s, 5′-CH_3), 4.28 (1H, s, 12-CH); 4.69–4.82 (2H, m, 6′-CH_2); 5.71 (1H, s, 14-CH); 7.28 (1H, s, 2-CH); 6.00 (1H, br. s, OH); 12.01 (1H, br. s, NH). ^{13}C NMR spectrum, δ, ppm: 14.9 (C-3′); 16.8 (C-19); 17.1 (C-9), 18.2 (C-18); 20.2 (C-17); 20.5 (C-16); 22.0 (C-6); 28.3 (C-11); 32.2 (C-15); 33.2 (C-5); 36.3 (C-12); 36.9 (C-8); 38.3 (C-4b); 38.8 (C-10); 46.6 (C-7); 47.4 (C-10a); 49.8 (C-6b); 51.4 (C-6′); 52.0 (C-5′); 54.8 (C-10b); 74.4 (C-8′); 78.1 (C-7′); 103.3 (C-2); 108.0 (C-1′); 124.5 (C-3); 127.5 (C-13); 130.9 (C-1a); 133.6 (C-4a); 143.5 (C-4); 146.5 (C-1); 153.2 (C-14); 162.9 (C-2′); 165.6 (C-4′); 178.2 (C-20). Mass spectrum, m/z (I_{rel}, %): 543 [M]+ (100). Found, %: C 75.15; H 7.61; N 2.60. $C_{34}H_{41}NO_5$. Calculated, %: C 75.11; H 7.60; N 2.58.

Ethyl 7-propynyl 1-hydroxy-13-isopropyl-7,10a,2′-trimethyl- 5,6,6b,7,8,9,10,10a,10b, 11,12,13-dodecahydro-12,4b-ethenophenanthro [2,1-g]indole-7,1′-dicarboxylate (**46**). Yield 0.42 g (75%), mp 98–100°C, $[\alpha]_{D20}$ +2.9° (c 0.15, $CHCl_3$). ^{1}H NMR spectrum, δ, ppm (J, Hz): 0.72 (3H, s, 18-CH_3); 0.80–0.96 (1H, m, 10-CH_2); 1.02 (3H, d, J = 7.0, 17-CH_3); 1.05 (3H, d, J = 7.0, 16-CH_3); 1.21 (3H, s, 19-CH_3); 1.31–1.69 (10H, m, 6,8,9,10,11-CH_2,10b-CH); 1.77–1.80 (1H, m, 6b-CH); 1.98–2.18 (1H, m, 15-CH); 2.41–2.48 (1H, m, 9′-CH); 2.65–2.70 (1H, m, 5-CH_2); 2.73 (3H, s, 3′-CH_3); 2.95–3.05 (1H, m, 5-CH_2); 3.73 (3H, s, 6′-CH_3), 3.95 (1H, s, 12-CH); 4.35-4.42 (2H, m, 5′-CH_2); 4.75 (2H, br.s., 6′-CH_2); 5.70 (1H, s, 14-CH); 7.49 (1H, s, 2-CH); 9.19 (1H, br. s, OH); 12.01 (1H, br. s, NH). ^{13}C NMR spectrum, δ, ppm: 14.1 (C-3′); 14.2 (C-19); 14.5 (C-9), 15.9 (C-18); 16.5 (C-17); 16.9 (C-16); 20.2 (C-6); 21.4 (C-11); 22.7 (C-15); 28.3 (C-5); 29.7 (C-12); 32.2 (C-8); 33.2 (C-4b); 36.2 (C-10); 36.9 (C-7); 38.8 (C-10a); 49.9 (C-6b); 51.9 (C-7′); 54.9 (C-6′); 57.1 (C-10b); 60.0 (C-5′), 74.9 (C-9′); 79.3 (C-8′); 101.8 (C-2); 108.6 (C-1′); 124.5 (C-3); 127.6 (C-13); 132.8 (C-1a); 133.8 (C-4a); 144.0 (C-4); 148.6 (C-1); 153.2 (C-14); 162.8 (C-2′); 164.8 (C-4′); 179.5 (C-20). Mass spectrum, m/z (I_{rel}, %): 558 [M]+ (100). Found, %: C 75.30; H 7.75; N 2.55. $C_{35}H_{43}NO_5$. Calculated, %: C 75.37; H 7.77; N 2.51.

Propynyl 13-isopropyl-7,10a-dimethyl-1,4-di oxo-4,5,6,6a,7,8,9,10,10a,10b,11,12-dodecahydro-1H-4b,12-ethenochrysene-7-carboxylate (**48**). Yield 0.33 g (75%), mp 85-89°C, $[\alpha]_{D20}$ +26.9° (c 0.75, $CHCl_3$). ^{1}H NMR spectrum, δ, ppm (J, Hz): 0.69 (3H, s, 18-CH_3); 0.86–0.96 (1H, m, 10-CH_2); 1.02 (3H, d, J = 7.0, 17-CH_3); 1.05 (3H, d, J = 7.0, 16-CH_3); 1.21 (3H, s, 19-CH_3); 1.31–1.69 (10H, m, 6,8,9,10,11-CH_2, 10b-CH); 1.77–1.80 (1H, m, 6b-CH); 1.98–2.18 (1H, m, 15-CH); 2.41–2.48 (1H, m, 5-CH_2); 2.50 (1H, br. s, 3′-CH); 2.85–2.89 (1H, m, 5-CH_2); 4.13 (1H, s, 12-CH); 4.67 (2H, br. s, 1′-CH_2); 5.63 (1H, s, 14-CH); 6.45-6.56 (2H, m, 2-CH, 3-CH). ^{13}C NMR spectrum, δ, ppm: 16.4 (C-19); 16.8 (C-9), 17.1 (C-18); 20.2 (C-17); 20.6 (C-16); 21.7 (C-6); 27.1 (C-11); 31.5 (C-15); 31.9 (C-5); 36.2 (C-12); 36.4 (C-8); 38.6 (C-4b); 39.3 (C-10); 47.1 (C-7); 49.1 (C-10a); 49.4 (C-6b); 52.1 (C-1′); 54.8 (C-10b); 74.6 (C-3′); 77.9 (C-2′); 127.3 (C-13); 133.6 (C-2); 137.5 (C-3); 150.6 (C-4a); 151.1 (C-14); 152.8 (C-1a); 177.7 (C-20); 184.0 (C-1); 185.3 (C-2). Mass spectrum, m/z (I_{rel}, %): 446 [M]+ (100). Found, %: C 78.05; H 7.65. $C_{29}H_{34}O_4$. Calculated, %: C 78.00; H 7.67.

3.5. *Synthesis of Compounds* **47** *and* **50**

A mixture of 1 mmol of the compounds **42** or **49**, 20 mmol (1.3 mL) of acrylonitrile and 0.5 mL of 30% KOH per one hydroxyl groups, 0.5 mmol (0.11 g) of BTEAC, in 20 mL of dioxane was stirred for 2 h at room temperature. The mixture was poured into a mixture of ice with HCl, the precipitate was filtered off, washed with water until neutral pH, air dried, and extracted with methylene chloride (3 × 80 mL) with heating, the solution was filtered. The filtrate was evaporated under reduced pressure, and the residue was purified on a silica gel column, eluent hexane–ethyl acetate, 10:1.

Methyl 1-(8′-cyanoethoxy)-13-isopropyl-1′-(ethoxycarbonyl)-7,10a,2′-trimethyl -5,6,6b, 7,8,9,10,10a,10b,11,12,13-dodecahydro-12,4b-ethenophenanthro [2,1-g]indole-7-carboxylate (**47**). Yield 0.39 g (68%), mp 127–129 °C, $[\alpha]_{D20}$ +77.9° (c 0.10, $CHCl_3$). ^{1}H NMR spectrum, δ, ppm (J, Hz): 0.79 (3H, s, 18-CH_3); 0.86–0.99 (1H, m, 10-CH_2); 1.04 (3H, d, J = 7.0, 17-CH_3); 1.07 (3H, d, J = 7.0, 16-CH_3); 1.24 (3H, s, 19-CH_3); 1.27–1.69 (10H, m, 6,8,9,10,11-CH_2, 10b-CH); 1.77–1.80 (1H, m, 6b-CH); 1.98–2.18 (1H, m, 15-CH); 2.45–2.58 (1H, m, 5-CH_2); 2.76 (3H, s, 3′-CH_3); 2.89–2.91 (2H, m, 7′-CH_2); 2.95–3.15 (1H, m, 5-CH_2); 3.75 (3H, s, 21-CH_3), 3.93 (3H, s, 5′-CH_3), 4.28-4.30 (2H, m, 6′-CH_2); 4.37 (1H, s, 12-CH); 5.72 (1H, s, 14-CH); 7.28 (1H, s, 2-CH); 12.10 (1H, br. s, NH). ^{13}C NMR spectrum, δ, ppm: 14.9 (C-3′); 15.5 (C-21), 16.9 (C-19); 16.8 (C-9), 17.2 (C-18); 20.2 (C-17); 20.5 (C-16); 22.1 (C-6); 28.3 (C-11); 32.2 (C-15); 33.2 (C-5); 36.3 (C-12); 36.9 (C-8); 38.4 (C-4b); 38.9 (C-10); 46.7 (C-7); 47.4 (C-10a); 49.9 (C-6b); 51.3 (C-7′); 52.0 (C-5′); 54.9 (C-10b); 63.7 (C-6′), 100.9 (C-2); 108.3 (C-1′); 117.4 (C-8′); 124.6 (C-3); 127.6 (C-13); 133.4 (C-1a); 134.1 (C-4a); 144.0 (C-4); 148.4 (C-1); 153.2 (C-14); 163.0 (C-2′); 165.1 (C-4′); 179.6 (C-20). Mass spectrum, m/z (I_{rel}, %): 572 [M]+ (100). Found, %: C 73.45; H 7.75; N 4.91. $C_{35}H_{44}N_2O_5$. Calculated, %: C 73.40; H 7.74; N 4.89.

1,4-Bis(2′-cyanoethoxy)-13-isopropyl-7,10a-dimethyl-6,6a,7,8,9,10,10a,10b,11,12-decahydro-5H-4b,12-ethenochrysene-7-carboxylic acid (**50**). Yield 0.41 g (80%), mp 157–159 °C, $[\alpha]_{D20}$ + 63.6° (c 0.1, $CHCl_3$). ^{1}H NMR spectrum, δ, ppm (J, Hz): 0.79 (3H, s, 18-CH_3); 0.86–0.99 (1H, m, 10-CH_2); 1.08 (3H, d, J = 7.0, 17-CH_3); 1.10 (3H, d, J = 7.0, 16-CH_3); 1.29 (3H, s, 19-CH_3); 1.22–1.69 (10H, m, 6,8,9,10,11-CH_2, 10b-CH); 1.77–1.80 (1H, m, 6b-CH); 1.98–2.18 (1H, m, 15-CH); 2.45–2.58 (1H, m, 5-CH_2); 2.80–2.82 (4H, m, 2′-CH_2, 2″-CH_2); 2.95–3.00 (1H, m, 5-CH_2); 4.05-4.13 (4H, m, 1′-CH_2, 1″-CH_2); 4.24 (1H, s, 12-CH); 5.71 (1H, s, 14-CH); 6.38-6.47 (2H, m, 2-CH, 3-CH); 9.12 (1H, br. s, OH). ^{13}C NMR spectrum, δ, ppm: 16.4 (C-19); 16.8

(C-9), 17.2 (C-18); 19.9 (C-2′, C-2″); 20.2 (C-17); 20.5 (C-16); 22.1 (C-6); 28.3 (C-11); 32.2 (C-15); 33.2 (C-5); 36.3 (C-12); 36.9 (C-8); 38.4 (C-4b); 38.9 (C-10); 46.6 (C-7); 47.4 (C-10a); 49.9 (C-6b); 54.8 (C-10b); 64.4 (C-1′, C-1″); 111.2 (C-1); 114.4 (C-2); 117.6 (C-3′, C-3″), 128.6 (C-14); 135.2 (C-4a); 138.4 (C-1a); 145.1 (C-4); 146.0 (C-1); 151.7 (C-13); 185.6 (C-20). Mass spectrum, m/z (I_{rel}, %): 517 [M]+H (100). Found, %: C 74.35; H 7.81; N 5.40. $C_{32}H_{40}N_2O_4$. Calculated, %: C 74.39; H 7.80; N 5.42.

Data of the study α-Gly inhibition in vitro, kinetic, docking studies and ADMET profiling of compound **45**. as well as studies of anti-antioxidant, antimicrobial and cytotoxic activities, can be found in the Supplementary Material (Section S1).

4. Conclusions

The screening of a series of 50 semisynthetic derivatives of levopimaric acid revealed that, in contrast to the majority of previously reported diterpene α-GLy inhibitors, a lead diterpene indole with an alkyne substituent **45** was identified as a competitive inhibitor. As a consequence, one might hope for better translatability to animal and clinical settings, since the active site of yeast α-GLy and intestinal mammalian maltase–glucoamylase are conserved, while allosteric sites are likely to be different. In addition, compound **45** is anticipated to have low intestinal absorption that benefits high concentration of the drug in the target area and helps to avoid systemic exposure. Additional experiments are warranted to confirm antihyperglycemic properties of compound **45** in vivo. In the event of the efficacy and safety being confirmed, novel glucosidase inhibitors open a promising venue to antidiabetic agents able not only to ameliorate postprandial hyperglycemia, but also reduce secretory load on pancreatic beta-cells.

Supplementary Materials: The following supporting information can be downloaded at: https://www.mdpi.com/article/10.3390/ijms232113535/s1. Ref [63–71] are cited in Supplementary Materials.

Author Contributions: E.T.—draft preparation; E.T. and I.S. prepared compounds for screening; O.K. brought the idea, managed the research and prepared the manuscript; H.T.T.N., A.S., E.S., D.B. and A.S. (Alina Shevchenko) conducted biological experiments; D.B. and A.S. (Alexander Spasov) prepared the manuscript. All authors have read and agreed to the published version of the manuscript.

Funding: This work was supported by the Federal program (Russian Federation) N°. 1021062311392-9-1.4.1. Ha Thi Thu Nguyen thanks the Vietnam Academy of Science and Technology (VAST) for support in the screening of compounds against α-glucosidase (Project N°. QTRU01.05/20-21).

Institutional Review Board Statement: Not applicable.

Informed Consent Statement: Not applicable.

Data Availability Statement: Not applicable.

Conflicts of Interest: The authors declare no conflict of interest.

References

1. Tang, O.; Matsushita, K.; Coresh, J.; Sharrett, A.R.; McEvoy, J.W.; Windham, B.G.; Ballantyne, C.M.; Selvin, E. Mortality implications of prediabetes and diabetes in older adults. *Diabetes Care* **2020**, *43*, 382–388. [CrossRef]
2. Kitabchi, A.E.; Umpierrez, G.E.; Miles, J.M.; Fisher, J.N. Hyperglycemic crises in adult patients with diabetes. *Diabetes Care* **2009**, *32*, 1335–1343. [CrossRef]
3. Ceriello, A.; Nicolucci, A. Intensive glucose control and type 2 diabetes—15 years on. *N. Engl. J. Med.* **2019**, *381*, 1292–1293. [CrossRef]
4. Zheng, Y.; Ley, S.H.; Hu, F.B. Global aetiology and epidemiology of type 2 diabetes mellitus and its complications. *Nat. Rev. Endocrinol.* **2018**, *14*, 88–98. [CrossRef]
5. Lean, M.; McCombie, L.; McSorely, J. Trends in type 2 diabetes. *BMJ* **2019**, *366*, l5407. [CrossRef]
6. Williams, J.; Loeffler, M. Global trends in type 2 diabetes, 2007–2017. *JAMA* **2019**, *322*, 1542. [CrossRef]
7. Bischoff, H. Pharmacology of alpha-glucosidase inhibition. *Eur. J. Clin. Investig.* **1994**, *24*, 3–10.
8. Toeller, M. α-Glucosidase inhibitors in diabetes: Efficacy in NIDDM subjects. *Eur. J. Clin. Investig.* **1994**, *24*, 31–35. [CrossRef]

9. Soccio, R.E.; Chen, E.R.; Lazar, M.A. Thiazolidinediones and the promise of insulin sensitization in type 2 diabetes. *Cell Metab.* **2014**, *20*, 573–591. [CrossRef]
10. Ullah, A.; Khan, A.; Khan, I. Diabetes mellitus and oxidative stress—A concise review. *Pharm. J.* **2016**, *24*, 547–553. [CrossRef]
11. Lorenzati, B.; Zucco, C.; Miglietta, S.; Lamberti, F.; Bruno, G. Oral hypoglycemic drugs: Pathophysiological basis of their mechanism of action. *Pharmaceuticals* **2010**, *3*, 3005–3020. [CrossRef] [PubMed]
12. Marín-Peñalver, J.J.; Martín-Timón, I.; Sevillano-Collantes, C.; del Cañizo-Gómez, F.J. Type 2 diabetes and cardiovascular disease: Have all risk factors the same strength. *World J. Diabetes* **2016**, *7*, 354–395. [CrossRef] [PubMed]
13. Feliciano, A.S.; Gordaliza, M.; Salinero, M.A.; del Corral, J.M.M. Abietane Acids: Sources, Biological Activities, and Therapeutic Uses. *Planta Med.* **1993**, *59*, 485–490. [CrossRef] [PubMed]
14. Tretyakova, E.V.; Smirnova, I.E.; Kazakova, O.B.; Tolstikov, G.A.; Yavorskaya, N.P.; Golubeva, I.S.; Pugacheva, R.B.; Apryshko, G.N.; Poroikov, V.V. Synthesis and anticancer activity of quinopimaric and maleopimaric acids' derivatives. *Bioorg. Med. Chem.* **2014**, *22*, 6481–6489. [CrossRef] [PubMed]
15. Tretyakova, E.V.; Smirnova, I.E.; Salimova, E.V.; Odinokov, V.N. Synthesis and antiviral activity of maleopimaric and quinopimaric acids' derivatives. *Bioorg. Med. Chem.* **2015**, *23*, 6543–6550. [CrossRef]
16. Smirnova, I.E.; Tret'yakova, E.V.; Baev, D.S.; Kazakova, O.B. Synthetic modifications of abietane diterpene acids to potent antimicrobial agents. *Nat. Prod. Res.* **2021**, 1–9. [CrossRef]
17. Kim, E.; Kang, Y.-G.; Kim, Y.-J.; Lee, T.R.; Yoo, B.C.; Jo, M.; Kim, J.H.; Kim, J.H.; Kim, D.; Cho, J.Y. Dehydroabietic acid suppresses inflammatory response via suppression of Src-, Syk-, and TAK1-mediated pathways. *Int. J. Mol. Sci.* **2019**, *20*, 1593. [CrossRef]
18. Tret'yakova, E.V.; Salimova, E.V.; Parfenova, L.V. Synthesis, modification, and biological activity of propargylated methyl dihydroquinopimarates. *Nat. Prod. Res.* **2022**, *36*, 79–86. [CrossRef]
19. Goncalves, M.D.; Bortoleti, B.T.S.; Tomiotto-Pellissier, F.; Miranda-Sapla, M.M.; Assolini, J.P.; Carloto, A.C.M.; Carvalho, P.G.C.; Tudisco, E.T.; Urbano, A.; Ambrosio, S.R.; et al. Dehydroabietic acid isolated from Pinus elliottii exerts in vitro antileishmanial action by pro-oxidant effect, inducing ROS production in promastigote and downregulating Nrf2/ferritin expression in amastigote forms of Leishmania amazonensis. *Fitoterapia* **2018**, *128*, 224–232. [CrossRef]
20. González, M.A. Aromatic abietane diterpenoids: Their biological activity and synthesis. *Nat. Prod. Rep.* **2015**, *32*, 684–704. [CrossRef]
21. Etsassala, N.G.E.R.; Cupido, C.N.; Iwuoha, I.E.; Hussein, A.A. Abietane Diterpenes as Potential Candidates for the Management of Type 2 Diabetes. *Curr. Pharm. Des.* **2020**, *26*, 2885–2891. [CrossRef] [PubMed]
22. Nachar, A.; Saleem, A.; Arnason, J.T.; Haddad, P.S. Regulation of liver cell glucose homeostasis by dehydroabietic acid, abietic acid and squalene isolated from balsam fir (*Abies balsamea* (L.) Mill.) a plant of the Eastern James Bay Cree traditional pharmacopeia. *Phytochemistry* **2015**, *117*, 373–379. [CrossRef] [PubMed]
23. Song, H.M.; Li, X.; Liu, Y.Y.; Lu, W.-P.; Cui, Z.-H.; Zhou, L.; Yao, D.; Zhang, H.-M. Carnosic acid protects mice from high-fat diet-induced NAFLD by regulating MARCKS. *Int. J. Mol. Med.* **2018**, *42*, 193–207. [CrossRef] [PubMed]
24. Lipina, C.; Hundal, H.S. Carnosic acid stimulates glucose uptake in skeletal muscle cells via a PME-1/PP2A/PKB signalling axis. *Cell. Signal.* **2014**, *26*, 2343–2349. [CrossRef] [PubMed]
25. Christensen, K.B.; Jørgensen, M.; Kotowska, D.; Petersen, R.K.; Kristiansen, K.; Christensen, L.P. Activation of the nuclear receptor PPARγ by metabolites isolated from sage (*Salvia officinalis* L.). *J. Ethnopharmacol.* **2010**, *132*, 127–133. [CrossRef] [PubMed]
26. Xie, Z.; Gao, G.; Wang, H.; Lia, E.; Yuan, Y.; Xu, J.; Zhang, Z.; Wang, P.; Fu, Y.; Zeng, H.; et al. Dehydroabietic acid alleviates high fat diet-induced insulin resistance and hepatic steatosis through dual activation of PPAR-γ and PPAR-α. *Biomed. Pharmacother.* **2020**, *127*, 110155. [CrossRef]
27. Vlavcheski, F.; Baron, D.; Vlachogiannis, I.A.; MacPherson, R.E.K.; Tsiani, E. Carnosol increases skeletal muscle cell glucose uptake via AMPK-Dependent GLUT4 glucose transporter translocation. *Int. J. Mol. Sci.* **2018**, *19*, 1321. [CrossRef]
28. Samarghandian, S.; Borji, A.; Farkhondeh, T. Evaluation of antidiabetic activity of carnosol (phenolic diterpene in rosemary) in Streptozotocin-induced diabetic rats. *Cardiovasc. Hematol. Disord. Drug Targets* **2017**, *17*, 11–17. [CrossRef]
29. Cui, L.; Kim, M.O.; Seo, J.H.; Kim, I.S.; Kim, N.Y.; Lee, S.H.; Park, J.; Kim, J.; Lee, H.S. Abietane diterpenoids of Rosmarinus officinalis and their diacylglycerol acyltransferase-inhibitory activity. *Food Chem.* **2012**, *132*, 1775–1780. [CrossRef]
30. Yun, Y.S.; Noda, S.; Shigemori, G.; Kuriyama, R.; Takahashi, S.; Umemura, M.; Takahashi, Y.; Inoue, H. Phenolic diterpenes from rosemary suppress cAMP responsiveness of gluconeogenic gene promoters. *Phytother. Res.* **2013**, *27*, 906–910. [CrossRef]
31. Kubínová, R.; Pořízková, R.; Navrátilová, A.; Farsa, O.; Hanáková, Z.; Bačinská, A.; Čížek, A.; Valentová, M. Antimicrobial and enzyme inhibitory activities of the constituents of Plectranthus madagascariensis (Pers.) Benth. *J. Enzym. Inhib. Med. Chem.* **2014**, *29*, 749–752. [CrossRef] [PubMed]
32. Kim, D.H.; Paudel, P.; Yu, T.; Ngo, T.M.; Kim, J.A.; Jung, H.A.; Yokozawa, T.; Choi, J.S. Characterization of the inhibitory activity of natural tanshinones from *Salvia miltiorrhiza* roots on protein tyrosine phosphatase 1B. *Chem. Biol. Interact.* **2017**, *278*, 65–73. [CrossRef] [PubMed]
33. Jung, S.H.; Seol, H.J.; Jeon, S.J.; Son, K.H.; Lee, J.R. Insulin-sensitizing activities of tanshinones, diterpene compounds of the root of Salvia miltiorrhiza Bunge. *Phytomedicine* **2009**, *16*, 327–335. [CrossRef] [PubMed]
34. Kang, M.S.; Hirai, S.; Goto, T.; Kuroyanagi, K.; Kim, Y.-I.; Ohyama, K.; Uemura, T.; Lee, J.-Y.; Sakamoto, T.; Ezaki, Y.; et al. Dehydroabietic acid, a diterpene, improves diabetes and hyperlipidemia in obese diabetic KK-Ay mice. *Biofactors* **2009**, *35*, 442–448. [CrossRef] [PubMed]

35. Ou, J.; Huang, J.; Zhao, D.; Du, B.; Wang, M. Protective effect of rosmarinic acid and carnosic acid against streptozotocin-induced oxidation, glycation, inflammation and microbiota imbalance in diabetic rats. *Food Funct.* **2018**, *9*, 851–860. [CrossRef] [PubMed]
36. Wei, Y.; Gao, J.; Qin, L.; Xu, Y.; Wang, D.; Shi, H.; Xu, T.; Liu, T. Tanshinone I alleviates insulin resistance in type 2 diabetes mellitus rats through IRS-1 pathway. *Biomed. Pharmacother.* **2017**, *93*, 352–358. [CrossRef]
37. Allen, G.R., Jr. *Organic Reactions*; Wiley: New York, NY, USA, 1973; Volume 20, p. 338.
38. Tretyakova, E.V.; Yarmukhametova, L.R.; Salimova, E.V.; Kukovinets, O.S.; Parfenova, L.V. The Nenitzescu reaction in the synthesis of new abietane diterpene indoles. *Chem. Heterocycl. Compd.* **2020**, *56*, 1366–1369. [CrossRef]
39. Tran, C.-L.; Dao, T.-B.-N.; Tran, T.-N.; Mai, D.-T.; Tran, T.-M.-D.; Tran, N.-M.-A.; Dang, V.-S.; Vo, T.-X.; Duong, T.-H.; Sichaem, J. Alpha-Glucosidase Inhibitory Diterpenes from Euphorbia antiquorum Growing in Vietnam. *Molecules* **2021**, *26*, 2257. [CrossRef]
40. Ghosh, S.; Rangan, L. Molecular Docking and Inhibition Kinetics of α-glucosidase Activity by Labdane Diterpenes Isolated from Tora Seeds (*Alpinia nigra* B.L. Burtt.). *Appl. Biochem. Biotechnol.* **2015**, *175*, 1477–1489. [CrossRef]
41. Yang, X.-T.; Geng, C.-A.; Li, T.-Z.; Deng, Z.-T.; Chen, J.-J. Synthesis and biological evaluation of chepraecoxin A derivatives as α-glucosidase inhibitors. *Bioorg. Med. Chem. Lett.* **2020**, *30*, 127020. [CrossRef]
42. Loo, K.Y.; Leong, K.H.; Sivasothy, Y.; Ibrahim, H.; Awang, K. Molecular Insight and Mode of Inhibition of α-Glucosidase and α-Amylase by Pahangensin A from Alpinia pahangensis Ridl. *Chem. Biodivers.* **2019**, *16*, e1900032. [CrossRef] [PubMed]
43. Ma, Y.-Y.; Zhao, D.-G.; Zhang, R.; He, X.; Li, B.Q.; Zhang, X.-Z.; Wang, Z.; Zhang, K. Identification of bioactive compounds that contribute to the α-glucosidase inhibitory activity of rosemary. *Food Funct.* **2020**, *11*, 1692–1701. [CrossRef] [PubMed]
44. Bissantz, C.; Kuhn, B.; Stahl, M. A medicinal chemist's guide to molecular interactions. *J. Med. Chem.* **2010**, *53*, 5061–5084. [CrossRef] [PubMed]
45. Stein, R.L. *Kinetics of Enzyme Action: Essential Principles for Drug Hunters*; John Wiley & Sons: Hoboken, NJ, USA, 2011.
46. Dong, J.; Wang, N.-N.; Yao, Z.-J.; Zhang, L.; Cheng, Y.; Ouyang, D.; Lu, A.-P.; Cao, D.-S. ADMETlab: A platform for systematic ADMET evaluation based on a comprehensively collected ADMET database. *J. Cheminform.* **2018**, *10*, 29. [CrossRef] [PubMed]
47. Xiong, G.; Wu, Z.; Yi, J.; Fu, L.; Yang, Z.; Hsieh, C.; Yin, M.; Zeng, X.; Wu, C.; Lu, A.; et al. ADMETlab 2.0: An integrated online platform for accurate and comprehensive predictions of ADMET properties. *Nucleic Acids Res.* **2021**, *49*, W5–W14. [CrossRef] [PubMed]
48. Daina, A.; Michielin, O.; Zoete, V. SwissADME: A free web tool to evaluate pharmacokinetics, drug-likeness and medicinal chemistry friendliness of small molecules. *Sci. Rep.* **2017**, *7*, 42717. [CrossRef]
49. Banerjee, P.; Eckert, A.O.; Schrey, A.K.; Preissner, R. ProTox-II: A webserver for the prediction of toxicity of chemicals. *Nucleic Acids Res.* **2018**, *46*, W257–W263. [CrossRef]
50. Herz, W.; Blackstone, R.C.; Nair, M.G. Resin acids. XI Configuration and transformations of the levopimaric acid-p-benzoquinone adduct. *J. Org. Chem.* **1967**, *32*, 2992–2998. [CrossRef]
51. Smirnova, I.E.; Tret'yakova, E.V.; Flekhter, O.B.; Spirikhin, L.V.; Galin, F.Z.; Tolstikov, G.A.; Starikova, Z.A.; Korlyukov, A.A. Synthesis, structure, and acylation of dihydroquinopimaric acid hydroxyl derivatives. *Russ. J Org. Chem.* **2008**, *44*, 1598–1605. [CrossRef]
52. Smirnova, I.E.; Tret'yakova, E.V.; Kazakova, O.B.; Starikova, Z.A.; Fedyanin, I.V. Molecular and crystal structure of a new compound methyl-18R-13-isopropyl-10a,7-dimethyl-4-oxo-1-oxahexacyclo 12.4.0.$0^{5a,4a}$.$0^{13,12}$0.$0^{10a,6a}$]heneicosane-7-Carboxylate. *J. Struct. Chem.* **2009**, *50*, 378–380. [CrossRef]
53. Smirnova, I.E.; Tret'yakova, E.V.; Kazakova, O.B.; Suponitsky, K.Y. Molecular structure of methyl 20-isopropyl-15(e)-hydroxyimino-5,9-dimethyl-18-oxahexacyclo[12.4.0.$2^{2,13}$.$1^{18,20}$.$0^{5,10}$.$0^{4,13}$]heneicosane-9-carboxylate. *J. Struct. Chem.* **2010**, *51*, 1208–1210. [CrossRef]
54. Tretyakova, E.V.; Salimova, E.V.; Odinokov, V.N.; Dzhemilev, U.M. Synthesis of a Novel 1,2,4-Oxadiazole Diterpene from the Oxime of the Methyl Ester of 1β,13-Epoxydihydroquinopimaric Acid. *Nat. Prod. Commun.* **2016**, *11*, 23–24. [CrossRef] [PubMed]
55. Kazakova, O.B.; Tret'yakova, E.V.; Smirnova, I.E.; Spirikhin, L.V.; Tolstikov, G.A.; Chudov, I.V.; Bazekin, G.V.; Ismagilova, A.F. The synthesis and anti-inflammatory activity of quinopimaric acid derivatives. *Russ. J. Bioorg. Chem.* **2010**, *36*, 257–262. [CrossRef] [PubMed]
56. Tret'yakova, E.V.; Salimova, E.V.; Parfenova, L.V.; Odinokov, V.N. Synthesis and Modifications of Alkyne Derivatives of Dihydroquinopimaric, Maleopimaric, and Fumaropimaric Acids. *Russ. J. Org. Chem.* **2016**, *52*, 1496–1502. [CrossRef]
57. Flekhter, O.B.; Smirnova, I.E.; Tret'yakova, E.V.; Tolstikov, G.A.; Savinova, O.V.; Boreko, E.I. Synthesis of Dihydroquinopymaric Acid Conjugates with Amino Acids. *Russ. J. Bioorg. Chem.* **2009**, *35*, 385–390. [CrossRef]
58. Tretyakova, E.V.; Salimova, E.V.; Parfenova, L.V.; Yunusbaeva, M.M.; Dzhemileva, L.U.; D'yakonov, V.A.; Dzhemilev, U.M. Synthesis of New Dihydroquinopimaric Acid Analogs with Nitrile Groups as Apoptosis-Inducing Anticancer Agents. *Anti-Cancer Agents Med. Chem.* **2019**, *19*, 1172–1183. [CrossRef]
59. Smirnova, I.E.; Kazakova, O.B.; Tret'yakova, E.V.; Spirikhin, L.V.; Glukhov, I.V.; Nelyubina, Y.V. Regioselective Bromination of Quinopimaric Acid Derivatives. *Russ. J. Org. Chem.* **2010**, *46*, 1135–1139. [CrossRef]
60. Kazakova, O.B.; Tret'yakova, E.V.; Smirnova, I.E.; Nazyrov, T.I.; Kukovinets, O.S.; Tolstikov, G.A.; Suponitskii, K.Y. An efficient oxyfunctionalization of quinopimqric acid derivatives with ozone. *Nat. Prod. Commun.* **2013**, *28*. 293–296.
61. Smirnova, I.E.; Kazakova, O.B.; Tret'yakova, E.V.; Tolstikov, G.A.; Spirikhin, L.V. Synthesis of Heterocyclic Derivatives of Dihydroquinopimaric Acid. *Russ. J. Org. Chem.* **2011**, *47*, 1576–1580. [CrossRef]

62. Shul'ts, E.E.; Oleinikov, D.S.; Nechepurenko, I.V.; Shakirov, M.M.; Tolstikov, G.A. Synthetic transformations of higher terpenoids: XVIII. Synthesis of optically active 9,10-anthraquinone derivatives. *Russ. J. Org. Chem.* **2009**, *45*, 102–114. [CrossRef]
63. Ha, N.T.T.; van Cuong, P.; Tra, N.T.; Anh, l.T.; Cham, B.T.; Son, N.T. Chemical constituents from methanolic extract of Garcinia mackeaniana leaves and their antioxydant activity. *Vietnam. J. Sci. Technol.* **2020**, *58*, 411–418.
64. Ha, N.T.T.; van Cuong, P.; Anh, l.T.; Tra, N.T.; Cham, B.T.; Son, N.T. Antimicrobacterial xanthones from Garcinia mackeaniana leaves. *Vietnam J. Chem.* **2020**, *58*, 343–348. [CrossRef]
65. Mosmann, T. Rapid colorimetric assay for cellular growth and survival: Application to proliferation and cytotoxicity assays. *J. Immunol. Methods* **1983**, *65*, 55–63. [CrossRef]
66. Kim, Y.M.; Wang, M.H.; Rhee, H.I. A novel α-glucosidase inhibitor from pine bark. *Carbohydr. Res.* **2004**, *339*, 715–717. [CrossRef] [PubMed]
67. Li, T.; Zhang, X.D.; Song, Y.W.; Liu, J.W. A microplate-based screening method for α-glucosidase inhibitors. *Nat. Prod. Res. Dev.* **2005**, *10*, 1128–1134.
68. *MarvinSketch*, 18.8.0; ChemAxon Ltd.: Budapest, Hungary, 2018.
69. Spasov, A.A.; Babkov, D.A.; Osipov, D.V.; Klochkov, V.G.; Prilepskaya, D.R.; Demidov, M.R.; Osyanin, V.A.; Klimochkin, Y.N. Synthesis, in vitro and in vivo evaluation of 2-aryl-4H-chromene and 3-aryl-1H-benzo[f]chromene derivatives as novel α-glucosidase inhibitors. *Bioorg. Med. Chem. Lett.* **2019**, *29*, 119–123. [CrossRef]
70. Trott, O.; Olson, A.J. AutoDock Vina: Improving the speed and accuracy of docking with a new scoring function, efficient optimization, and multithreading. *J. Comput. Chem.* **2009**, *31*, 455–461. [CrossRef]
71. *Discovery Studio Visualizer*, 17.2.0.16349; Dassault Systemes Biovia Corp.: San Diego, CA, USA, 2016.

International Journal of
Molecular Sciences

MDPI

Review

Reviewing the Modern Therapeutical Options and the Outcomes of Sacubitril/Valsartan in Heart Failure

Diana-Carina Iovanovici [1], Simona Gabriela Bungau [1,2,*], Cosmin Mihai Vesa [3], Madalina Moisi [3], Elena Emilia Babes [4,*], Delia Mirela Tit [1,2], Tunde Horvath [2], Tapan Behl [5] and Marius Rus [4]

1 Doctoral School of Biomedical Sciences, Faculty of Medicine and Pharmacy, University of Oradea, 410087 Oradea, Romania
2 Department of Pharmacy, Faculty of Medicine and Pharmacy, University of Oradea, 410028 Oradea, Romania
3 Department of Preclinical Disciplines, Faculty of Medicine and Pharmacy, University of Oradea, 410087 Oradea, Romania
4 Department of Medical Disciplines, Faculty of Medicine and Pharmacy, University of Oradea, 410087 Oradea, Romania
5 School of Health Sciences &Technology, University of Petroleum and Energy Studies, Bidholi, Dehradun 248007, India
* Correspondence: sbungau@uoradea.ro (S.G.B.); eebabes@uoradea.ro (E.E.B.)

Abstract: Sacubitril/valsartan (S/V) is a pharmaceutical strategy that increases natriuretic peptide levels by inhibiting neprilysin and regulating the renin-angiotensin-aldosterone pathway, blocking AT1 receptors. The data for this innovative medication are mainly based on the PARADIGM-HF study, which included heart failure with reduced ejection fraction (HFrEF)-diagnosed patients and indicated a major improvement in morbidity and mortality when S/V is administrated compared to enalapril. A large part of the observed favorable results is related to significant reverse cardiac remodeling confirmed in two prospective trials, PROVE-HF and EVALUATE-HF. Furthermore, according to a subgroup analysis from the PARAGON-HF research, S/V shows benefits in HFrEF and in many subjects having preserved ejection fraction (HFpEF), which indicated a decrease in HF hospitalizations among those with a left ventricular ejection fraction (LVEF) < 57%. This review examines the proven benefits of S/V and highlights continuing research in treating individuals with varied HF characteristics. The article analyses published data regarding both the safeness and efficacy of S/V in patients with HF, including decreases in mortality and hospitalization, increased quality of life, and reversible heart remodeling. These benefits led to the HF guidelines recommendations updating and inclusion of S/V combinations a key component of HFrEF treatment.

Keywords: sacubitril/valsartan; mortality; morbidity; heart failure; ejection fraction

Citation: Iovanovici, D.-C.; Bungau, S.G.; Vesa, C.M.; Moisi, M.; Babes, E.E.; Tit, D.M.; Horvath, T.; Behl, T.; Rus, M. Reviewing the Modern Therapeutical Options and the Outcomes of Sacubitril/Valsartan in Heart Failure. *Int. J. Mol. Sci.* **2022**, *23*, 11336. https://doi.org/10.3390/ijms231911336

Academic Editor: Raffaele Marfella

Received: 17 August 2022
Accepted: 22 September 2022
Published: 26 September 2022

1. Introduction

Heart failure (HF) is a prevalent disease these days, having a variety of etiologies. This syndrome involves the ventricle's structure and function and affects patients' quality of life, which means that work capacity, effort tolerance, sleep and psychosocial profile are altered. Even though HF is debilitating, and deadly, continuous research has developed more effective therapies. HF is the final stage of most types of heart disease. As a result, established risk factors play a crucial role in developing HF. High blood pressure, metabolic syndrome, low physical activity, dyslipidemia, and smoking [1,2] have all been tied to incident HF, either through coronary disease [3] or through conditions associated with HF, such as type 2 diabetes mellitus (T2DM) [4], chronic kidney disease [5] or overweight [6], which are widely known to be implicated in the genesis of HF via multiple pathways. HF describes symptoms and signs caused by cardiac abnormalities. HF most used terminology is the left ventricular ejection fraction (LVEF). HF with preserved ejection fraction (HFpEF) is described as HF with normal LVEF ($\geq$50%) and HF with reduced LVEF ($\leq$40%) as HF with reduced ejection fraction (HFrEF). Subjects that are HF-diagnosed, also having an LVEF of 40 to 49%, are known to have mildly reduced ejection fraction (HFmrEF) [7–9].

Decompensated heart failure (DHF) can be defined as an exacerbation of a chronic HF or an acute condition [10]. DHF usually appears in patients pre-diagnosed with HF and is characterized by signs and symptoms that are not tolerable and imply rapid therapeutic intervention [11]. However, DHF can arise de novo when it is triggered, among other causes, by complications of acute myocardial infarction (e.g., rupture of the chordae tendinae, acute mitral regurgitation, etc.), pulmonary embolism, and arrhythmias. This type of DHF is also known as acute heart failure (AHF).

The persistent burden of HF has lately been highlighted by data on cardiovascular mortality in the United States [12,13]. HF is significantly more common in older age groups, with a prevalence of 4.3% among 65–70 year-olds, and is expected to rise rapidly until 2030 when the incidence of HF might reach 8.5%. HF is usually included in elderly cardiovascular syndromes, which have a built-in burden of multiple chronic conditions and frailty, considerably exacerbating the disease's personal and social costs. In addition, people of color with HF, especially women, have a disproportionately high impairment rate [14].

The New York Heart Association categorizes the relationship between dyspnea symptoms and physical activity: class I: no symptoms; class II: minor symptoms when engaged in regular physical activity; class III: patients still have no symptoms at rest but occur at a lower-than-normal activity; and class IV: extreme breathlessness even when patients are resting.

Drug therapy is gradually introduced according to the symptoms and stages of HF. Stage A (high risk, no symptoms) focuses on treating risk factors and comorbidities. Stages B (structural heart disease, symptoms missing) and C (structural heart disease, positive symptoms) require drug therapy. If bundle branch block is present, it should be considered cardiac resynchronization; if acute myocardial infarction is a problem, revascularization (PCI and CABG) must be performed. Finally, refractory symptoms require intervention in stage D: VAD (ventricular assisted device) and transplantation. Since neurohormonal involvement in HF has been recognized, there has been increased attention to the renin-angiotensin-aldosterone pathway (RAAS) and sympathetic activation. Thus, by suppressing the two, a decrease in mortality and rehospitalizations was demonstrated; moreover, some beta-blockers (BB) (carvedilol, nebivolol, prolonged-release metoprolol, bisoprolol) have also been shown to improve left ventricular function. Drug classes such as mineralocorticoid receptor antagonists (MRAs), angiotensin-converting enzyme inhibitors (ACEI)/angiotensin receptor blockers (ARBs) showed improved prognosis when combined [15–18].

Since the 2000s, a new range of drugs has been introduced in HF therapy [19–22], some of them having a spectacular evolution in terms of formulation [23]. A relatively new drug class that has made its presence felt in cardiology is co-transporter 2 (SGLT2) inhibitors, which has been proven as being effective in treating HFrEF, even if patients do not have DM; thus, doctors are encouraged to add this class of drugs (if they are not contraindicated or intolerated) to the treatment plan, among a beta-blocker ACEI, MRA, and an Angiotensin Receptor-Neprilysin Inhibitor (ARNI), to diminish the cardiovascular death risk or exacerbating HF. ARNI is known as combination of sacubitril and valsartan.

Sacubitril and valsartan (S/V) formed this new drug class—ARNI, because ACEI and sacubitril combined produced significant angioedema. *Neprilysin* is an endopeptidase that degrades natriuretic peptides (NPs) and other endogenous vasoactive peptides. Sacubitril inhibits neprilysin, which raises the quantities of these peptides, and counteracts the opposite effect of neurohormonal overactivation [24]. The clinical efficacy of ARNI in HFrEF was proven in the prospective comparison of ARNI with ACEI to determine impact on global mortality and morbidity in HF (PARADIGM-HF) trial published in 2014 [25]. Experimental investigations have shown that inhibiting the RAAS and neprilysin simultaneously can reduce neurohormonal activation [26]. ACEI alone was inferior to ARNI in decreasing the hospitalization and the risk of death, in patients diagnosed with HFrEF in a double-blind large RCT [27].

MiRNAs are small molecules that can be identified at the blood level and are potential biomarkers to monitor in cardiovascular pathology. more precisely in heart failure [28]. These molecules are involved in cardiac adaptation processes. Some mechanisms involved in damaging the heart are fibrosis, hypertrophy, and apoptosis. These alterations at the cardiac level correspond to changes at the molecular level, so it is possible that in the future these genes will also be used as therapeutic targets. Published data [29] showed that ARNI increases the level of miRNA-18 and miRNA-145, which offers some protection against myocardial remodeling and oxidative stress at the cardiomyocyte level. Increased levels of miRNA-181 are associated with myocardial hypertrophy and fibrosis. ARNI has been shown to reduce the level of miRNA-181 [29].

In patients who have a poor therapeutic effect of cardiac resynchronization therapy with defibrillator (CRTd) and a worse prognosis, ARNI leads to significant improvement in clinical symptoms, cardiac pump, and reduction in NYHA class [29]. These effects ARNI-induced in CRTd patients lead to reduction in hospitalizations [29]. Notably, these effects are due to the regression of reverse cardiac remodeling via the modulation of microRNAs expression [29]. Indeed, the microRNAs are implied in the control of cardiac adaptive processes in CRTd patients [28], ARNI already being a common practice for HFrEF patients [25,30–32].

This study would like to present the modern therapeutic options, adapted according to the pathophysiological mechanism of the diseases, in HFrEF and HFpEF, and to emphasize the role of sacubitril/valsartan in heart failure therapy based on the data provided by the large trials. The review also aims to raise awareness to the medical public, starting from the general practitioner to the cardiologist, about the benefits of inclusion of S/V in the complex management of HF that led to its inclusion in the cardiology guidelines for heart failure.

2. The Pathophysiology of Heart Failure

Cardiac dysfunction, both structural and functional, causes decreased cardiac output and increased intracardiac pressures, which dictate the signs and symptoms of HF [7]. Cardiac injury, including myocyte cell loss, myocardial deformity, fibrosis, LV gradual dilation, and changes in ventricular shape, leads to cardiac remodeling, an imbalance in the demand/supply of oxygen from the heart, and altered contractility. In addition, arrhythmias also cause loss of heart pump function and systolic dysfunction [33]. Moreover, vasoconstrictor, pro-thrombotic and pro-inflammatory factors also contribute to cardiac injury, by altering diastole (both atrial and ventricular), so relaxation and filling of the cavities are no longer possible [34].

Considering only the ejection fraction, HF was divided into two major categories: HFpEF and HFrEF, which helped to make the diagnosis more accessible to establish the therapeutic course as quickly as possible, having an essential predictive value. Besides the two categories which consider EF < 40% and > 50%, a third category should be mentioned (that covers the gray area of 40–50%), namely HFmrEF. When more than one variable is considered, research has shown that myocardial dysfunction can be global, both systolic and diastolic, and that fibrosis, cardiomyocyte cell loss, oxidative stress, and coronary heart disease contribute at the onset of HF too. Moreover, the negative involvement of the RAAS in varying degrees in both types of HF has been shown, but the NPs counteracts the effects of RAAS by vasodilation, decreased wall thickness and inflammation of the heart, but also by its action on the nervous system [35]. Figure 1 demonstrates the relationship between the mechanisms described and HFrEF occurrence.

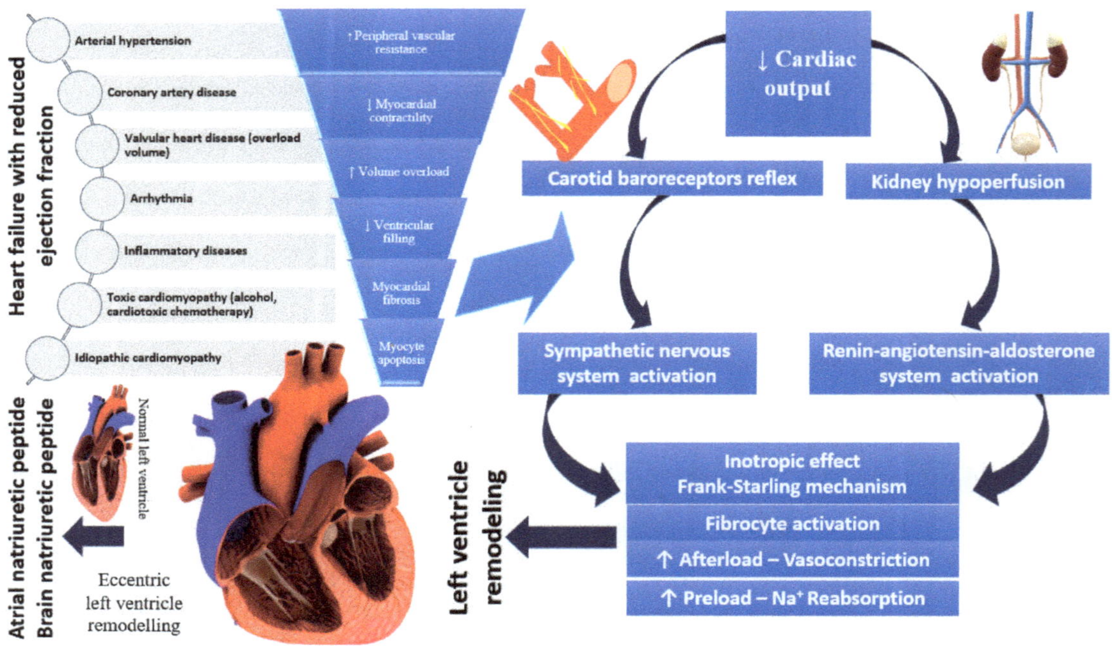

Figure 1. Pathophysiology of heart failure with reduced ejection fraction.

Evidence-based therapy improves symptoms and prognosis in HFrEF, and less in HFpEF. These variations underscore the importance of understanding the pathophysiological differences between HFrEF and HFpEF, which may influence therapeutic targets. The rising incidence and high mortality rates are standard features of both. It has been noticed that differences in giant spring titin, fibrosis, endothelial malfunction, and inflammation, as well as cardiomyocyte hypertrophy and apoptosis, vary in HF pathology. Cardiomyocyte hypertrophy, intercellular fibrosis, abnormal cardiomyocyte relaxation, and inflammation are all characteristics of HFpEF, resulting in the LV's inability to relax adequately. Most of the time, HFpEF is associated with other chronic diseases, which can lead to its aggravation, and implicitly to an increase in the hospitalization rate. Non-cardiomyocytes are made up of about 60% endothelial cells, and endothelial dysfunction, which can be recognized early in cardiovascular disease, being less common in HFrEF vs. HFpEF. Several adaptive mechanisms can cause endothelial dysfunction in response to low cardiac output, such as vasoconstriction, nitric oxide imbalance, enhanced oxidative stress, neurohormonal activation and energy bioavailability [36]. Figure 2 illustrates the pathophysiology of HFpEF.

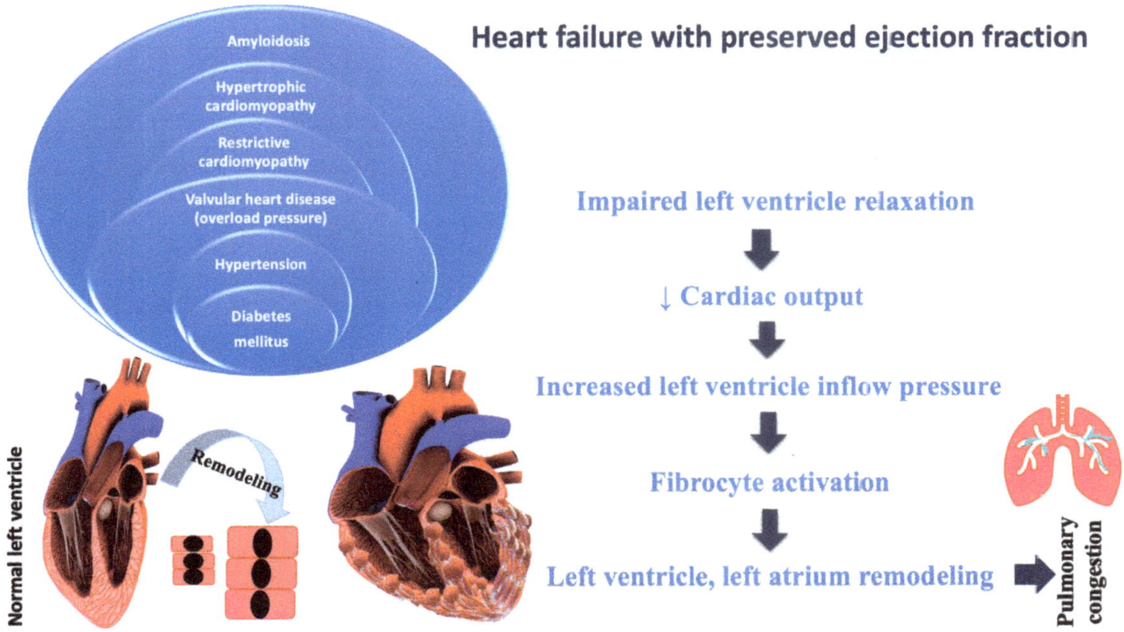

Figure 2. Pathophysiology of heart failure with preserved ejection fraction.

3. Diagnosis of Heart Failure and Types of Heart Failure

The HF presence is suggested by various signs and symptoms such as shortness of breath, cough, disrupted sleep, exercise intolerance, edema, and fatigue; in addition, displacement apex shock and increased jugular venous pressure may also be present. However, these variables are not always sufficient for the definite diagnosis of HF, because they can appear in other disorders as well (i.e., kidney failure, chronic obstructive pulmonary disease (COPD), and obesity) [7].

Thus, in addition to clinical evaluation and routine laboratory tests, specific laboratory tests are sometimes required—brain natriuretic peptides (BNP) and N-terminal proBNP (70% sensitivity, 99% specificity, respectively, 99% sensitivity, 85% specificity) [37]. The two cardiac biomarkers, BNP and proBNP, are secreted mainly by the ventricles but also by the atria. In the treatment of HF, BNP and NT-proBNP are known as having clinical importance in prognostic/diagnostic indicators. For example, BNP levels < 100 pg/mL had 90% predictive negative value during the diagnosis of HF in patients with acute dyspnea. In comparison, values > 500 pg/mL have >80% predictive positive value [38]. LVEF is a widely used phenotypic criterion for HF diagnosis. Both HFrEF (EF $\leq$ 40%) and HFpEF (EF $\geq$50%) are primary HF subtypes with distinct pathophysiology, etiology, and therapy outcomes. In addition, the proportions of the two phenotypes with specific risk factors differ. HFpEF, for example, is defined by female gender and advanced age [39–43].

Comorbidities, numerous risk factors, and pre-existing illnesses contribute to HF, harming heart's function and structure. The first choice of HF medications aimed to improve quality of life by reducing morbidity and death in HFrEF while reducing symptoms and slowing disease progression. However, to date, no HF therapy has been reported to reverse (constantly and permanently) the evolution of structural and functional degeneration of the heart [7,44]. Despite breakthroughs in treatment, the prognosis in HF remains poor. Patients with HFpEF exhibit symptoms and signs of HF, and proof of cardiac dysfunction as a source of symptoms [2]. HFpEF has the same clinical symptoms as typical HF, including HFrEF [45]. HFpEF refers to those having HF signs/symptoms of HF, or cardiac abnormalities (LV diastolic dysfunction/increased left ventricular filling pressures, and/or elevated NPs, as well as an LVEF above 50%) [46,47].

4. Pharmacologic Therapy for Heart Failure

4.1. Treatment in HFrEF

The decrease in cardiac output causes an inadequate circulating volume that will be ameliorated, at the onset of HF, by the compensatory mechanisms through neurohormonal involvement: sympathetic activation, RAAS, and release of antidiuretic (ADH). Chronic activation of these mechanisms will cause vasoconstriction and fluid retention, depletion of catecholamines, and a weak response to the action of circulating catecholamines, contributing to myocardial hypertrophy and cardiac remodeling, which are present in HF [48–50]. ACEI impacts significantly the neurohormonal state of HF subjects by interfering with the RAAS by limiting angiotensin I (ATI) to convert to angiotensin II (ATII), causing vascular relaxation, decreased vasoconstriction and vascular resistance [51–54].

Low levels of ATII promote the elimination of Na from the body, decreased vasoconstriction and blood pressure (BP). These effects are due to decreased sympathetic activity, ADH production and aldosterone. Low preload and afterload results from low venous and arterial pressure lead to improved ventricular filling and better blood ejection. ACEI can help prevent ventricular remodeling by limiting cardiac hypertrophy and myocardial fibrosis and reducing cardiomyocyte death by acting at the cellular level. ACEI have been demonstrated to have positive benefits in chronic HF [55–58].

In patients with HFrEF, ACEI enhance symptoms, life quality, and physical function. When compared to placebo, ACEI reduced death by 23% and HFrEF-related mortality or hospitalization by 35%, according to an analysis of 32 randomized clinical studies in people with HFrEF. ACEI improve survival and lower the risk of HF and coronary events in patients with a reduced LVEF, but no HF [59–61].

Angiotensin receptor blockers (ARBs) suppress the RAAS by blocking angiotensin II from binding to its receptor, preventing constriction of the blood vessels and aldosterone release; ARBs do not inhibit kininase, which lowers cough compared to ACE inhibitors. Therefore, ARBs should be used to minimize morbidity and fatality in those patients which cannot be administered ACEI (considering their side effects) [62], or in patients in whom ARNI is not feasible, according to the 2022 ACCF/AHA/HFSA guidelines [63]. Furthermore, due to the danger of cross-reaction, ARBs should be administrated with caution in the case of the subjects having in their medical history of angioedema induced by ACEI [64].

According to the guidelines, in patients having HFrEF NYHA class II/III, ARB medication should be replaced with an ARNI [65]. The Candesartan in Heart Failure (CHARM Alternative) study compared candesartan to placebo and found that candesartan improved cardiovascular outcomes compared to placebo, including cardiovascular death or hospital readmission. In comparison to 40% of placebo patients, only 33% of candesartan patients died of cardiovascular cause or were hospitalized for HF [66].

RAAS inhibitors have been shown to have cardioprotective effects. In the study published by Marfella et al. in 2022, the effect of RAAS blockers was analyzed in heart transplant patients with/without T2DM. When the research started, no significant differences were observed in terms of myocardial fibrosis, but at one year of follow-up, there were differences between patients who did not have T2DM and those who were diagnosed with T2DM, and more than that, differences were observed in patients who had a more rigorous glycemic control. It should be clarified that all patients followed a similar therapy with ACEI or ARB. The study evaluated the involvement of Ang 1–7 and Ang 1–9 molecules responsible for the antifibrotic effects at the cardiac myocyte level and observed that their levels are higher in patients without T2DM and in patients with better glycemic control [67].

Beta-blockers bind to beta-adrenoceptors and prevent adrenaline and noradrenaline from binding to these receptors, reducing the SNS's functions. Fundamental studies have shown that giving a BB to symptomatic patients with decreased LVEF reduces mortality and morbidity. In the SENIORS, MERIT-HF, COPERNICUS, and Cardiac Insufficiency Bisoprolol Study (CIBIS)-II trials, this was proved with nebivolol, carvedilol, bisoprolol, and controlled-release metoprolol [68–71].

Because of detrimental inotropic effects, it is not indicated that BB be started during an HF exacerbation; instead, this class is recommended when the patient is volume-stable [22,72]. Only certain BBs reduce mortality and rehospitalizations because BBs do not show a class effect. The effects of bisoprolol and metoprolol CR/XL were compared to placebo and resulted that there has been a reduction of all generating causes of mortality, hospitalizations, and even NYHA functional status [71,73].

Mineralocorticoid receptor antagonists work by inhibiting the mineralocorticoid receptor, which counteracts the effects of aldosterone since MRAs provide diuretic properties and can contribute to a fluid balance. The main beneficial effects are observed in decreasing mortality, respectively, the number and duration of hospitalizations due to HF (explainable by both decreasing the risk of hypokalemia, prevention of myocardial/renal fibrosis caused by excess aldosterone, etc.) [74–76]. Spironolactone side effects (e.g., gynecomastia) are caused by the affinity of spironolactone for glucocorticoid, progesterone, and androgen receptors. Eplerenone does not determine gynecomastia, so it is a better choice, but a higher dose is needed, because of the lower affinity [77–79].

Sodium-glucose cotransporter-2 inhibitors (SGLT2i) consider a newer class of drugs, with antidiabetic effect, that improve natriuresis and osmotic diuresis, while increasing the excretion of glucose in the urine to reduce blood glucose levels.

The Dapagliflozin and Prevention of Adverse-outcomes in Heart Failure Trial (DAPA-HF) randomized trial with dapagliflozin was designed in the HFrEF group, even if patients did not have DM [80].

When compared to placebo, the study main purpose was to see how adding 10 mg of dapagliflozin to the optimal treatment for HFrEF affects the primary endpoint's occurrence, aggravation of HF (decompensated HF) and cardiovascular mortality. The dapagliflozin-treated group exceeded the placebo-treated group in every parameter tested throughout the course of an average 18-month follow-up. In addition, concerning the first exacerbating HF event and death from cardiac or other causes, the group using dapagliflozin (as treatment) had a reduced risk. Empagliflozin improves physical performance and life quality, while lowering the risk of HF hospitalization in the Empagliflozin Outcome Trial in Patients with Chronic Heart Failure with Reduced Ejection Fraction (EMPEROR-Reduced study) [81].

In HFrEF patients, the two approaches joined (inhibition of NP breakdown and blockage of the RAA) proving to be the most effective together [82]. In real-life and contemporary research, ARNI has established itself as a first-line drug in the treatment of HFrEF. Its favorable effect on cardiac remodeling has also been confirmed. The major objective of the Rationale and Methods of a Prospective Study of Biomarkers, Symptom Improvement, and Ventricular Remodeling During Sacubitril/Valsartan Therapy for Heart Failure (PROVE-HF) trial was to find a relationship between changes of cardiac remodeling and NT-proBNP concentration values [83]. In the PRIME trial (Pharmacological Reduction of Functional, Ischemic Mitral Regurgitation), another study found that valsartan alone was inferior to ARNI in standard therapy, in reverse remodeling and downsizing functional mitral regurgitation (FMR) in a group of participants diagnosed with HF with an EF $\approx$ 25–50%, symptomatic (NYHA II-III) and significant FMR lasting more than 6 months. After 12 months, in the group treated with ARNI, a reduction in EROA (effective regurgitant orifice area) was noticed to be more significant than in the group treated with valsartan; this was the main indicator of FMR improvement. Reducing the effective regurgitant orifice area determined a reduction in the end systolic/diastolic volume of the LV, in both groups, S/V, respectively, valsartan. The S/V group outperformed the valsartan group in regurgitant volume (mean difference, 7.3 mL). Additionally, the S/V group had a substantially higher decline in diastolic LV function—E/e'. These data support S/V role in cardiac reverse remodeling [84].

New information concerning the mechanisms of this drug class in HFrEF was revealed after researchers analyzed the effect on inflammation, functional ability, and peripheral vascular activity. A group of patients with HFrEF (LVEF~28%) were administered S/V and investigated prospectively. Patients were evaluated at the beginning of treatment, then monthly for up to three months, and a decrease in pro-inflammatory markers, improvement in functional capacity and peripheral vascular activity were observed [85].

4.2. Treatment in HFpEF

The development and progression of HFpEF are linked to abnormal activation of the RAAS. In addition, the RAAS reduced LV diastolic performance (increasing myocardial/arterial stiffness) and caused LV hypertrophy [86,87]. As a result, multiple randomized clinical trials have assessed the RAAS blockade's prognostic value (CHARM-preserved, PEPCHF, I-PRESERVED). Unfortunately, the results were not as optimistic as expected: candesartan and perindopril reduced hospitalization rates due to HF and symptoms, but irbesartan did not reduce hospitalizations and patients' quality of life did not improve in the long term [88–91]. Due to their effectiveness in HFrEF, attempts have been made to initiate BB in treating HFpEF. Given that there is also a ventricular filling defect in HFpEF due to adrenergic stimulation, BB may be helpful in decreasing the adrenergic response and HR and improving exercise tolerance [92,93].

Aldosterone has implications for the development of myocardial fibrosis that will cause adverse effects on the heart muscle. MRAs act on aldosterone receptors and are useful in treating HF regardless of the left ventricular ejection fraction. The ALDO-HF study proved the usefulness of this class in HFpEF from a cardiac point of view but not showing improvements in symptoms or in the life's quality [94]. The TOPCAT-HF trial studied the effects of spironolactone compared to placebo, but the results were not encouraging. In the group treated with spironolactone, the unwanted effects did not take long to appear, so hyperkalemia and renal failure were present to a higher degree. However, this trial opens the way for other studies regarding HFpEF therapy, depending on the etiology [95].

ARNI combines RAAS blockade and endogenous natriuretic peptide pathway upregulation. In HFrEF, S/V is an emerging and game-changing disease-modifying drug. Inhibition of neprilysin increases endogenous natriuretic/vasoactive peptides, such as cGMP, which are reduced in HFpEF and linked to myocardial stiffness. In a phase II study of patients with HFpEF, S/V was linked to LA reverse remodeling, a more considerable reduction in NP levels, and symptoms improvement [96,97]. Patients with HFpEF may benefit from this discovery. To determine the efficacy of S/V in HFpEF, the Prospective Comparison of ARNI with ARB Global Outcomes in HF with Preserved Ejection Fraction (PARAGON-HF) trial was conducted. Subjects having LVEF > 45% and NYHA II-IV were included in the study. The study's findings were not as statistically significant as expected, but it was found that S/V is more effective among women [98–100].

5. Sacubitril/Valsartan Therapy

5.1. Sacubitril/Valsartan in HFrEF—Evidence of Efficacy in Clinical Trials

In the last three decades, HF treatment has experienced an unparalleled evolution, based on evidence, with an increase in the quality of living and the survival of patients suffering from HF. International guidelines support this progress, especially the ESC Guide. Compared to the guide published in 2016, the 2021 guide focuses on patient involvement in managing the treatment of their disease, of course with a multidisciplinary team. Along with conventional therapy (ACEI, MRAs, beta-blockers), ARNI and SGLT2i are added to the therapeutic regimen of subjects with HFrEF. These medication classes have both independent and additive therapeutic effects [46,101].

In the PARADIGM-HF, patients with HFrEF and NYHA class II, III, or IV HF have been randomly selected to be administered either S/V or enalapril, in addition to standard therapy. The study aimed to identify the variations in mortality rates from CV causes. The study ended earlier in agreement with predetermined rules because the threshold for an overwhelming advantage with S/V had been crossed. A total of 17% of patients treated with S/V and 19.8% with enalapril died, but S/V reduced the rehospitalization rate due to HF by 21% and improved symptoms. Hypotension and angioedema were more pronounced in the ARNI group, but fewer patients developed coughs, electrolyte imbalances, and kidney damage [25]. A blinded independent committee conducted and adjudicated a meticulous and highly extensive investigation of the mode of death subsequent to the first release of PARADIGM-HF. The mortality causes were divided into non-CV or CV deaths. In CV deaths, it was observed that S/V reduced the death rate by 20% compared to the control group treated with enalapril [102].

In the PIONEER-HF trial, subjects suffering from HFrEF hospitalized due to IC decompensation were randomized into two groups: one was treated with ARNI and another with enalapril. Shortly afterward, a more considerable reduction in NT-proBNP levels was observed in the group treated with ARNI. The decrease in NT proBNP, which indicates neurohormonal activation and hemodynamic stress, was linked to a decrease in high-sensitivity cardiac troponin, indicating myocardial damage. Compared to enalapril, S/V subjects were unlikely to be re-hospitalized for HF at eight weeks. This trial was the first to show that starting S/V in the hospital was tolerable and safe [103].

The goal of the TITRATION trial was to offer information on how to start and up titrate S/V in people with chronic HFrEF. TITRATION included patients who had never been treated before or had varying levels of ACEI/ARB pre-treatment [33,42]. Patients were divided into two groups: one for concentrated titration, where the dose was increased in three weeks, and one for conservative titration, where the dose was increased in six weeks. By measuring the patient's tolerance to treatment, therapeutic success was obtained in both groups analyzed. Patients who have not received any ACEI/ARBs previously can tolerate S/V if introduced gradually, according to the TITRATION trial. It was observed that if the higher dose is not tolerated in the beginning, a down titration can be beneficial; this technique can allow the physicians to reach the target dose in time. The TITRATION trial reveals that S/V can be titrated in most of the subjects in three weeks, excluding the patients naïve to ACEI/ARBs therapy [104].

PRIME study aimed to show that S/V has a favorable effect on remodeling in HFrEF patients. In the PRIME study patients suffer from chronic functional mitral regurgitation due to left ventricular malfunction and reduced EF. These subjects received guideline-directed medical therapy in a double-blind experiment. Valsartan or S/V were given to the patients. The S/V group developed a decrease in EROA, thus proving that the drug has a better effect on heart remodeling than the valsartan group; additionally, an improvement in the regurgitant volume was noticed. Some other noticeable results show that the S/V group had a higher decrease in LV end-diastolic volume index, but on partial mitral leaflet closure area, other measures of the LV, or changes in blood pressure, it was not a remarkable difference. PRIME is considered a modest trial being the first to demonstrate the reverse remodeling impact of S/V in people with HFrEF, associated with suffering from FMR [84].

In the PARADIGM-HF study, a lower NT-proBNP level was linked to a better result in patients using S/V. The PROVE-HF wanted to investigate this further because NT-proBNP lowering during guideline-directed healing therapy has already been associated with cardiac remodeling reversal. PROVE-HF was open-label research in which 794 individuals suffering from chronic HFrEF were distributed to S/V and had their echocardiogram carried out before starting treatment, six months later, and 1 year later. At each study visit, the concentration of NT-proBNP was assessed. Following the completion of all research procedures, at a core laboratory, echocardiograms were blindly analyzed temporally and clinically. After starting S/V, the researchers found a significant 37 percent reduction in NT-proBNP and reverse cardiac remodeling intimately tied to it. At the starting point, the LVEF was ~28%, but after 12 months of treatment, it increased by 9.4%, with a significant improvement in other patients. Additionally, more variables were modified: diastolic function, measured by E/e' proportion, improved, and the mass index of LV and volumes of LV and LA reduced. These results are maintained in patients who were newly diagnosed with HF or those who had not previously been treated with ACE inhibitors or ARBs or patients who did not achieve the target dose of S/V. The PROVE-HF research adds to the growing body of evidence supporting reversible cardiac remodeling and the drop of NT-proBNP levels when ARNI is administered [105].

Whether aortic impedance can contribute (from the pathophysiologic point of view) in patients diagnosed with HFrEF and treated with ARNI was one of the questions answered in the EVALUATE-HF trial. Patients were divided into two groups: one that received S/V and the other one, enalapril. It resulted that the S/V group showed a decreased aortic impedance, while the valsartan's group was increased; still, the dissimilarity was not eloquent. Compared to the enalapril group, the S/V group exhibited considerably lessened NT-proBNP levels and a significant reduction in numerous echocardiographic parameters (LVED, SVI, LAVI mitral E/e′ index). An exploratory secondary goal was to show a substantial increase in the general summary score for the 12-module Kansas City Cardiomyopathy Questionnaire (KCCQ). Despite three months of S/V medication, these findings demonstrate a clear remodeling benefit compared to usual care. The EVALUATE-HF research found that, while no considerable improvement was observed when administering S/V in aortic impedance, in contrast with enalapril, it did show more substantial cardiac reverse remodeling and improved quality of life [106].

A summary of the results obtained from the above presented clinical trials is presented in the Table 1.

Table 1. Clinical trials main information.

Trial Acronym [Ref], Setting	Study Design Patients' no./Inclusion Criteria Period	Results
PARADIGM-HF [25], ambulatory preceded	S/V (target dose 97/103 mg × 2/day) vs. enalapril; a multicenter, prospective, randomized clinical trial 8442/NYHA II–IV (EF ≤ 40%) follow-up of 27 months	• CVD mortality ↓20% • HF hospitalization ↓21%
TITRATION [104], ambulatory	S/V clinical trial, multicenter, prospective, randomized/ 498/NYHA II–IV (EF ≤ 35%)/ 16-week study period	• 76% of patients achieved and maintained S/V without modifying the dose • 77.8% of patients reached the appropriate S/V dosage, • 84.3% with a prolonged titration • Safety was equal
PRIME HF [84], ambulatory	S/V vs. valsartan in a multicenter, prospective, randomized clinical study 118/HFrEF (EF < 50%) 12-month study period	• reduction in EROA • significant changes in regurgitant volume and LVEDV • no significant changes in LVESV and inadequate mitral leaflet closure area.
EVALUATE-HF [106], ambulatory	S/V vs. enalapril in a multicenter, prospective, randomized clinical trial 464/ HFrEF (EF ≤ 40%) S/V or enalapril were assigned at random. 12-week study period	• difference in aortic characteristic impedance—not substantial • changes in LAVI, LVEDV, LVESV, mitral E/e′ ratio • EF increased by 1.9% in S/V group
PROVE-HF [105], ambulatory	S/V clinical trial, multicenter, prospective, open label 794/HFrEF (EF ≤ 40%) 12-month study period	• reduction in NT-proBNP • changes in LVEDV, LVESV, LAVI and E/e′ ration • reversal cardiac remodeling • low NT-proBNP, not attain target dose, new-onset HF, or not taking an ACEI/ARB at the time of enrolling—similar outcomes.

Table 1. *Cont.*

Trial Acronym [Ref], Setting	Study Design Patients' no./Inclusion Criteria Period	Results
PIONEER-HF [107], in-hospital	S/V was compared to enalapril in a multicenter, prospective, randomized clinical trial 881/NYHA II–IV (EF $\leq$ 40%) 8-week study period	• 47% against a 25% drop in NT-proBNP • S/V is safe in AHF or new-onset HF • recurrent HF hospitalizations—reduced • no significant difference between the two groups regarding renal function, hypotension, and hyperkalemia
TRANSITION [108], in-hospital	A multicenter, prospective, randomized clinical trial 1.002/HFrEF (EF $\leq$ 40%)	• patients who reached the target dose of S/V was similar • S/V safe and well-tolerated in patients with AHF and new-onset HF

AHF, acute heart failure; EF, ejection fraction; HF, heart failure; HFrEF, HF with reduced ejection fraction; LVEDV, left ventricle end-diastolic volume; LVESV, left ventricle end-systolic volume; NYHA, New York Heart Association; S/V, sacubitril/valsartan; LAVI, left atrial volume index; ↓ reducing.

Associations between S/V post-discharge adherence (PDC) and clinical outcomes after readmissions for HFrEF revealed that 32.9% patients who received ARNI therapy were compliant when discharged (PDC $\geq$ 80%), but 67.1% were not compliant to the treatment (PDC < 80%). Between groups, baseline attributes were evenly distributed. Subjects with PDC $\geq$ 80% presented considerably decreased adjusted risk of all-cause readmission to hospital and mortality after 90 days and one year compared to patients with PDC < 80%. At one year, patients saw a significant reduction in rehospitalization for every five-percentage point rise in PDC. Finally, the trial proved that compliance to the therapy with S/V, 90 days after discharge was related to diminished rehospitalization rates and death in patients hospitalized for HFrEF and discharged on S/V. In HFrEF, more work is needed to increase adherence to S/V and other guideline-directed medical therapy [109].

5.2. Sacubitril/Valsartan in the Treatment of HFpEF

The researchers assigned HF patients with LVEF > 45%, NYHA II-IV, high NP values, and structural heart disease to receive target doses of S/V and valsartan (97/203 mg $\times$ 2/day, respectively 160 mg $\times$ 2/day). Renal function, safety, functional class, KCCQ score, CV death, and HF hospitalizations were monitored. The results favor S/V except for hospitalizations and CV deaths, where there was no statistically significant difference [98].

A randomized, parallel-group, double-blind clinical experiment included subjects with HF, LVEF > 40%, high NT-proBNP concentrations, structural heart disease, and bad quality of life registered from 396 facilities in 32 countries. The aim of the research was to compare the results of S/V on NT-proBNP levels, 6-min walk test (6MWT) and life quality in subjects suffering from chronic HF and LVEF greater than 40% to background medication-based individualized comparators. The trial was completed by 87.1% of patients. At the baseline, the levels of NT-proBNP were comparative in both groups (786 pg/mL—S/V group, respectively, 760 pg/mL—comparative group). After 12 weeks, the group that received S/V had a considerable reduction in NT-proBNP levels compared to the other group. A big difference was not noticed regarding the 6MWT after 24 weeks (at the baseline 9.7 m and at the end 12.2 m) nor in NYHA class and KCCQ score (12.3 vs. 11.8, respectively 23.6% vs. 24.0% of patients). S/V treatment determined a more notable decrease in plasma NT-proBNP peptide concentrations compared to standard RAAS inhibitor treatment or placebo. However, it did not significantly enhance the 6MWT in subjects suffering from HF and a LVEF > 40% [110].

A systematic review on four randomized controlled studies of S/V for HFpEF patients showed a reduced rate of HF hospitalization in HFpEF patients when compared to the control group. S/V did not exhibit any clear benefits in cardiovascular mortality, mortality of all causes, or improvement in the NYHA class. Even though S/V was connected to an increased risk of symptomatic hypotension, there were no signs of deteriorating renal function or hyperkalemia. Except for the hospitalization rate, in which the S/V treatment group favored HFpEF patients, the study found no differences between the two groups compared to valsartan or personalized medical therapy [111].

Previous research has validated S/V's efficacy in treating HFrEF. However, the role of S/V in HFpEF is still an area of research. S/V is a combination therapy that includes sacubitril and valsartan and functions as neprilysin inhibitor (ARNI) and a first-generation angiotensin receptor blocker. Using animal models, the effect of S/V was evaluated in a high sodium diet that causes HFpEF and vascular damage, in addition to the primary mechanism. Thus, a considerable amount of salt (68 mg/kg) was administered intragastrically before S/V treatment. According to the findings of functional tests, it has been observed that a high sodium diet leads to lesions in the heart and vascular system. These effects were reversed by S/V administration; in addition, S/V had an antifibrotic impact by decreasing the amount of type 1 and type 3 collagen and decreasing the MMP3/SMAD3 ratio. The most plausible explanation for the favorable effects on HFpEF is the action of S/V on the TGF1/SMAD3 signaling pathway. Following this experiment, S/V proved its therapeutic potential in HFpEF, reversing the effects of HFpEF induced by a high sodium diet. However, this study had some drawbacks: to corroborate the findings, the study did not identify the agonists or antagonists of the TGF-/Smad signaling pathway; instead, it merely examined the biochemical effects of a high-salt diet and S/V. HFpEF is still controversial in high salt-induced rat models even after thorough reports by the team of Y Sakata/M Hori. For the thorough verification of these data, in vitro experiments are also necessary [112]. S/V has been given a pioneering expanded indication by the FDA, making it the first medicine in the US to be approved for chronic HF that is not defined by ejection fraction.

5.3. *Sacubitril/Valsartan in Advanced Heart Failure*

According to estimates, 1% to 10% of people with heart failure have advanced heart failure [113]. However, the prevalence is growing due to the rising number of people with heart failure and improved survival and treatment [114]. Advanced heart failure is characterized by low ejection fraction (LVEF $\leq$ 30%), right ventricle failure, congenital or valve abnormalities that are not operable, high values of the cardiac biomarkers and low to very low quality of life of the patients (e.g., NYHA III/IV, inability to fulfill 300 m without symptoms). Therefore, this stage of HF can be classified as stage D according to ACC/AHA classification. In addition, advanced HF involves repeated interventions by the medical staff and more aggressive therapy to manage the severe overlapping symptoms [46,115]. Researchers analyzed the medical records of all severe HF patients who were assessed at their center for heart transplant therapy. They looked at individuals who had started ARNI medication and had their hemodynamics checked before and after six months. At six months after initiating ARNI medication, the first noticed result was pulmonary pressures and filling pressures variety. Systolic pulmonary artery pressure (32 mm Hg vs. 25 mm Hg) in addition to average pulmonary artery pressure (20 mm Hg vs. 17 mm Hg) were significantly lower six months after commencing ARNI. Because of improvement, 23% patients were removed from the heart transplant list, while four new patients were added to the list. After almost two years, three patients were treated with a left ventricular assist device, while six patients had their hearts transplanted. S/V was found to be safe and effective in lowering filling pressures and pulmonary pressures in subjects suffering from advanced HF [116]. Advanced HF in patients with reduced EF NYHA IV was not an inclusion criterion in the large clinical trials PARADIGM-HF and PIONEER-HF, so in these patients, the effectiveness of S/V in their treatment could not be confirmed. To assess the efficacy and impacts of S/V, the randomized clinical trial LIFE (LCZ696 In Hospitalized Advanced Heart Failure) reached an answer [117]. Therefore, the LIFE trial tried to objectify the efficacy and effects of S/V through the changes made to NT-proBNP levels from the beginning of the study to 24 weeks of therapy. The study included 335 patients with this pathology who were administered S/V or valsartan (target dose—200 mg × 2/day, respectively, 160 mg × 2/day). The trial showed that there are no significant differences between the two groups [118]. Moreover, in patients whose tolerability was not verified to one of the ACEI/ARB classes before initiating S/V, it was proven that 50% of patients would have an intolerance to S/V. Therefore, this study concludes that the benefit of S/V in patients with HFrEF in an advanced stage must be weighed against the risk [117].

6. Ongoing Research with Sacubitril/Valsartan in the Treatment of Heart Failure

The first study (PARAGLIDE-HF) is a multicenter, randomized, and double-blinded clinical trial that is expected to enroll 450 participants. Following hospitalization for acute decompensated HF, this study will assess the effects of S/V compared to valsartan monotherapy on clinical outcomes, safety, tolerability and NT-proBNP levels in decompensated HFpEF patients that were stabilized, and the treatment with S/V will be instituted at the time or in 30 days after patient stabilization [119].

A total of 50 people will be enrolled in the second study. It will look at the effects of S/V on the autonomic cardiac nerve system in HF patients by measuring HR variability [120].

The third study will assess the effects of S/V vs. valsartan on cardiac oxygen demand and cardiac work efficiency in 60 participants with NYHA II-III HFrEF after six weeks of therapy [121].

The fourth trial will enroll 48 patients and will examine S/V vs. placebo in terms of neurohormonal activation and physical function in people presenting right ventricular dysfunction (moderate/severe) and those that have present NYHA II-III symptoms [122].

The researchers will enroll 100 patients in the fifth study, which will be a phase 3 randomized controlled trial to see if S/V can reverse heart hypertrophy and fibrosis [123].

The SHORT trial, the sixth study, is a randomized experiment on HF patients with LVEF < 35%. Its goal is to see if a quicker protocol translates to faster optimization and a higher degree of optimization than the current standard protocol [124].

A multicenter prospective randomized study will assess the effect of in-hospital initiation of S/V on the NT-proBNP concentrations in patients admitted due to AHF (PREMIER) and is estimated to include 400 participants [125].

7. Current Guidelines Recommendations

Recent ACC AHA 2022 HF recommendations indicate the use of ARNI as first line therapy to lessen morbidity and fatality in subjects suffering from HFrEF and those that present NYHA class II/III symptoms. Subjects with HFrEF, NYHA II-III that are under treatment with ACEI/ARBs should be switched to ARNI due to favorable morbidity and mortality results. ARNI is indicated as a de novo medication in subjects suffering from AHF and chronic symptomatic HFrEF, given the improved prognosis, decreased NTproBNP value, and cardiac remodeling [44].

From the perspective of mortality and morbidity, none of therapies described in this review have been shown efficient in HFpEF; it is crucial to understand the mechanisms underlying this condition. However, the PARAGON-HF study revealed a lower readmissions rate in those patients with HF and LVEF < 57%, and a systematic review that included PARADIGM-HF and PARAGON-HF trials showed a decrease in CV mortality and HF hospitalization in patients that had an LVEF lower than the normal interval [46]. FDA granted S/V, an indication to treat patients with HFpEF. Management of HF according to latest guideline recommendations is illustrated in the Figure 3.

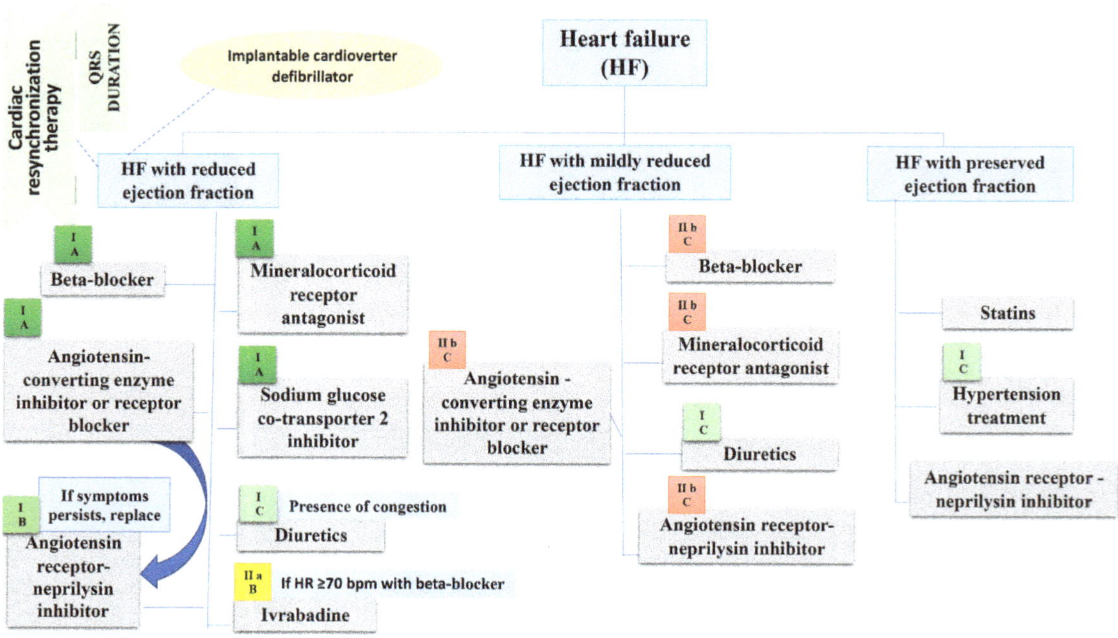

Figure 3. Management of heart failure according to the European Society of Cardiology guideline. Beside each therapeutic agent there is the class and level of recommendation mentioned in the European Society of Cardiology guideline. In HF with preserved ejection fraction the levels of recommendation for drugs included in the figure are not clearly specified because some trials were not published by the time the guideline has been released, so we cannot consider them class I treatment options. b.p.m., beats per minute; HR, Heart rate.

8. Conclusions

S/V is a breakthrough drug in HF therapy. Compelling benefits of ARNI therapy in HFrEF were demonstrated in PARADIGM HF with a reduction in CV mortality and hospital admission for HF, and a decrease in all-cause mortality. Sudden cardiac death was reduced by 22%, as were all-cause and HF readmissions and AHF. S/V reduced symptoms and physical constraints linked to HF, as assessed by the KCCQ, and this advantage expanded to almost all areas of the score when examined individually. There is an expanding body of evidence supporting S/V's role in cardiac reverse remodeling process correlated with a significant reduction of NT-proBNP level in EVALUATE HF and PROVE HF. The PRIME study demonstrates for the first time the result of S/V treatment on reverse remodeling in HFrEF subjects with functional mitral regurgitation. S/V decreased the rate of cardiovascular mortality and readmissions in subjects suffering from HFpEF in PARAGON-HF, though the results are not statistically significant. In addition, advantages for life quality and renal function have been reported, but also from a clinical perspective. S/V treatment reduced the symptoms of HFpEF produced by a high-salt diet, most likely by decreasing fibrosis, highlighting S/V's therapeutic promise for HFpEF. For patients admitted to hospital for HFrEF and discharged on S/V, increased adherence to therapy was connected to considerably lower readmission rates and death.

Funding: This research received no external funding.

Institutional Review Board Statement: Not applicable.

Informed Consent Statement: Not applicable.

Data Availability Statement: Not applicable.

Conflicts of Interest: The authors declare no conflict of interest.

References

1. Ursoniu, S.; Mikhailidis, D.P.; Serban, M.C.; Penson, P.; Toth, P.P.; Ridker, P.M.; Ray, K.K.; Kees Hovingh, G.; Kastelein, J.J.; Hernandez, A.V.; et al. The Effect of Statins on Cardiovascular Outcomes by Smoking Status: A Systematic Review and Meta-Analysis of Randomized Controlled Trials. *Pharmacol. Res.* **2017**, *122*, 105–117. [CrossRef] [PubMed]
2. Gheorghe, G.; Toth, P.P.; Bungau, S.; Behl, T.; Ilie, M.; Stoian, A.P.; Bratu, O.G.; Bacalbasa, N.; Rus, M.; Diaconu, C.C. Cardiovascular Risk and Statin Therapy Considerations in Women. *Diagnostics* **2020**, *10*, 483. [CrossRef]
3. Babes, E.E.; Bustea, C.; Behl, T.; Abdel-Daim, M.M.; Nechifor, A.C.; Stoicescu, M.; Brisc, C.M.; Moisi, M.; Gitea, D.; Iovanovici, D.C.; et al. Acute coronary syndromes in diabetic patients, outcome, revascularization, and antithrombotic therapy. *Biomed. Pharmacother.* **2022**, *148*, 112772. [CrossRef] [PubMed]
4. Vesa, C.M.; Popa, L.; Popa, A.R.; Rus, M.; Zaha, A.A.; Bungau, S.; Tit, D.M.; Corb Aron, R.A.; Zaha, D.C. Current Data Regarding the Relationship between Type 2 Diabetes Mellitus and Cardiovascular Risk Factors. *Diagnostics* **2020**, *10*, 314. [CrossRef] [PubMed]
5. Moisi, M.I.; Bungau, S.G.; Vesa, C.M.; Diaconu, C.C.; Behl, T.; Stoicescu, M.; Toma, M.M.; Bustea, C.; Sava, C.; Popescu, M.I. Framing Cause-Effect Relationship of Acute Coronary Syndrome in Patients with Chronic Kidney Disease. *Diagnostics* **2021**, *11*, 1518. [CrossRef] [PubMed]
6. Remus Popa, A.; Fratila, O.; Rus, M.; Anca Corb Aron, R.; Mihai Vesa, C.; Pantis, C.C.; Diaconu, C.; Bratu, O.; Bungau, S.; Nemeth, S. Risk factors for adiposity in the urban population and influence on the prevalence of overweight and obesity. *Exp. Ther. Med.* **2020**, *20*, 129–133. [CrossRef] [PubMed]
7. Ponikowski, P.; Voors, A.A.; Anker, S.D.; Bueno, H.; Cleland, J.G.; Coats, A.J.; Falk, V.; González-Juanatey, J.R.; Harjola, V.P.; Jankowska, E.A.; et al. 2016 ESC Guidelines for the diagnosis and treatment of acute and chronic heart failure: The Task Force for the diagnosis and treatment of acute and chronic heart failure of the European Society of Cardiology (ESC). Developed with the special contribution of the Heart Failure Association (HFA) of the ESC. *Eur. J. Heart Fail.* **2016**, *18*, 891–975. [CrossRef]
8. Kim, M.S.; Lee, J.H.; Kim, E.J.; Park, D.G.; Park, S.J.; Park, J.J.; Shin, M.S.; Yoo, B.S.; Youn, J.C.; Lee, S.E.; et al. Korean Guidelines for Diagnosis and Management of Chronic Heart Failure. *Korean Circ. J.* **2017**, *47*, 555–643. [CrossRef]
9. Youn, J.C.; Han, S.; Ryu, K.H. Temporal Trends of Hospitalized Patients with Heart Failure in Korea. *Korean Circ. J.* **2017**, *47*, 16–24. [CrossRef]
10. Montera, M.W.; Pereira, S.B.; Colafranceschi, A.S.; Almeida, D.R.; Tinoco, E.M.; Rocha, R.M.; Moura, L.A.; Réa-Neto, Á.; Mangini, S.; Braga, F.G.; et al. Summary of the II Brazilian Guideline update on Acute Heart Failure 2009/2011. *Arq. Bras. Cardiol.* **2012**, *98*, 375–383. [CrossRef]
11. Bocchi, E.A.; Vilas-Boas, F.; Perrone, S.; Caamaño, A.G.; Clausell, N.; Moreira Mda, C.; Thierer, J.; Grancelli, H.O.; Serrano Junior, C.V.; Albuquerque, D.; et al. I Latin American Guidelines for the Assessment and Management of Decompensated Heart Failure. *Arq. Bras. Cardiol.* **2005**, *85*, 41–94.
12. Sidney, S.; Quesenberry, C.P.; Jaffe, M.G.; Sorel, M.; Nguyen-Huynh, M.N.; Kushi, L.H.; Go, A.S.; Rana, J.S. Recent Trends in Cardiovascular Mortality in the United States and Public Health Goals. *JAMA Cardiol.* **2016**, *1*, 594–599. [CrossRef]
13. Sidney, S.; Quesenberry, C.P.; Jaffe, M.G.; Sorel, M.; Go, A.S.; Rana, J.S. Heterogeneity in national U.S. mortality trends within heart disease subgroups, 2000–2015. *BMC Cardiovasc. Disord.* **2017**, *17*, 192. [CrossRef]
14. Van Nuys, K.E.; Xie, Z.; Tysinger, B.; Hlatky, M.A.; Goldman, D.P. Innovation in Heart Failure Treatment: Life Expectancy, Disability, and Health Disparities. *JACC Heart Fail.* **2018**, *6*, 401–409. [CrossRef]
15. Flather, M.D.; Yusuf, S.; Køber, L.; Pfeffer, M.; Hall, A.; Murray, G.; Torp-Pedersen, C.; Ball, S.; Pogue, J.; Moyé, L.; et al. Long-term ACE-inhibitor therapy in patients with heart failure or left-ventricular dysfunction: A systematic overview of data from individual patients. ACE-Inhibitor Myocardial Infarction Collaborative Group. *Lancet* **2000**, *355*, 1575–1581. [CrossRef]
16. Bristow, M.R.; Gilbert, E.M.; Abraham, W.T.; Adams, K.F.; Fowler, M.B.; Hershberger, R.E.; Kubo, S.H.; Narahara, K.A.; Ingersoll, H.; Krueger, S.; et al. Carvedilol produces dose-related improvements in left ventricular function and survival in subjects with chronic heart failure. MOCHA Investigators. *Circulation* **1996**, *94*, 2807–2816. [CrossRef]
17. Heidenreich, P.A.; Lee, T.T.; Massie, B.M. Effect of beta-blockade on mortality in patients with heart failure: A meta-analysis of randomized clinical trials. *J. Am. Coll. Cardiol.* **1997**, *30*, 27–34. [CrossRef]
18. Choi, K.H.; Lee, G.Y.; Choi, J.O.; Jeon, E.S.; Lee, H.Y.; Lee, S.E.; Kim, J.J.; Chae, S.C.; Baek, S.H.; Kang, S.M.; et al. The mortality benefit of carvedilol versus bisoprolol in patients with heart failure with reduced ejection fraction. *Korean J. Intern. Med.* **2019**, *34*, 1030–1039. [CrossRef]
19. Felker, G.M.; Mentz, R.J.; Adams, K.F.; Cole, R.T.; Egnaczyk, G.F.; Patel, C.B.; Fiuzat, M.; Gregory, D.; Wedge, P.; O'Connor, C.M.; et al. Tolvaptan in Patients Hospitalized with Acute Heart Failure: Rationale and Design of the TACTICS and the SECRET of CHF Trials. *Circ. Heart Fail.* **2015**, *8*, 997–1005. [CrossRef]
20. Konstam, M.A.; Kiernan, M.; Chandler, A.; Dhingra, R.; Mody, F.V.; Eisen, H.; Haught, W.H.; Wagoner, L.; Gupta, D.; Patten, R.; et al. Short-Term Effects of Tolvaptan in Patients with Acute Heart Failure and Volume Overload. *J. Am. Coll. Cardiol.* **2017**, *69*, 1409–1419. [CrossRef]

21. Park, G.H.; Lee, C.M.; Song, J.W.; Jung, M.C.; Kim, J.K.; Song, Y.R.; Kim, H.J.; Kim, S.G. Comparison of tolvaptan treatment between patients with the SIADH and congestive heart failure: A single-center experience. *Korean J. Intern. Med.* **2018**, *33*, 561–567. [CrossRef]
22. Swedberg, K.; Komajda, M.; Böhm, M.; Borer, J.S.; Ford, I.; Dubost-Brama, A.; Lerebours, G.; Tavazzi, L.; SHIFT Investigators. Ivabradine and outcomes in chronic heart failure (SHIFT): A randomised placebo-controlled study. *Lancet* **2010**, *376*, 875–885. [CrossRef]
23. Sabir, F.; Barani, M.; Mukhtar, M.; Rahdar, A.; Cucchiarini, M.; Zafar, M.N.; Behl, T.; Bungau, S. Nanodiagnosis and Nanotreatment of Cardiovascular Diseases: An Overview. *Chemosensors* **2021**, *9*, 67. [CrossRef]
24. Berliner, D.; Bauersachs, J. Current Drug Therapy in Chronic Heart Failure: The New Guidelines of the European Society of Cardiology (ESC). *Korean Circ. J.* **2017**, *47*, 543–554. [CrossRef]
25. McMurray, J.J.; Packer, M.; Desai, A.S.; Gong, J.; Lefkowitz, M.P.; Rizkala, A.R.; Rouleau, J.L.; Shi, V.C.; Solomon, S.D.; Swedberg, K.; et al. Angiotensin-neprilysin inhibition versus enalapril in heart failure. *N. Engl. J. Med.* **2014**, *371*, 993–1004. [CrossRef]
26. Singh, J.S.; Lang, C.C. Angiotensin receptor-neprilysin inhibitors: Clinical potential in heart failure and beyond. *Vasc. Health Risk Manag.* **2015**, *11*, 283–295. [CrossRef]
27. McMurray, J.J.; Packer, M.; Desai, A.S.; Gong, J.; Lefkowitz, M.; Rizkala, A.R.; Rouleau, J.L.; Shi, V.C.; Solomon, S.D.; Swedberg, K.; et al. Baseline characteristics and treatment of patients in prospective comparison of ARNI with ACEI to determine impact on global mortality and morbidity in heart failure trial (PARADIGM-HF). *Eur. J. Heart Fail.* **2014**, *16*, 817–825. [CrossRef]
28. Sardu, C.; Marfella, R.; Santulli, G.; Paolisso, G. Functional role of miRNA in cardiac resynchronization therapy. *Pharmacogenomics* **2014**, *15*, 1159–1168. [CrossRef]
29. Sardu, C.; Massetti, M.; Scisciola, L.; Trotta, M.C.; Santamaria, M.; Volpicelli, M.; Ducceschi, V.; Signoriello, G.; D'Onofrio, N.; Marfella, L.; et al. Angiotensin receptor/Neprilysin inhibitor effects in CRTd non-responders: From epigenetic to clinical beside. *Pharmacol. Res.* **2022**, *182*, 106303. [CrossRef]
30. Jering, K.S.; Claggett, B.; Pfeffer, M.A.; Granger, C.; Køber, L.; Lewis, E.F.; Maggioni, A.P.; Mann, D.; McMurray, J.J.V.; Rouleau, J.L.; et al. Prospective ARNI vs. ACE inhibitor trial to DetermIne Superiority in reducing heart failure Events after Myocardial Infarction (PARADISE-MI): Design and baseline characteristics. *Eur. J. Heart Fail.* **2021**, *23*, 1040–1048. [CrossRef]
31. Solomon, S.D.; McMurray, J.J.V.; PARAGON-HF Steering Committee and Investigators. Angiotensin-Neprilysin Inhibition in Heart Failure with Preserved Ejection Fraction. Reply. *N. Engl. J. Med.* **2020**, *382*, 1182–1183. [CrossRef] [PubMed]
32. Wijkman, M.O.; Claggett, B.; Vaduganathan, M.; Cunningham, J.W.; Rørth, R.; Jackson, A.; Packer, M.; Zile, M.; Rouleau, J.; Swedberg, K.; et al. Effects of sacubitril/valsartan on glycemia in patients with diabetes and heart failure: The PARAGON-HF and PARADIGM-HF trials. *Cardiovasc. Diabetol.* **2022**, *21*, 110. [CrossRef] [PubMed]
33. Katz, A.M.; Rolett, E.L. Heart failure: When form fails to follow function. *Eur. Heart J.* **2016**, *37*, 449–454. [CrossRef] [PubMed]
34. Packer, M.; Lam, C.S.P.; Lund, L.H.; Maurer, M.S.; Borlaug, B.A. Characterization of the inflammatory-metabolic phenotype of heart failure with a preserved ejection fraction: A hypothesis to explain influence of sex on the evolution and potential treatment of the disease. *Eur. J. Heart Fail.* **2020**, *22*, 1551–1567. [CrossRef]
35. McMurray, J.J. Neprilysin inhibition to treat heart failure: A tale of science, serendipity, and second chances. *Eur. J. Heart Fail.* **2015**, *17*, 242–247. [CrossRef]
36. Simmonds, S.J.; Cuijpers, I.; Heymans, S.; Jones, E.A.V. Cellular and Molecular Differences between HFpEF and HFrEF: A Step Ahead in an Improved Pathological Understanding. *Cells* **2020**, *9*, 242. [CrossRef]
37. Dassanayaka, S.; Jones, S.P. Recent Developments in Heart Failure. *Circ. Res.* **2015**, *117*, e58–e63. [CrossRef]
38. Maries, L.; Manitiu, I. Diagnostic and prognostic values of B-type natriuretic peptides (BNP) and N-terminal fragment brain natriuretic peptides (NT-pro-BNP). *Cardiovasc. J. Afr.* **2013**, *24*, 286–289. [CrossRef]
39. Tschöpe, C.; Birner, C.; Böhm, M.; Bruder, O.; Frantz, S.; Luchner, A.; Maier, L.; Störk, S.; Kherad, B.; Laufs, U. Heart failure with preserved ejection fraction: Current management and future strategies: Expert opinion on the behalf of the Nucleus of the "Heart Failure Working Group" of the German Society of Cardiology (DKG). *Clin. Res. Cardiol.* **2018**, *107*, 1–19. [CrossRef]
40. Ferrari, R.; Böhm, M.; Cleland, J.G.; Paulus, W.J.; Pieske, B.; Rapezzi, C.; Tavazzi, L. Heart failure with preserved ejection fraction: Uncertainties and dilemmas. *Eur. J. Heart Fail.* **2015**, *17*, 665–671. [CrossRef]
41. Loffredo, F.S.; Nikolova, A.P.; Pancoast, J.R.; Lee, R.T. Heart failure with preserved ejection fraction: Molecular pathways of the aging myocardium. *Circ. Res.* **2014**, *115*, 97–107. [CrossRef]
42. de Boer, R.A.; Nayor, M.; de Filippi, C.R.; Enserro, D.; Bhambhani, V.; Kizer, J.R.; Blaha, M.J.; Brouwers, F.P.; Cushman, M.; Lima, J.A.C.; et al. Association of Cardiovascular Biomarkers with Incident Heart Failure With Preserved and Reduced Ejection Fraction. *JAMA Cardiol.* **2018**, *3*, 215–224. [CrossRef]
43. Steinmann, E.; Brunner-La Rocca, H.P.; Maeder, M.T.; Kaufmann, B.A.; Pfisterer, M.; Rickenbacher, P. Is the clinical presentation of chronic heart failure different in elderly versus younger patients and those with preserved versus reduced ejection fraction? *Eur. J. Intern. Med.* **2018**, *57*, 61–69. [CrossRef]
44. Yancy, C.W.; Jessup, M.; Bozkurt, B.; Butler, J.; Casey, D.E.; Colvin, M.M.; Drazner, M.H.; Filippatos, G.S.; Fonarow, G.C.; Givertz, M.M.; et al. 2017 ACC/AHA/HFSA Focused Update of the 2013 ACCF/AHA Guideline for the Management of Heart Failure: A Report of the American College of Cardiology/American Heart Association Task Force on Clinical Practice Guidelines and the Heart Failure Society of America. *J. Am. Coll. Cardiol.* **2017**, *70*, 776–803. [CrossRef]

45. Solomon, S.D.; Rizkala, A.R.; Lefkowitz, M.P.; Shi, V.C.; Gong, J.; Anavekar, N.; Anker, S.D.; Arango, J.L.; Arenas, J.L.; Atar, D.; et al. Baseline Characteristics of Patients with Heart Failure and Preserved Ejection Fraction in the PARAGON-HF Trial. *Circ. Heart Fail.* **2018**, *11*, e004962. [CrossRef]
46. McDonagh, T.A.; Metra, M.; Adamo, M.; Gardner, R.S.; Baumbach, A.; Böhm, M.; Burri, H.; Butler, J.; Čelutkiene, J.; Chioncel, O.; et al. 2021 ESC Guidelines for the diagnosis and treatment of acute and chronic heart failure: Developed by the Task Force for the diagnosis and treatment of acute and chronic heart failure of the European Society of Cardiology (ESC) With the special contribution of the Heart Failure Association (HFA) of the ESC. *Eur. Heart J.* **2021**, *42*, 3599–3726. [CrossRef]
47. Reddy, Y.N.V.; Carter, R.E.; Obokata, M.; Redfield, M.M.; Borlaug, B.A. A Simple, Evidence-Based Approach to Help Guide Diagnosis of Heart Failure with Preserved Ejection Fraction. *Circ.* **2018**, *138*, 861–870. [CrossRef]
48. Michel, F.S.; Magubane, M.; Mokotedi, L.; Norton, G.R.; Woodiwiss, A.J. Sex-Specific Effects of Adrenergic-Induced Left Ventricular Remodeling in Spontaneously Hypertensive Rats. *J. Card Fail.* **2017**, *23*, 161–168 [CrossRef]
49. Gibbs, M.; Veliotes, D.G.; Anamourlis, C.; Badenhorst, D.; Osadchii, O.; Norton, G.R.; Woodiwiss, A.J. Chronic beta-adrenoreceptor activation increases cardiac cavity size through chamber remodeling and not via modifications in myocardial material properties. *Am. J. Physiol. Heart Circ. Physiol.* **2004**, *287*, H2762–H2767. [CrossRef]
50. Bolam, H.; Morton, G.; Kalra, P.R. Drug therapies in chronic heart failure: A focus on reduced ejection fraction. *Clin. Med.* **2018**, *18*, 138–145. [CrossRef]
51. Group, C.T.S. Effects of enalapril on mortality in severe congestive heart failure. Results of the Cooperative North Scandinavian Enalapril Survival Study (CONSENSUS). *N. Engl. J. Med.* **1987**, *316*, 1429–1435. [CrossRef]
52. Cohn, J.N.; Archibald, D.G.; Ziesche, S.; Franciosa, J.A.; Harston, W.E.; Tristani, F.E.; Dunkman, W.B.; Jacobs, W.; Francis, G.S.; Flohr, K.H. Effect of vasodilator therapy on mortality in chronic congestive heart failure. Results of a Veterans Administration Cooperative Study. *N. Engl. J. Med.* **1986**, *314*, 1547–1552. [CrossRef] [PubMed]
53. Yusuf, S.; Pitt, B.; Davis, C.E.; Hood, W.B.; Cohn, J.N.; SHIFT Investigators. Effect of enalapril on survival in patients with reduced left ventricular ejection fractions and congestive heart failure. *N. Engl. J. Med.* **1991**, *325*, 293–302. [CrossRef] [PubMed]
54. Garg, R.; Yusuf, S. Overview of randomized trials of angiotensin-converting enzyme inhibitors on mortality and morbidity in patients with heart failure. Collaborative Group on ACE Inhibitor Trials. *JAMA* **1995**, *273*, 1450–1456. [CrossRef]
55. Frigerio, M.; Roubina, E. Drugs for left ventricular remodeling in heart failure. *Am. J. Cardiol.* **2005**, *96*, 10L–18L. [CrossRef] [PubMed]
56. DiBianco, R. ACE inhibitors in the treatment of heart failure. *Clin. Cardiol.* **1990**, *13*, VII32–VII38. [CrossRef] [PubMed]
57. Pahor, M.; Bernabei, R.; Sgadari, A.; Gambassi, G.; Lo Giudice, P.; Pacifici, L.; Ramacci, M.T.; Lagrasta, C.; Olivetti, G.; Carbonin, P. Enalapril prevents cardiac fibrosis and arrhythmias in hypertensive rats. *Hypertension* **1991**, *18*, 148–157. [CrossRef] [PubMed]
58. Goussev, A.; Sharov, V.G.; Shimoyama, H.; Tanimura, M.; Lesch, M.; Goldstein, S.; Sabbah, H.N. Effects of ACE inhibition on cardiomyocyte apoptosis in dogs with heart failure. *Am. J. Physiol.* **1998**, *275*, H626–H631 [CrossRef]
59. Pfeffer, M.A.; Braunwald, E.; Moyé, L.A.; Basta, L.; Brown, E.J.; Cuddy, T.E.; Davis, B.R.; Geltman, E.M.; Goldman, S.; Flaker, G.C. Effect of captopril on mortality and morbidity in patients with left ventricular dysfunction after myocardial infarction. Results of the survival and ventricular enlargement trial. The SAVE Investigators. *N. Engl. J. Med.* **1992**, *327*, 669–677. [CrossRef]
60. Yusuf, S.; Pitt, B.; Davis, C.E.; Hood, W.B.; Cohn, J.N.; SHIFT Investigators. Effect of enalapril on mortality and the development of heart failure in asymptomatic patients with reduced left ventricular ejection fractions. *N. Engl. J. Med.* **1992**, *327*, 685–691. [CrossRef]
61. Køber, L.; Torp-Pedersen, C.; Carlsen, J.E.; Bagger, H.; Eliasen, P.; Lyngborg, K.; Videbaek, J.; Cole, D.S.; Auclert, L.; Pauly, N.C. A clinical trial of the angiotensin-converting-enzyme inhibitor trandolapril in patients with left ventricular dysfunction after myocardial infarction. Trandolapril Cardiac Evaluation (TRACE) Study Group. *N. Engl. J. Med.* **1995**, *333*, 1670–1676. [CrossRef]
62. Yılmaz, İ. Angiotensin-Converting Enzyme Inhibitors Induce Cough. *Turk. Thorac. J.* **2019**, *20*, 36–42. [CrossRef]
63. Heidenreich, P.A.; Bozkurt, B.; Aguilar, D.; Allen, L.A.; Byun, J.J.; Colvin, M.M.; Deswal, A.; Drazner, M.H.; Dunlay, S.M.; Evers, L.R.; et al. 2022 AHA/ACC/HFSA Guideline for the Management of Heart Failure: A Report of the American College of Cardiology/American Heart Association Joint Committee on Clinical Practice Guidelines. *Circulation* **2022**, *145*, e895–e1032. [CrossRef]
64. Brown, T.; Gonzalez, J.; Monteleone, C. Angiotensin-converting enzyme inhibitor-induced angioedema: A review of the literature. *J. Clin. Hypertens.* **2017**, *19*, 1377–1382. [CrossRef]
65. Yancy, C.W.; Jessup, M.; Bozkurt, B.; Butler, J.; Casey, D.E.; Colvin, M.M.; Drazner, M.H.; Filippatos, G.; Fonarow, G.C.; Givertz, M.M.; et al. 2016 ACC/AHA/HFSA Focused Update on New Pharmacological Therapy for Heart Failure: An Update of the 2013 ACCF/AHA Guideline for the Management of Heart Failure: A Report of the American College of Cardiology/American Heart Association Task Force on Clinical Practice Guidelines and the Heart Failure Society of America. *Circulation* **2016**, *134*, e282–e293. [CrossRef]
66. Granger, C.B.; McMurray, J.J.; Yusuf, S.; Held, P.; Michelson, E.L.; Olofsson, B.; Ostergren, J.; Pfeffer, M.A.; Swedberg, K.; CHARM Investigators and Committees. Effects of candesartan in patients with chronic heart failure and reduced left-ventricular systolic function intolerant to angiotensin-converting-enzyme inhibitors: The CHARM-Alternative trial. *Lancet* **2003**, *362*, 772–776. [CrossRef]

67. Marfella, R.; D'Onofrio, N.; Mansueto, G.; Grimaldi, V.; Trotta, M.C.; Sardu, C.; Sasso, F.C.; Scisciola, L.; Amarelli, C.; Esposito, S.; et al. Glycated ACE2 reduces anti-remodeling effects of renin-angiotensin system inhibition in human diabetic hearts. *Cardiovasc. Diabetol.* **2022**, *21*, 146. [CrossRef]
68. The Cardiac Insufficiency Bisoprolol Study II (CIBIS-II): A randomised trial. *Lancet* **1999**, *353*, 9–13. [CrossRef]
69. Flather, M.D.; Shibata, M.C.; Coats, A.J.; Van Veldhuisen, D.J.; Parkhomenko, A.; Borbola, J.; Cohen-Solal, A.; Dumitrascu, D.; Ferrari, R.; Lechat, P.; et al. Randomized trial to determine the effect of nebivolol on mortality and cardiovascular hospital admission in elderly patients with heart failure (SENIORS). *Eur. Heart J.* **2005**, *26*, 215–225. [CrossRef]
70. Krum, H.; Roecker, E.B.; Mohacsi, P.; Rouleau, J.L.; Tendera, M.; Coats, A.J.; Katus, H.A.; Fowler, M.B.; Packer, M.; Carvedilol Prospective Randomized Cumulative Survival (COPERNICUS) Study Group. Effects of initiating carvedilol in patients with severe chronic heart failure: Results from the COPERNICUS Study. *JAMA* **2003**, *289*, 712–718. [CrossRef]
71. Effect of metoprolol CR/XL in chronic heart failure: Metoprolol CR/XL Randomised Intervention Trial in Congestive Heart Failure (MERIT-HF). *Lancet* **1999**, *353*, 2001–2007. [CrossRef]
72. Meuwese, C.L.; Kirkels, J.H.; de Jonge, N.; Nathoe, H.M.; Doevendans, P.A.; Klöpping, C. Beta-blocker therapy in unstable severe heart failure, evidence or experience? *Neth. Heart J.* **2013**, *21*, 3–5. [CrossRef]
73. CIBIS Investigators and Committees. A randomized trial of beta-blockade in heart failure: The Cardiac Insufficiency Bisoprolol Study (CIBIS). *Circulation* **1994**, *90*, 1765–1773. [CrossRef]
74. Zannad, F.; McMurray, J.J.; Krum, H.; van Veldhuisen, D.J.; Swedberg, K.; Shi, H.; Vincent, J.; Pocock, S.J.; Pitt, B.; Group, E.-H.S. Eplerenone in patients with systolic heart failure and mild symptoms. *N. Engl. J. Med.* **2011**, *364*, 11–21. [CrossRef]
75. Pitt, B.; Zannad, F.; Remme, W.J.; Cody, R.; Castaigne, A.; Perez, A.; Palensky, J.; Wittes, J. The effect of spironolactone on morbidity and mortality in patients with severe heart failure. Randomized Aldactone Evaluation Study Investigators. *N. Engl. J. Med.* **1999**, *341*, 709–717. [CrossRef]
76. Chaplin, S. Chronic Heart Failure in Adults: Diagnosis and Management. *Prescriber* **2019**, *30*, 16–18. [CrossRef]
77. Pitt, B.; Remme, W.; Zannad, F.; Neaton, J.; Martinez, F.; Roniker, B.; Bittman, R.; Hurley, S.; Kleiman, J.; Gatlin, M.; et al. Eplerenone, a selective aldosterone blocker, in patients with left ventricular dysfunction after myocardial infarction. *N. Engl. J. Med.* **2003**, *348*, 1309–1321. [CrossRef]
78. Struthers, A.; Krum, H.; Williams, G.H. A comparison of the aldosterone-blocking agents eplerenone and spironolactone. *Clin. Cardiol.* **2008**, *31*, 153–158. [CrossRef]
79. Bloch, M.J.; Basile, J.N. Spironolactone is more effective than eplerenone at lowering blood pressure in patients with primary aldosteronism. *J. Clin. Hypertens.* **2011**, *13*, 629–631. [CrossRef] [PubMed]
80. McMurray, J.J.V.; DeMets, D.L.; Inzucchi, S.E.; Køber, L.; Kosiborod, M.N.; Langkilde, A.M.; Martinez, F.A.; Bengtsson, O.; Ponikowski, P.; Sabatine, M.S.; et al. A trial to evaluate the effect of the sodium-glucose co-transporter 2 inhibitor dapagliflozin on morbidity and mortality in patients with heart failure and reduced left ventricular ejection fraction (DAPA-HF). *Eur. J. Heart Fail.* **2019**, *21*, 665–675. [CrossRef]
81. McMurray, J.J.V.; Solomon, S.D.; Inzucchi, S.E.; Køber, L.; Kosiborod, M.N.; Martinez, F.A.; Ponikowski, P.; Sabatine, M.S.; Anand, I.S.; Bělohlávek, J.; et al. Dapagliflozin in Patients with Heart Failure and Reduced Ejection Fraction. *N. Engl. J. Med.* **2019**, *381*, 1995–2008. [CrossRef] [PubMed]
82. Bayés-Genís, A. Neprilysin in Heart Failure: From Oblivion to Center Stage. *JACC Heart Fail.* **2015**, *3*, 637–640. [CrossRef] [PubMed]
83. Januzzi, J.L.; Butler, J.; Fombu, E.; Maisel, A.; McCague, K.; Piña, I.L.; Prescott, M.F.; Riebman, J.B.; Solomon, S. Rationale and methods of the Prospective Study of Biomarkers, Symptom Improvement, and Ventricular Remodeling During Sacubitril/Valsartan Therapy for Heart Failure (PROVE-HF). *Am. Heart J.* **2018**, *199*, 130–136. [CrossRef] [PubMed]
84. Kang, D.H.; Park, S.J.; Shin, S.H.; Hong, G.R.; Lee, S.; Kim, M.S.; Yun, S.C.; Song, J.M.; Park, S.W.; Kim, J.J. Angiotensin Receptor Neprilysin Inhibitor for Functional Mitral Regurgitation. *Circulation* **2019**, *139*, 1354–1365. [CrossRef] [PubMed]
85. Bunsawat, K.; Ratchford, S.M.; Alpenglow, J.K.; Park, S.H.; Jarrett, C.L.; Stehlik, J.; Smith, A.S.; Richardson, R.S.; Wray, D.W. Sacubitril-valsartan improves conduit vessel function and functional capacity and reduces inflammation in heart failure with reduced ejection fraction. *J. Appl. Physiol.* **2021**, *130*, 256–268. [CrossRef]
86. Jia, G.; Aroor, A.R.; Hill, M.A.; Sowers, J.R. Role of Renin-Angiotensin-Aldosterone System Activation in Promoting Cardiovascular Fibrosis and Stiffness. *Hypertension* **2018**, *72*, 537–548. [CrossRef]
87. Pugliese, N.R.; Masi, S.; Taddei, S. The renin-angiotensin-aldosterone system: A crossroad from arterial hypertension to heart failure. *Heart Fail. Rev.* **2020**, *25*, 31–42. [CrossRef]
88. Cleland, J.G.; Tendera, M.; Adamus, J.; Freemantle, N.; Polonski, L.; Taylor, J.; PEP-CHF Investigators. The perindopril in elderly people with chronic heart failure (PEP-CHF) study. *Eur. Heart J.* **2006**, *27*, 2338–2345. [CrossRef]
89. Massie, B.M.; Carson, P.E.; McMurray, J.J.; Komajda, M.; McKelvie, R.; Zile, M.R.; Anderson, S.; Donovan, M.; Iverson, E.; Staiger, C.; et al. Irbesartan in patients with heart failure and preserved ejection fraction. *N. Engl. J. Med.* **2008**, *359*, 2456–2467. [CrossRef]
90. Rector, T.S.; Carson, P.E.; Anand, I.S.; McMurray, J.J.; Zile, M.R.; McKelvie, R.S.; Komajda, M.; Kuskowski, M.; Massie, B.M.; I-PRESERVE Trial Investigators. Assessment of long-term effects of irbesartan on heart failure with preserved ejection fraction as measured by the minnesota living with heart failure questionnaire in the irbesartan in heart failure with preserved systolic function (I-PRESERVE) trial. *Circ. Heart Fail.* **2012**, *5*, 217–225. [CrossRef]

91. Yusuf, S.; Pfeffer, M.A.; Swedberg, K.; Granger, C.B.; Held, P.; McMurray, J.J.; Michelson, E.L.; Olofsson, B.; Ostergren, J.; CHARM Investigators and Committees. Effects of candesartan in patients with chronic heart failure and preserved left-ventricular ejection fraction: The CHARM-Preserved Trial. *Lancet* **2003**, *362*, 777–781. [CrossRef]
92. Simpson, J.; Castagno, D.; Doughty, R.N.; Poppe, K.K.; Earle, N.; Squire, I.; Richards, M.; Andersson, B.; Ezekowitz, J.A.; Komajda, M.; et al. Is heart rate a risk marker in patients with chronic heart failure and concomitant atrial fibrillation? Results from the MAGGIC meta-analysis. *Eur. J. Heart Fail.* **2015**, *17*, 1182–1191. [CrossRef]
93. Yanagihara, K.; Kinugasa, Y.; Sugihara, S.; Hirai, M.; Yamada, K.; Ishida, K.; Kato, M.; Yamamoto, K. Discharge use of carvedilol is associated with higher survival in Japanese elderly patients with heart failure regardless of left ventricular ejection fraction. *J. Cardiovasc. Pharmacol.* **2013**, *62*, 485–490. [CrossRef]
94. Edelmann, F.; Wachter, R.; Schmidt, A.G.; Kraigher-Krainer, E.; Colantonio, C.; Kamke, W.; Duvinage, A.; Stahrenberg, R.; Durstewitz, K.; Löffler, M.; et al. Effect of spironolactone on diastolic function and exercise capacity in patients with heart failure with preserved ejection fraction: The Aldo-DHF randomized controlled trial. *JAMA* **2013**, *309*, 781–791. [CrossRef]
95. Pitt, B.; Pfeffer, M.A.; Assmann, S.F.; Boineau, R.; Anand, I.S.; Claggett, B.; Clausell, N.; Desai, A.S.; Diaz, R.; Fleg, J.L.; et al. Spironolactone for heart failure with preserved ejection fraction. *N. Engl. J. Med.* **2014**, *370*, 1383–1392. [CrossRef]
96. Greenberg, B. Angiotensin Receptor-Neprilysin Inhibition (ARNI) in Heart Failure. *Int J. Heart Fail.* **2020**, *2*, 73–90. [CrossRef]
97. Solomon, S.D.; Zile, M.; Pieske, B.; Voors, A.; Shah, A.; Kraigher-Krainer, E.; Shi, V.; Bransford, T.; Takeuchi, M.; Gong, J.; et al. The angiotensin receptor neprilysin inhibitor LCZ696 in heart failure with preserved ejection fraction: A phase 2 double-blind randomised controlled trial. *Lancet* **2012**, *380*, 1387–1395. [CrossRef]
98. Solomon, S.D.; McMurray, J.J.V.; Anand, I.S.; Ge, J.; Lam, C.S.P.; Maggioni, A.P.; Martinez, F.; Packer, M.; Pfeffer, M.A.; Pieske, B.; et al. Angiotensin-Neprilysin Inhibition in Heart Failure with Preserved Ejection Fraction. *N. Engl. J. Med.* **2019**, *381*, 1609–1620. [CrossRef]
99. McMurray, J.J.V.; Jackson, A.M.; Lam, C.S.P.; Redfield, M.M.; Anand, I.S.; Ge, J.; Lefkowitz, M.P.; Maggioni, A.P.; Martinez, F.; Packer, M.; et al. Effects of Sacubitril-Valsartan Versus Valsartan in Women Compared with Men with Heart Failure and Preserved Ejection Fraction: Insights From PARAGON-HF. *Circulation* **2020**, *141*, 338–351. [CrossRef]
100. Regitz-Zagrosek, V. Sex and Gender Differences in Heart Failure. *Int J. Heart Fail.* **2020**, *2*, 157–181. [CrossRef]
101. Rotariu, D.; Babes, E.E.; Tit, D.M.; Moisi, M.; Bustea, C.; Stoicescu, M.; Radu, A.-F.; Vesa, C.M.; Behl, T.; Bungau, A.F.; et al. Oxidative stress—Complex pathological issues concerning the hallmark of cardiovascular and metabolic disorders. *Biomed. Pharmacother.* **2022**, *152*, 113238. [CrossRef]
102. Desai, A.S.; McMurray, J.J.; Packer, M.; Swedberg, K.; Rouleau, J.L.; Chen, F.; Gong, J.; Rizkala, A.R.; Brahimi, A.; Claggett, B.; et al. Effect of the angiotensin-receptor-neprilysin inhibitor LCZ696 compared with enalapril on mode of death in heart failure patients. *Eur. Heart J.* **2015**, *36*, 1990–1997. [CrossRef]
103. Velazquez, E.J.; Morrow, D.A.; DeVore, A.D.; Duffy, C.I.; Ambrosy, A.P.; McCague, K.; Rocha, R.; Braunwald, E.; PIONEER-HF Investigators. Angiotensin-Neprilysin Inhibition in Acute Decompensated Heart Failure. *N. Engl. J. Med.* **2019**, *380*, 539–548. [CrossRef]
104. Senni, M.; McMurray, J.J.; Wachter, R.; McIntyre, H.F.; Reyes, A.; Majercak, I.; Andreka, P.; Shehova-Yankova, N.; Anand, I.; Yilmaz, M.B.; et al. Initiating sacubitril/valsartan (LCZ696) in heart failure: Results of TITRATION, a double-blind, randomized comparison of two uptitration regimens. *Eur. J. Heart Fail.* **2016**, *18*, 1193–1202. [CrossRef]
105. Januzzi, J.L.; Prescott, M.F.; Butler, J.; Felker, G.M.; Maisel, A.S.; McCague, K.; Camacho, A.; Piña, I.L.; Rocha, R.A.; Shah, A.M.; et al. Association of Change in N-Terminal Pro-B-Type Natriuretic Peptide Following Initiation of Sacubitril-Valsartan Treatment With Cardiac Structure and Function in Patients With Heart Failure With Reduced Ejection Fraction. *JAMA* **2019**, *322*, 1085–1095. [CrossRef]
106. Desai, A.S.; Solomon, S.D.; Shah, A.M.; Claggett, B.L.; Fang, J.C.; Izzo, J.; McCague, K.; Abbas, C.A.; Rocha, R.; Mitchell, G.F.; et al. Effect of Sacubitril-Valsartan vs Enalapril on Aortic Stiffness in Patients with Heart Failure and Reduced Ejection Fraction: A Randomized Clinical Trial. *JAMA* **2019**, *322*, 1077–1084. [CrossRef]
107. Velazquez, E.J.; Morrow, D.A.; DeVore, A.D.; Ambrosy, A.P.; Duffy, C.I.; McCague, K.; Hernandez, A.F.; Rocha, R.A.; Braunwald, E. Rationale and design of the comParIson of sacubitril/valsartaN versus Enalapril on Effect on nt-pRo-bnp in patients stabilized from an acute Heart Failure episode (PIONEER-HF) trial. *Am. Heart J.* **2018**, *198*, 145–151. [CrossRef]
108. Wachter, R.; Senni, M.; Belohlavek, J.; Straburzynska-Migaj, E.; Witte, K.K.; Kobalava, Z.; Fonseca, C.; Goncalvesova, E.; Cavusoglu, Y.; Fernandez, A.; et al. Initiation of sacubitril/valsartan in haemodynamically stabilised heart failure patients in hospital or early after discharge: Primary results of the randomised TRANSITION study. *Eur. J. Heart Fail.* **2019**, *21*, 998–1007. [CrossRef]
109. Carnicelli, A.P.; Li, Z.; Greiner, M.A.; Lippmann, S.J.; Greene, S.J.; Mentz, R.J.; Hardy, N.C.; Blumer, V.; Shen, X.; Yancy, C.W.; et al. Sacubitril/Valsartan Adherence and Postdischarge Outcomes Among Patients Hospitalized for Heart Failure with Reduced Ejection Fraction. *JACC Heart Fail.* **2021**, *9*, 876–886. [CrossRef]
110. Pieske, B.; Wachter, R.; Shah, S.J.; Baldridge, A.; Szeczoedy, P.; Ibram, G.; Shi, V.; Zhao, Z.; Cowie, M.R.; PARALLAX Investigators and Committee members; et al. Effect of Sacubitril/Valsartan vs Standard Medical Therapies on Plasma NT-proBNP Concentration and Submaximal Exercise Capacity in Patients with Heart Failure and Preserved Ejection Fraction: The PARALLAX Randomized Clinical Trial. *JAMA* **2021**, *326*, 1919–1929. [CrossRef] [PubMed]
111. Yuheng, J.; Yanyan, L.; Song, Z.; Yafang, Z.; Xiaowei, M.; Jiayan, Z. The effects of sacubitril/valsartan on heart failure with preserved ejection fraction: A meta-analysis. *Acta Cardiol.* **2021**, 1–9. [CrossRef] [PubMed]

112. Zhang, W.; Liu, J.; Fu, Y.; Ji, H.; Fang, Z.; Zhou, W.; Fan, H.; Zhang, Y.; Liao, Y.; Yang, T.; et al. Sacubitril/Valsartan Reduces Fibrosis and Alleviates High-Salt Diet-Induced HFpEF in Rats. *Front. Pharmacol.* **2020**, *11*, 600953. [CrossRef] [PubMed]
113. Bjork, J.B.; Alton, K.K.; Georgiopoulou, V.V.; Butler, J.; Kalogeropoulos, A.P. Defining Advanced Heart Failure: A Systematic Review of Criteria Used in Clinical Trials. *J. Card Fail.* **2016**, *22*, 569–577. [CrossRef] [PubMed]
114. Xanthakis, V.; Enserro, D.M.; Larson, M.G.; Wollert, K.C.; Januzzi, J.L.; Levy, D.; Aragam, J.; Benjamin, E.J.; Cheng, S.; Wang, T.J.; et al. Prevalence, Neurohormonal Correlates, and Prognosis of Heart Failure Stages in the Community. *JACC Heart Fail.* **2016**, *4*, 808–815. [CrossRef]
115. Crespo-Leiro, M.G.; Metra, M.; Lund, L.H.; Milicic, D.; Costanzo, M.R.; Filippatos, G.; Gustafsson, F.; Tsui, S.; Barge-Caballero, E.; De Jonge, N.; et al. Advanced heart failure: A position statement of the Heart Failure Association of the European Society of Cardiology. *Eur. J. Heart Fail.* **2018**, *20*, 1505–1535. [CrossRef]
116. Gentile, P.; Cantone, R.; Perna, E.; Ammirati, E.; Varrenti, M.; D'Angelo, L.; Verde, A.; Foti, G.; Masciocco, G.; Garascia, A.; et al. Haemodynamic effects of sacubitril/valsartan in advanced heart failure. *ESC Heart Fail.* **2022**, *9*, 894–904. [CrossRef]
117. Samsky Marc, D.; Sen, S. Sacubitril/Valsartan in Advanced Heart Failure. *JACC Heart Fail.* **2022**, *10*, 457–458. [CrossRef]
118. Mann, D.L.; Givertz, M.M.; Vader, J.M.; Starling, R.C.; Shah, P.; McNulty, S.E.; Anstrom, K.J.; Margulies, K.B.; Kiernan, M.S.; Mahr, C.; et al. Effect of Treatment with Sacubitril/Valsartan in Patients With Advanced Heart Failure and Reduced Ejection Fraction: A Randomized Clinical Trial. *JAMA Cardiol.* **2022**, *7*, 17–25. [CrossRef]
119. Changes in NT-proBNP, Safety, and Tolerability in HFpEF Patients with a WHF Event (HFpEF Decompensation) Who Have Been Stabilized and Initiated at the Time of or within 30 Days Post-Decompensation (PARAGLIDE-HF). Available online: https://clinicaltrials.gov/ct2/show/NCT03988634 (accessed on 19 January 2022).
120. Influence of Sacubitril/Valsartan on Autonomic Cardiac Nervous System in Heart Failure Patients: An Exploratory Study. Available online: https://clinicaltrials.gov/ct2/show/NCT04587947 (accessed on 11 January 2022).
121. The Effects of Sacubitril/Valsartan on Cardiac Oxygen Consumption and Efficiency of Cardiac Work in Heart Failure Patients (TurkuPET). Available online: https://clinicaltrials.gov/ct2/show/NCT03300427 (accessed on 12 January 2022).
122. ARNI Versus plAcebo in Patients with Congenital sYStemic Right Ventricle Heart Failure (PARACYS-RV). Available online: https://clinicaltrials.gov/ct2/show/record/NCT05117736 (accessed on 21 January 2022).
123. Role of ARNi in Ventricular Remodeling in Hypertensive LVH (REVERSE-LVH). Available online: https://clinicaltrials.gov/ct2/show/NCT03553810 (accessed on 16 February 2022).
124. Efficacy of a Streamlined Heart Failure Optimization Protocol (SHORT). Available online: https://clinicaltrials.gov/ct2/show/NCT05021419 (accessed on 24 January 2022).
125. Program of Angiotensin-Neprilysin Inhibition in Admitted Patients with Worsening Heart Failure (PREMIER). Available online: https://clinicaltrials.gov/ct2/show/NCT05164653 (accessed on 26 March 2022).

International Journal of
Molecular Sciences

MDPI

Article

Nicotinamide Mononucleotide Prevents Retinal Dysfunction in a Mouse Model of Retinal Ischemia/Reperfusion Injury

Deokho Lee [1,2], Yohei Tomita [1,2], Yukihiro Miwa [1,2,3], Ari Shinojima [1,2], Norimitsu Ban [2], Shintaro Yamaguchi [4], Ken Nishioka [4], Kazuno Negishi [2], Jun Yoshino [4] and Toshihide Kurihara [1,2,*]

1 Laboratory of Photobiology, Keio University School of Medicine, Shinjuku-ku, Tokyo 160-8582, Japan
2 Department of Ophthalmology, Keio University School of Medicine, Shinjuku-ku, Tokyo 160-8582, Japan
3 Aichi Animal Eye Clinic, Nagoya 466-0827, Japan
4 Department of Internal Medicine, Keio University School of Medicine, Shinjuku-ku, Tokyo 160-8582, Japan
* Correspondence: kurihara@z8.keio.jp

Abstract: Retinal ischemia/reperfusion (I/R) injury can cause severe vision impairment. Retinal I/R injury is associated with pathological increases in reactive oxygen species and inflammation, resulting in retinal neuronal cell death. To date, effective therapies have not been developed. Nicotinamide mononucleotide (NMN), a key nicotinamide adenine dinucleotide (NAD^+) intermediate, has been shown to exert neuroprotection for retinal diseases. However, it remains unclear whether NMN can prevent retinal I/R injury. Thus, we aimed to determine whether NMN therapy is useful for retinal I/R injury-induced retinal degeneration. One day after NMN intraperitoneal (IP) injection, adult mice were subjected to retinal I/R injury. Then, the mice were injected with NMN once every day for three days. Electroretinography and immunohistochemistry were used to measure retinal functional alterations and retinal inflammation, respectively. The protective effect of NMN administration was further examined using a retinal cell line, 661W, under $CoCl_2$-induced oxidative stress conditions. NMN IP injection significantly suppressed retinal functional damage, as well as inflammation. NMN treatment showed protective effects against oxidative stress-induced cell death. The antioxidant pathway (*Nrf2* and *Hmox-1*) was activated by NMN treatment. In conclusion, NMN could be a promising preventive neuroprotective drug for ischemic retinopathy.

Keywords: nicotinamide mononucleotide; retinal ischemia/reperfusion; oxidative stress; neuroprotection; inflammation

Citation: Lee, D.; Tomita, Y.; Miwa, Y.; Shinojima, A.; Ban, N.; Yamaguchi, S.; Nishioka, K.; Negishi, K.; Yoshino, J.; Kurihara, T. Nicotinamide Mononucleotide Prevents Retinal Dysfunction in a Mouse Model of Retinal Ischemia/Reperfusion Injury. *Int. J. Mol. Sci.* **2022**, *23*, 11228. https://doi.org/10.3390/ijms231911228

Academic Editors: Simona Gabriela Bungau and Vesa Cosmin

Received: 11 September 2022
Accepted: 22 September 2022
Published: 23 September 2022

1. Introduction

Retinal ischemia/reperfusion (I/R) injury is involved in various retinal ischemic diseases, including diabetic retinopathy, glaucoma, and vascular ischemic retinopathy. Retinal I/R injury can cause pathological events, such as the induction of reactive oxygen species and retinal inflammation, ultimately leading to retinal neuronal cell death [1–3]. As effective therapeutics have not yet been found or developed for retinal I/R injury, research on searching for promising neuroprotective drugs to prevent and/or suppress retinal I/R injury has been attempted at the preclinical stage [4,5].

Nicotinamide mononucleotide (NMN) is one of the major intermediates of nicotinamide adenine dinucleotide (NAD^+), a crucial co-enzyme for various cellular redox metabolisms, including cellular proliferation, DNA repair, and cell death/survival [6,7]. In this aspect, boosting NAD^+ biosynthesis has been nominated as beneficial for age-related metabolic disorders and diseases, including obesity, insulin resistance, and diabetes [8]. When it comes to the eye, recent studies have demonstrated that NMN administration could protect against photoreceptor cell damage in rodent models of light-induced retinopathy or retinal detachment [9,10]. However, little is known about the preventive role of NMN for retinal I/R injury in mice.

Thus, in the present study, we aimed to determine whether NMN treatment could exert retinal protection in a mouse model of retinal I/R injury induced by acute transient elevation of intraocular pressure. Furthermore, we attempted to investigate how NMN treatment could show therapeutic effects using a retinal 661W in vitro system.

2. Results

2.1. NMN Treatment Prevents Retinal Dysfunction in a Mouse Model of Retinal I/R Injury Induced by Acute Elevation of Intraocular Pressure

To examine whether NMN treatment could prevent retinal dysfunction in a mouse model of retinal I/R injury, NMN was intraperitoneally injected into the mice one day before retinal I/R injury (Figure 1A). After injury, NMN was continuously intraperitoneally injected into the mice every day according to our experimental scheme. The dose of NMN was determined based on previous studies that evaluated various effects of NMN treatment [9–12]. We found that the amplitude of b-wave significantly decreased five days after retinal I/R injury, and its reduction was lessened by NMN treatment (Figure 1B).

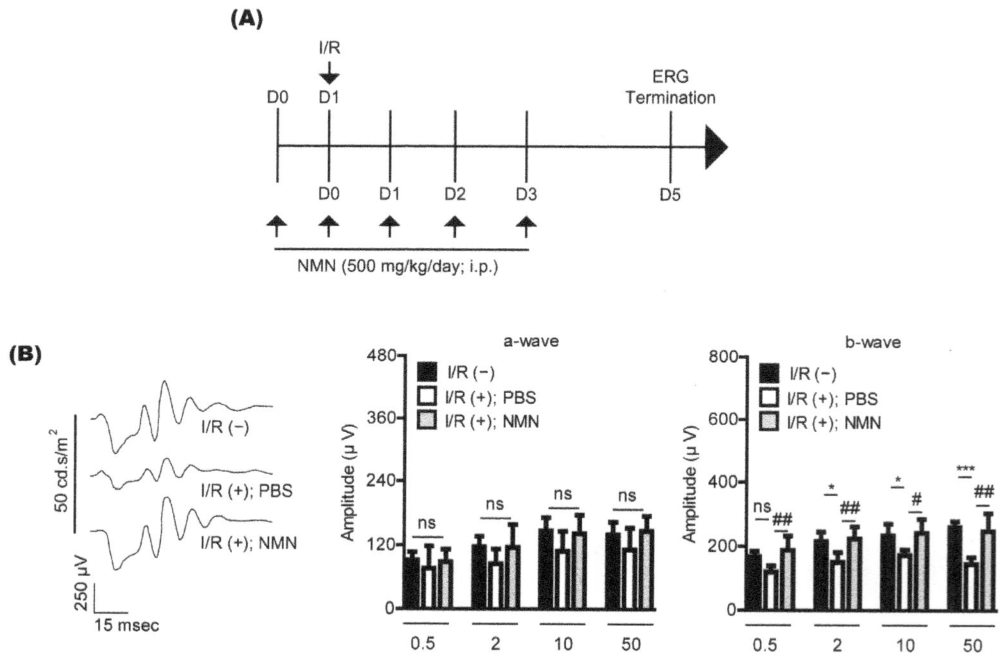

Figure 1. Retinal functional changes by nicotinamide mononucleotide (NMN) treatment in mice. (**A**) The schematic illustration shows whole experimental plans for NMN injection, retinal ischemia/reperfusion (I/R) injury, and the termination day of the whole experiment. i.p., intraperitoneal; D, day; ERG, electroretinography. (**B**) Representative waveforms (50 cd·s/m^2) of ERG (n = 5–6 per group) demonstrated that the ERG amplitudes decreased 5 days after retinal I/R injury. NMN injection suppressed reductions in the ERG amplitude (especially b-wave), flashed with various intensities (0.5, 2, 10, or 50 cd·s/m^2). * $p < 0.05$, *** $p < 0.001$, # $p < 0.05$, and ## $p < 0.01$. ns, not significant. The data were analyzed using one-way ANOVA followed by a Bonferroni post-hoc test. The data are drawn as the mean ± standard deviation.

2.2. NMN Treatment Reduces Retinal Inflammation in a Mouse Model of Retinal I/R Injury Induced by Acute Elevation of Intraocular Pressure

Previously, we found pathologic increases in isolectin GS-IB4 (IB4)-positive inflammatory cells in the retina five days after retinal I/R injury [13]. Therefore, we examined whether the likelihood of its occurrence could be lessened by NMN treatment (Figure 2A,B). In our current system, the number of IB4-positive inflammatory cells in the ischemic retina was markedly reduced by NMN treatment.

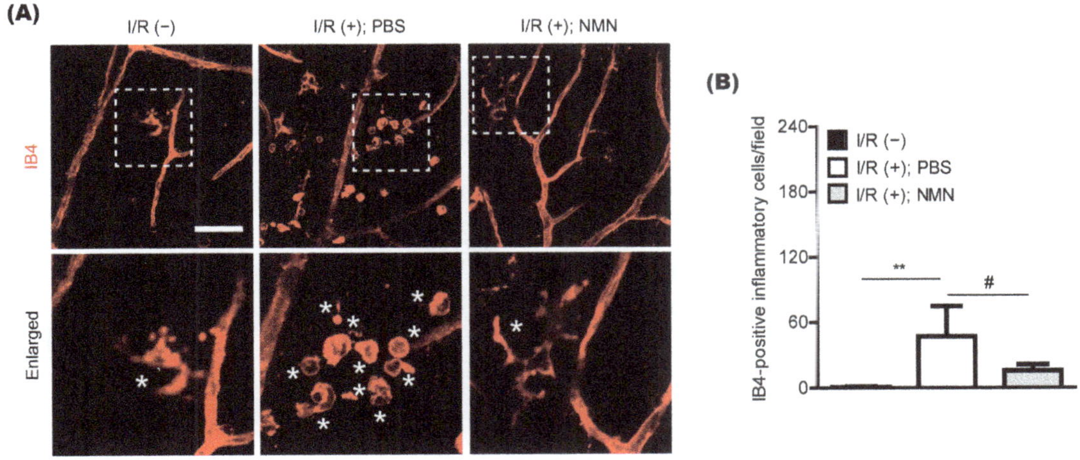

Figure 2. General modulation of retinal inflammation by nicotinamide mononucleotide (NMN) treatment in mice. (**A**,**B**) Representative pictures and quantitative analyses (n = 5–6 per group) demonstrated that the number of retinal I/R injury-induced IB4 positive inflammatory cells in the retina was reduced by NMN injection 5 days after injury. Scale bar: 50 μm. White boxes, enlarged pictures; Stars (*) in enlarged pictures, IB4 positive inflammatory cells. ** $p < 0.01$ and # $p < 0.05$. One-way ANOVA followed by a Bonferroni post-hoc test was used for the data analysis. Bar graphs are shown as the mean ± standard deviation. IB4, isolectin GS-IB4 from *Griffonia simplicifolia*.

2.3. *NMN Treatment Exerts Neuroprotection in Retinal 661W Cells under $CoCl_2$-Induced Oxidative Stress Conditions*

To further determine whether NMN treatment could show protective effects in retinal neuronal cells, we applied an in vitro system to our current study (Figure 3). 661W cells, a mouse-derived photoreceptor cell line possessing features of retinal ganglion precursor-like cells, were used, as they are generally used as one of the in vitro models for studying retinal degeneration [14–16]. Previously, we found that $CoCl_2$-induced oxidative stress could cause 661W cell death [17]. In our current system, we also found 661W cell death by $CoCl_2$-induced oxidative stress, analyzed with TUNEL assay (Figure 3A,B). NMN treatment significantly reduced oxidative stress-induced 661W cell death. This outcome was further confirmed with MTT assay (Figure 3C).

2.4. *NMN Treatment Upregulates the Antioxidant Genes in Retinal 661W Cells under $CoCl_2$-Induced Oxidative Stress Conditions*

NMN treatment has been shown to increase antioxidant genes such as *Nrf2* and *Hmox-1* in various cell types [9–11,18,19]. Thus, we applied its concept to our current system (Figure 4). Under the same conditions of $CoCl_2$-induced oxidative stress above, we found that the expression of *Nrf2* and *Hmox-1* mRNA was markedly increased by NMN treatment in 661W cells (Figure 4A,B).

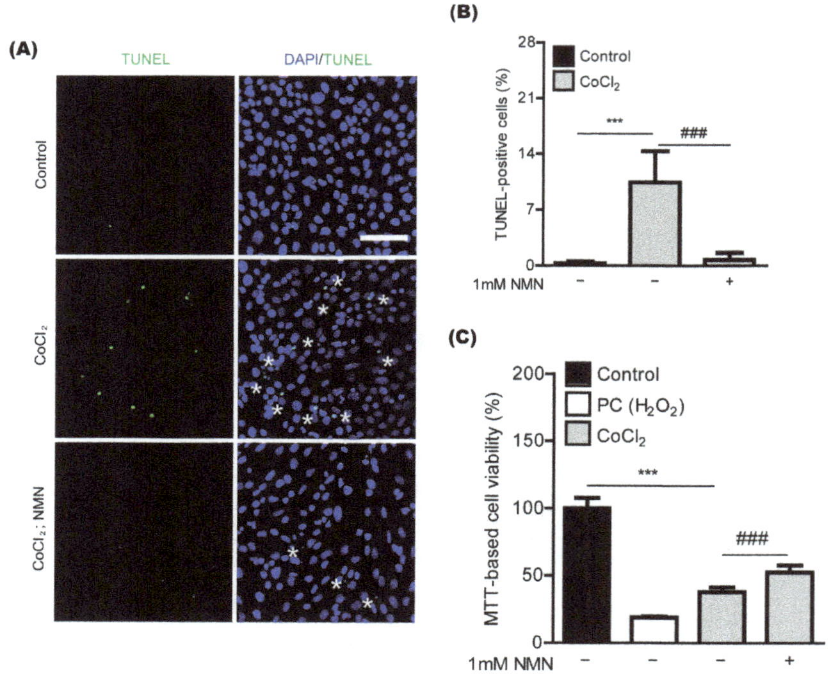

Figure 3. Neuroprotection in vitro by nicotinamide mononucleotide (NMN) treatment. (**A**,**B**) Representative pictures and quantitative analyses (n = 5 per group) demonstrated that the amount of oxidative stress-induced 661W cell death stained by TUNEL assay was reduced by 1 mM NMN treatment after 24 h of 400 µM of $CoCl_2$ incubation. Scale bar: 100 µm. DAPI (blue); TUNEL (green); Stars (*) in pictures, TUNEL positive cells. *** $p < 0.001$ and ### $p < 0.001$. One-way ANOVA followed by a Bonferroni post-hoc test was used for the data analysis. Graphs are depicted as the mean ± standard deviation. (**C**) Quantitative analyses (n = 9 per group) demonstrated that MTT-based 661W cell viability increased by 1 mM NMN treatment after 10 h of 400 µM of $CoCl_2$ incubation. *** $p < 0.001$ and ### $p < 0.001$. One-way ANOVA followed by a Bonferroni post-hoc test was used for the data analysis. Bar graphs are shown as the mean ± standard deviation. PC, positive control (250 µM of H_2O_2).

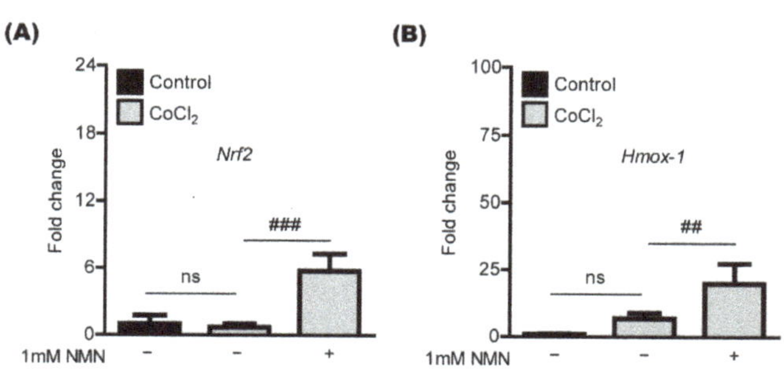

Figure 4. Antioxidant gene regulation by nicotinamide mononucleotide (NMN) treatment. (**A**,**B**) Quantitative analyses (n = 5 per group) demonstrated that NMN treatment (6 h) increases mRNA expression in the antioxidant genes (*Nrf2* and *Hmox-1*) in 661W cells under 400 µM $CoCl_2$-induced oxidative stress conditions. ## $p < 0.01$ and ### $p < 0.001$. ns, not significant. The data were analyzed using one-way ANOVA followed by a Bonferroni post-hoc test and drawn as the mean ± standard deviation.

3. Discussion

The current study demonstrated that retinal neuronal cells could be damaged by retinal I/R injury, and consecutive NMN treatment could show preventive effects against such injury. The therapeutic effects of NMN were further examined using the 661W cell line. Oxidative stress-induced 661W cell death was markedly decreased by NMN treatment, with increases in antioxidant gene expression (*Nrf2* and *Hmox-1*). Although NMN therapy has gradually been reported to be beneficial for eye diseases [20,21], as far as we know, this report is the first to expand the role of NMN in a mouse model of retinal I/R injury induced by acute increases in intraocular pressure.

NMN is a key intermediate of NAD^+, an important metabolic redox co-enzyme in most eukaryotic cells. NMN supplements could be involved in various biological processes such as aging, cell growth, cell death and protection, and DNA repair [22,23]. To date, the therapeutic outcomes of treating NMN or targeting its related NAD^+ pathways have been unraveled in various experimental models of metabolic diseases and disorders, including diabetes, obesity, ischemia/reperfusion injury, heart failure, vascular dysfunction, hemorrhage, cognitive dysfunction, kidney injury, and alcoholic liver disease [22–26]. This indicates that NMN and its related NAD^+ pathways might be crucial for regulating multiple cellular functions in the body.

Based on our current data, retinal I/R-induced retinal dysfunction and oxidative stress-induced cell death were reduced by NMN treatment. Other groups have reported similar effects. Chen et al. demonstrated that NMN supplementation reduces photoreceptor cell loss in the early phase of retinal detachment [10]. They further found that NMN treatment exerts neuroprotection in 661W cells under tBuOOH-induced oxidative stress conditions (tBuOOH; one of the most widely used inducers of oxidative stress [27–29]). Lin et al. showed that NMN treatment protects against retinal dysfunction in mice lacking nicotinamide phosphoribosyltransferase (*Nampt*; a rate-limiting enzyme in the NAD^+ salvage biosynthesis pathway [30,31]) and mice with light-induced retinopathy [9]. They further found that NAMPT inhibitor FK866-induced cell death was decreased by NMN treatment in 661W cells. In this regard, our current results are consistent with those of other previous reports, which implies that NMN could have therapeutic effects against retinal cell damage.

Previous reports from us and others have suggested that reductions in pathologic retinal inflammatory cells could be beneficial for retinal neuronal protection in a murine model of retinal I/R injury [13,32,33]. The therapeutic strategy of modulating pathologic inflammatory cells for neuroprotection can be seen in brain injury, rather than just retinal I/R injury [34–36]. Our current system detected reductions in pathologic retinal inflammatory cells in the NMN-treated retinas. Chen et al. also found that NMN supplementation could suppress retinal inflammation in retinal detachment [10]. Wei et al. demonstrated that NMN treatment could suppress neuroinflammation to exert neuroprotection in intracerebral hemorrhage [18]. Although the way in which NMN treatment is involved in modulating inflammation requires further investigation, its effect might contribute to retinal protection against retinal I/R injury.

NMN is widely known to be associated with antioxidants to protect cells against free radicals [37]. Based on previous reports, the antioxidant function of NMN has been linked to the *Nrf2/Hmox-1;Ho-1* antioxidant signaling pathway [9–11,18,19,38]. Pu et al. demonstrated that NMN treatment could increase cell viability and restore tight junctions in human corneal epithelial cells through the *Nrf2/Hmox-1* pathway [19]. Luo et al. showed that NMN administration could restore redox homeostasis in a murine model of oxidative stress-induced liver injury through the *Nrf2* pathway [39]. Wei et al. showed that NMN treatment could attenuate brain damage induced by intracerebral hemorrhage through the *Nrf2/Hmox-1* pathway [18]. Chen et al. demonstrated that NMN treatment could increase HO-1 protein expression in 661W cells under tBuOOH-induced oxidative stress conditions [10]. Taken together, our current data also support the notion that the therapeutic role of NMN could be involved in the *Nrf2/Hmox-1* antioxidant signaling pathway in various eukaryotic cells.

In summary, we applied the promising NMN therapy to retinal I/R injury in the current study. We further found that NMN treatment could protect against oxidative stress-induced retinal cell death and could upregulate the antioxidant pathway (*Nrf2* and *Hmox-1*) in retinal cells. Although more studies (such as unraveling the clear mode of action of NMN treatment in ischemic eyes, including the retina, choroid, and retinal pigment epithelium) are needed, we suggest a promising NMN therapy for ischemic retinopathy based on our short current summary.

4. Materials and Methods

4.1. Animal and Retinal Ischemia/Reperfusion (I/R) Injury

The mouse experimental processes of the Ethics Committee on Animal Research of Keio University School of Medicine (#16017), the ARVO Statement for the Use of Animals in Ophthalmic and Vision Research, and the International Standards of Animal Care and Use, Animal Research: Reporting in Vivo Experiments were followed. Adult male mice (C57BL/6, 6–8 weeks old) were bought from CLEA Japan (Tokyo, Japan). After mouse randomization, retinal I/R injury was induced in their eyes, as described in our previous paper [13]. Briefly, anterior chamber cannulation was performed using a 35-gauge needle and high intraocular pressure was maintained by PBS solution for 40 min. After injury, the mice were recovered with hot pads to maintain their body temperature and then moved back to their cages for further experiments.

4.2. Electroretinography (ERG)

General scotopic ERG was performed as described in our previous retinal I/R injury study [13]. After 24 h of dark adaptation, the mice were placed on a Ganzfeld dome table under dimly lit conditions. After pupil dilation followed by anesthesia (midazolam, medetomidine, and butorphanol tartrate, called MMB [40]), active ERG electrodes, connected to a PuREC acquisition system (MAYO, Inazawa, Japan), were softly contacted with the mouse cornea. Then, the general scotopic ERG amplitudes (a- and b-waves) were obtained at standard flash intensities.

4.3. Immunohistochemistry (IHC)

IHC was performed as described in our previous retinal I/R injury study [13]. Briefly, after paraformaldehyde (4% PFA) fixation for more than 3 h, the mouse eyeballs were moved to a Petri dish with cold PBS. After the retinas were collected from the eyeballs, the retinal samples were flat-mounted using micro-scissors and incubated in IB4 solution for 24 h. After washing with cold PBS several times, the retinal samples were mounted and examined by an LSM710 fluorescent microscope (Carl Zeiss, Jena, Germany).

4.4. Cell Culture

Retinal 661W cells were generally cultured in DMEM (Cat #08456-36, Nacalai Tesque, Kyoto, Japan) media with 10% FBS and 1% streptomycin–penicillin as described in our previous papers [41,42]. The cell culture was maintained under atmospheric conditions containing 5% CO_2 at 37 °C.

4.5. Terminal Deoxynucleotidyl Transferase dUTP Nick end Labeling (TUNEL) and MTT Assays

A TUNEL assay was conducted as described in our previous papers [13,43]. We basically followed the manufacturer's manual instructions (in situ Apoptosis Detection Kit, Cat #MK500, Takara Bio, Japan). PFA (4%)-fixed 661W cells were subjected to permeabilization using a permeabilization buffer supplied from the assay kit for 5 min on ice. A labeling reaction mixture with TdT enzymes was added to the cells for 1 h. Then, the DAPI solution was incubated for 1 min to stain the nuclei. The samples were mounted and investigated using an LSM710 fluorescent microscope (Carl Zeiss, Jena, Germany).

An MTT assay was also performed as described in our previous report [41]. The MTT solution (Cat #M2128, Sigma, St. Louis, MO, USA) was given to each well for 2 h. After removing the supernatant, DMSO was added to each well. Then, color absorbance was measured and determined using a microplate reader (Synergy HT Multi-Mode, Winooski, VT, USA).

4.6. Quantitative PCR (qPCR)

All steps for quantitative PCR (qPCR) were described in our previous reports [42,44]. Briefly, RNA extraction, cDNA synthesis, and qPCR were performed with each kit (Qiagen, Velno, Netherlands; TOYOBO, Osaka, Japan; Applied Biosystems, Waltham, MA, USA, respectively). The primer information used in our current study is outlined in Table 1. The general $\Delta\Delta CT$ calculation method was applied for our qPCR analysis.

Table 1. Primer list.

Name	Direction	Sequence (5′→3′)	Accession Number
Hprt	Forward	TCAGTCAACGGGGGACATAAA	NM_013556.2
	Reverse	GGGGCTGTACTGCTTAACCAG	
Hmox-1	Forward	CACTCTGGAGATGACACCTGAG	NM_010442.2
	Reverse	GTGTTCCTCTGTCAGCATCACC	
Nrf2	Forward	TAGATGACCATGAGTCGCTTGC	NM_010902.4
	Reverse	GCCAAACTTGCTCCATGTCC	

4.7. Statistical Analysis

All values in our current data were depicted as the mean ± standard deviation. Statistical significance was determined using one-way ANOVA followed by a Bonferroni post-hoc test. Statistical significance was considered when the p-value < 0.05.

Author Contributions: D.L. designed the whole study, performed the experiments, analyzed the data, and wrote the manuscript. Y.T., A.S., N.B., Y.M., S.Y., K.N. (Ken Nishioka) and J.Y. provided experimental support. K.N. (Kazuno Negishi) reviewed and revised the manuscript. T.K. supervised the entire project. All authors have read and agreed to the published version of the manuscript.

Funding: The current research was supported by Grants-in-Aid for Scientific Research (KAKENHI, number 15K10881, and 18K09424) from the Ministry of Education, Culture, Sports, Science and Technology (MEXT) to T.K. and JST SPRING (number JPMJSP2123) to D.L.

Institutional Review Board Statement: All mouse procedures of the Ethics Committee on Animal Research of the Keio University School of Medicine (#16017), the ARVO Statement for the Use of Animals in Ophthalmic and Vision Research, and the International Standards of Animal Care and Use, Animal Research: Reporting in Vivo Experiments guidelines were followed.

Informed Consent Statement: Not applicable.

Data Availability Statement: The current study's data are available upon request from the corresponding author.

Conflicts of Interest: Y.M. is employed by Aichi Animal Eye Clinic. The other authors declare no conflicts of interest.

References

1. Hangai, M.; Yoshimura, N.; Hiroi, K.; Mandai, M.; Honda, Y. Inducible nitric oxide synthase in retinal ischemia-reperfusion injury. *Exp. Eye Res.* **1996**, *63*, 501–509. [CrossRef]
2. Neufeld, A.H.; Kawai, S.; Das, S.; Vora, S.; Gachie, E.; Connor, J.R.; Manning, P.T. Loss of retinal ganglion cells following retinal ischemia: The role of inducible nitric oxide synthase. *Exp. Eye Res.* **2002**, *75*, 521–528. [CrossRef]
3. Wei, Y.; Gong, J.; Yoshida, T.; Eberhart, C.G.; Xu, Z.; Kombairaju, P.; Sporn, M.B.; Handa, J.T.; Duh, E.J. Nrf2 has a protective role against neuronal and capillary degeneration in retinal ischemia-reperfusion injury. *Free. Radic. Biol. Med.* **2011**, *51*, 216–224. [CrossRef]

4. Minhas, G.; Morishita, R.; Anand, A. Preclinical models to investigate retinal ischemia: Advances and drawbacks. *Front. Neurol.* **2012**, *3*, 75. [CrossRef]
5. Shah, M.; Cabrera-Ghayouri, S.; Christie, L.A.; Held, K.S.; Viswanath, V. Translational Preclinical Pharmacologic Disease Model for Ophthalmic Drug Development. *Pharm. Res.* **2019**, *36*, 58. [CrossRef]
6. Shade, C. The Science Behind NMN-A Stable, Reliable NAD+Activator and Anti-Aging Molecule. *Integr. Med.* **2020**, *19*, 12–14.
7. Braidy, N.; Berg, J.; Clement, J.; Khorshidi, F.; Poljak, A.; Jayasena, T.; Grant, R.; Sachdev, P. Role of Nicotinamide Adenine Dinucleotide and Related Precursors as Therapeutic Targets for Age-Related Degenerative Diseases: Rationale, Biochemistry, Pharmacokinetics, and Outcomes. *Antioxid. Redox Signal.* **2019**, *30*, 251–294. [CrossRef]
8. Nadeeshani, H.; Li, J.; Ying, T.; Zhang, B.; Lu, J. Nicotinamide mononucleotide (NMN) as an anti-aging health product—Promises and safety concerns. *J. Adv. Res.* **2022**, *37*, 267–278. [CrossRef]
9. Lin, J.B.; Kubota, S.; Ban, N.; Yoshida, M.; Santeford, A.; Sene, A.; Nakamura, R.; Zapata, N.; Kubota, M.; Tsubota, K.; et al. NAMPT-Mediated NAD(+) Biosynthesis Is Essential for Vision in Mice. *Cell Rep.* **2016**, *17*, 69–85. [CrossRef]
10. Chen, X.; Amorim, J.A.; Moustafa, G.A.; Lee, J.J.; Yu, Z.; Ishihara, K.; Iesato, Y.; Barbisan, P.; Ueta, T.; Togka, K.A.; et al. Neuroprotective effects and mechanisms of action of nicotinamide mononucleotide (NMN) in a photoreceptor degenerative model of retinal detachment. *Aging* **2020**, *12*, 24504–24521. [CrossRef]
11. Yoshino, J.; Mills, K.F.; Yoon, M.J.; Imai, S. Nicotinamide mononucleotide, a key NAD(+) intermediate, treats the pathophysiology of diet- and age-induced diabetes in mice. *Cell Metab.* **2011**, *14*, 528–536. [CrossRef] [PubMed]
12. Ramanathan, C.; Lackie, T.; Williams, D.H.; Simone, P.S.; Zhang, Y.; Bloomer, R.J. Oral Administration of Nicotinamide Mononucleotide Increases Nicotinamide Adenine Dinucleotide Level in an Animal Brain. *Nutrients* **2022**, *14*, 300. [CrossRef]
13. Lee, D.; Nakai, A.; Miwa, Y.; Tomita, Y.; Kunimi, H.; Chen, J.; Ikeda, S.I.; Tsubota, K.; Negishi, K.; Kurihara, T. Retinal degeneration induced in a mouse model of ischemia-reperfusion injury and its management by pemafibrate treatment. *FASEB J. Off. Publ. Fed. Am. Soc. Exp. Biol.* **2022**, *36*, e22497. [CrossRef]
14. Wheway, G.; Nazlamova, L.; Turner, D.; Cross, S. 661W Photoreceptor Cell Line as a Cell Model for Studying Retinal Ciliopathies. *Front. Genet.* **2019**, *10*, 308. [CrossRef] [PubMed]
15. Sayyad, Z.; Sirohi, K.; Radha, V.; Swarup, G. 661W is a retinal ganglion precursor-like cell line in which glaucoma-associated optineurin mutants induce cell death selectively. *Sci. Rep.* **2017**, *7*, 16855. [CrossRef] [PubMed]
16. Thompson, A.F.; Crowe, M.E.; Lieven, C.J.; Levin, L.A. Induction of Neuronal Morphology in the 661W Cone Photoreceptor Cell Line with Staurosporine. *PLoS ONE* **2015**, *10*, e0145270. [CrossRef]
17. Kunimi, H.; Lee, D.; Ibuki, M.; Katada, Y.; Negishi, K.; Tsubota, K.; Kurihara, T. Inhibition of the HIF-1α/BNIP3 pathway has a retinal neuroprotective effect. *FASEB J. Off. Publ. Fed. Am. Soc. Exp. Biol.* **2021**, *35*, e21829. [CrossRef]
18. Wei, C.C.; Kong, Y.Y.; Li, G.Q.; Guan, Y.F.; Wang, P.; Miao, C.Y. Nicotinamide mononucleotide attenuates brain injury after intracerebral hemorrhage by activating Nrf2/HO-1 signaling pathway. *Sci. Rep.* **2017**, *7*, 717. [CrossRef] [PubMed]
19. Pu, Q.; Guo, X.X.; Hu, J.J.; Li, A.L.; Li, G.G.; Li, X.Y. Nicotinamide mononucleotide increases cell viability and restores tight junctions in high-glucose-treated human corneal epithelial cells via the SIRT1/Nrf2/HO-1 pathway. *Biomed. Pharmacother. Biomed. Pharmacother.* **2022**, *147*, 112659. [CrossRef]
20. Westenskow, P.D. Nicotinamide: A novel treatment for age-related macular degeneration? *Stem Cell Investig.* **2017**, *4*, 86. [CrossRef]
21. Pîrvu, A.S.; Andrei, A.M.; Stănciulescu, E.C.; Baniță, I.M.; Pisoschi, C.G.; Jurja, S.; Ciuluvica, R. NAD(+) metabolism and retinal degeneration (Review). *Exp. Ther. Med.* **2021**, *22*, 670. [CrossRef] [PubMed]
22. Covarrubias, A.J.; Perrone, R.; Grozio, A.; Verdin, E. NAD(+) metabolism and its roles in cellular processes during ageing. *Nat. Reviews. Mol. Cell Biol.* **2021**, *22*, 119–141. [CrossRef]
23. Hong, W.; Mo, F.; Zhang, Z.; Huang, M.; Wei, X. Nicotinamide Mononucleotide: A Promising Molecule for Therapy of Diverse Diseases by Targeting NAD+ Metabolism. *Front. Cell Dev. Biol.* **2020**, *8*, 246. [CrossRef] [PubMed]
24. Lautrup, S.; Sinclair, D.A.; Mattson, M.P.; Fang, E.F. NAD(+) in Brain Aging and Neurodegenerative Disorders. *Cell Metab.* **2019**, *30*, 630–655. [CrossRef]
25. Yoshino, J.; Baur, J.A.; Imai, S.I. NAD(+) Intermediates: The Biology and Therapeutic Potential of NMN and NR. *Cell Metab.* **2018**, *27*, 513–528. [CrossRef]
26. Sasaki, Y.; Kakita, H.; Kubota, S.; Sene, A.; Lee, T.J.; Ban, N.; Dong, Z.; Lin, J.B.; Boye, S.L.; DiAntonio, A.; et al. SARM1 depletion rescues NMNAT1-dependent photoreceptor cell death and retinal degeneration. *Elife* **2020**, *9*, e62027. [CrossRef]
27. Wenz, C.; Faust, D.; Linz, B.; Turmann, C.; Nikolova, T.; Bertin, J.; Gough, P.; Wipf, P.; Schröder, A.S.; Krautwald, S.; et al. t-BuOOH induces ferroptosis in human and murine cell lines. *Arch. Toxicol.* **2018**, *92*, 759–775. [CrossRef]
28. Takayama, F.; Egashira, T.; Yamanaka, Y. Protective effect of Ninjin-yoei-to on damage to isolated hepatocytes following transient exposure to tert-butyl hydroperoxide. *Jpn. J. Pharmacol.* **2001**, *85*, 227–233. [CrossRef]
29. Baysal, E.; Sullivan, S.G.; Stern, A. Prooxidant and antioxidant effects of ascorbate on tBuOOH-induced erythrocyte membrane damage. *Int. J. Biochem.* **1989**, *21*, 1109–1113. [CrossRef]
30. Audrito, V.; Messana, V.G.; Deaglio, S. NAMPT and NAPRT: Two Metabolic Enzymes with Key Roles in Inflammation. *Front. Oncol.* **2020**, *10*, 358. [CrossRef]
31. Gardell, S.J.; Hopf, M.; Khan, A.; Dispagna, M.; Hampton Sessions, E.; Falter, R.; Kapoor, N.; Brooks, J.; Culver, J.; Petucci, C.; et al. Boosting NAD(+) with a small molecule that activates NAMPT. *Nat. Commun.* **2019**, *10*, 3241. [CrossRef] [PubMed]

32. Abcouwer, S.F.; Shanmugam, S.; Muthusamy, A.; Lin, C.M.; Kong, D.; Hager, H.; Liu, X.; Antonetti, D.A. Inflammatory resolution and vascular barrier restoration after retinal ischemia reperfusion injury. *J. Neuroinflammation* **2021**, *18*, 186. [CrossRef] [PubMed]
33. Huang, R.; Liang, S.; Fang, L.; Wu, M.; Cheng, H.; Mi, X.; Ding, Y. Low-dose minocycline mediated neuroprotection on retinal ischemia-reperfusion injury of mice. *Mol. Vis.* **2018**, *24*, 367–378. [PubMed]
34. Celorrio, M.; Shumilov, K.; Payne, C.; Vadivelu, S.; Friess, S.H. Acute minocycline administration reduces brain injury and improves long-term functional outcomes after delayed hypoxemia following traumatic brain injury. *Acta Neuropathol. Commun.* **2022**, *10*, 10. [CrossRef] [PubMed]
35. Schimmel, S.J.; Acosta, S.; Lozano, D. Neuroinflammation in traumatic brain injury: A chronic response to an acute injury. *Brain Circ.* **2017**, *3*, 135–142. [CrossRef]
36. Begemann, M.; Leon, M.; van der Horn, H.J.; van der Naalt, J.; Sommer, I. Drugs with anti-inflammatory effects to improve outcome of traumatic brain injury: A meta-analysis. *Sci. Rep.* **2020**, *10*, 16179. [CrossRef]
37. Yamaura, K.; Mifune, Y.; Inui, A.; Nishimoto, H.; Kurosawa, T.; Mukohara, S.; Hoshino, Y.; Niikura, T.; Kuroda, R. Antioxidant effect of nicotinamide mononucleotide in tendinopathy. *BMC Musculoskelet. Disord.* **2022**, *23*, 249. [CrossRef]
38. Liu, X.; Dilxat, T.; Shi, Q.; Qiu, T.; Lin, J. The combination of nicotinamide mononucleotide and lycopene prevents cognitive impairment and attenuates oxidative damage in D-galactose induced aging models via Keap1-Nrf2 signaling. *Gene* **2022**, *822*, 146348. [CrossRef]
39. Luo, C.; Ding, W.; Yang, C.; Zhang, W.; Liu, X.; Deng, H. Nicotinamide Mononucleotide Administration Restores Redox Homeostasis via the Sirt3-Nrf2 Axis and Protects Aged Mice from Oxidative Stress-Induced Liver Injury. *J. Proteome Res.* **2022**, *21*, 1759–1770. [CrossRef]
40. Miwa, Y.; Tsubota, K.; Kurihara, T. Effect of midazolam, medetomidine, and butorphanol tartrate combination anesthetic on electroretinograms of mice. *Mol. Vis.* **2019**, *25*, 645–653.
41. Ibuki, M.; Lee, D.; Shinojima, A.; Miwa, Y.; Tsubota, K.; Kurihara, T. Rice Bran and Vitamin B6 Suppress Pathological Neovascularization in a Murine Model of Age-Related Macular Degeneration as Novel HIF Inhibitors. *Int. J. Mol. Sci.* **2020**, *21*, 8940. [CrossRef] [PubMed]
42. Lee, D.; Miwa, Y.; Wu, J.; Shoda, C.; Jeong, H.; Kawagishi, H.; Tsubota, K.; Kurihara, T. A Fairy Chemical Suppresses Retinal Angiogenesis as a HIF Inhibitor. *Biomolecules* **2020**, *10*, 1405. [CrossRef]
43. Lee, D.; Jeong, H.; Miwa, Y.; Shinojima, A.; Katada, Y.; Tsubota, K.; Kurihara, T. Retinal dysfunction induced in a mouse model of unilateral common carotid artery occlusion. *PeerJ* **2021**, *9*, e11665. [CrossRef]
44. Lee, D.; Tomita, Y.; Jeong, H.; Miwa, Y.; Tsubota, K.; Negishi, K.; Kurihara, T. Pemafibrate Prevents Retinal Dysfunction in a Mouse Model of Unilateral Common Carotid Artery Occlusion. *Int. J. Mol. Sci.* **2021**, *22*, 9408. [CrossRef]

International Journal of
Molecular Sciences

MDPI

Article

Enhancement of Sphingomyelinase-Induced Endothelial Nitric Oxide Synthase-Mediated Vasorelaxation in a Murine Model of Type 2 Diabetes

Éva Ruisanchez [1,2,†], Anna Janovicz [1,2,†], Rita Cecília Panta [1], Levente Kiss [3], Adrienn Párkányi [1], Zsuzsa Straky [1], Dávid Korda [1], Károly Liliom [4], Gábor Tigyi [1,5] and Zoltán Benyó [1,2,*]

1 Institute of Translational Medicine, Semmelweis University, H-1094 Budapest, Hungary
2 Eötvös Loránd Research Network and Semmelweis University (ELKH-SE) Cerebrovascular and Neurocognitive Disorders Research Group, H-1052 Budapest, Hungary
3 Department of Physiology, Semmelweis University, H-1094 Budapest, Hungary
4 Institute of Biophysics and Radiation Biology, Semmelweis University, H-1094 Budapest, Hungary
5 Department of Physiology, University of Tennessee Health Science Center, Memphis, TN 38163, USA
* Correspondence: benyo.zoltan@med.semmelweis-univ.hu
† These authors contributed equally to this work.

Abstract: Sphingolipids are important biological mediators both in health and disease. We investigated the vascular effects of enhanced sphingomyelinase (SMase) activity in a mouse model of type 2 diabetes mellitus (T2DM) to gain an understanding of the signaling pathways involved. Myography was used to measure changes in the tone of the thoracic aorta after administration of 0.2 U/mL neutral SMase in the presence or absence of the thromboxane prostanoid (TP) receptor antagonist SQ 29,548 and the nitric oxide synthase (NOS) inhibitor L-NAME. In precontracted aortic segments of non-diabetic mice, SMase induced transient contraction and subsequent weak relaxation, whereas vessels of diabetic ($Lepr^{db}/Lepr^{db}$, referred to as db/db) mice showed marked relaxation. In the presence of the TP receptor antagonist, SMase induced enhanced relaxation in both groups, which was 3-fold stronger in the vessels of db/db mice as compared to controls and could not be abolished by ceramidase or sphingosine-kinase inhibitors. Co-administration of the NOS inhibitor L-NAME abolished vasorelaxation in both groups. Our results indicate dual vasoactive effects of SMase: TP-mediated vasoconstriction and NO-mediated vasorelaxation. Surprisingly, in spite of the general endothelial dysfunction in T2DM, the endothelial NOS-mediated vasorelaxant effect of SMase was markedly enhanced.

Keywords: sphingolipids; sphingomyelinase; vasorelaxation; endothelial nitric oxide synthase; type 2 diabetes; thromboxane prostanoid receptor

Citation: Ruisanchez, É.; Janovicz, A.; Panta, R.C.; Kiss, L.; Párkányi, A.; Straky, Z.; Korda, D.; Liliom, K.; Tigyi, G.; Benyó, Z. Enhancement of Sphingomyelinase-Induced Endothelial Nitric Oxide Synthase-Mediated Vasorelaxation in a Murine Model of Type 2 Diabetes. *Int. J. Mol. Sci.* **2023**, *24*, 8375. https://doi.org/10.3390/ijms24098375

Academic Editors: Simona Gabriela Bungau and Cosmin Mihai Vesa

Received: 28 February 2023
Revised: 30 April 2023
Accepted: 4 May 2023
Published: 6 May 2023

1. Introduction

Sphingolipids, derived from sphingomyelin metabolism, have been implicated as important mediators in the physiology and pathophysiology of the cardiovascular system [1–6]. Sphingomyelinase (SMase) catalyzes the conversion of sphingomyelin to ceramide, which is the precursor of other sphingolipid mediators, including ceramide-1-phosphate (C1P), sphingosine (Sph), and sphingosine-1-phosphate (S1P) [7]. The majority of S1P-induced biological effects are mediated by G-protein-coupled receptors (GPCRs), termed $S1P_{1-5}$ [8]. Other sphingolipid mediators may exert biological effects by directly interacting with membrane or intracellular protein targets, independently of the activation of S1P receptors [5,9–11].

Based on the optimal pH for their catalytic activity, SMase isoforms can be divided into three groups: alkaline, acidic, and neutral [12]. The expression and known functions of alkaline SMases are mostly restricted to the gastrointestinal system, whereas acidic and neutral SMases are more widely expressed and involved in physiological and pathophysiological reactions in many systems, including the cardiovascular system. In the vasculature, SMases are implicated in the regulation of vascular tone and permeability as well as in causing atherosclerotic lesions and vascular wall remodeling [13]. Interestingly, neutral SMase has been reported to induce a wide range of changes in the vascular tone, depending on the species, vessel type, and experimental conditions (Table 1). Taken into account the large number of biologically active mediators (including ceramides, C1P, Sph, and S1P) that can be generated both extra- and intracellularly upon triggering the sphingolipid biosynthesis by neutral SMase, the diversity of vascular effects is not unexpected.

Table 1. Reported vasoactive effects of neutral SMase.

Species	Vessel	Vasoactive Effects	Proposed Mechanism	Refs.
Yorkshire pig	Coronary artery	Transient endothelium-dependent contraction followed by endothelium-dependent relaxation	Vasoconstriction: prostanoid(s) Vasorelaxation: NO	[14]
Sprague-Dawley rat	Thoracic aorta	Endothelium-independent relaxation	Inhibition of protein kinase C (PKC)	[15,16]
Wistar rat	Thoracic aorta	Partly endothelium-independent relaxation	Endothelium-mediated components are independent of NO or prostanoids Non-endothelial components are independent of PKC	[17]
Mongrel dog	Basilar artery	Endothelium-independent contraction	Activation of VDCC and PKC	[18]
Wistar rat	Pial venule (60–70 μm in diameter)	Constriction and spasm	Activation of VDCC, PKC, and MAP kinase	[19]
Wistar rat	Thoracic aorta	Endothelium-independent relaxation	Inhibition of both Ca^{2+}-dependent and Ca^{2+}-independent (RhoA-/Rho kinase-mediated) contractile pathways	[20]
Cow	Coronary artery	Endothelium-dependent relaxation	Ca^{2+}-independent eNOS activation, involving phosphorylation on serine 1179 and dissociation of eNOS from plasma membrane caveolae	[21]
Wistar rat	Pulmonary artery	Endothelium-independent contraction	Activation of VDCC, PKCζ, and Rho kinase	[22]
Wistar-Kyoto (WKY) and *Spontaneously hypertensive rat* (SHR)	Carotid artery	SHR: strong endothelium-dependent contraction WKY: weak endothelium-dependent contraction	Vasoconstriction is mediated by PLA_2- and COX2-mediated TXA_2 release and attenuated by NO	[23–25]

NO, nitric oxide; PKC, protein kinase C; VDCC, voltage-gated calcium channel; MAP, mitogen activated protein; eNOS, endothelial NO synthase; PLA2, phospholipase A2; TXA2, thromboxane A2; COX, cyclooxygenase.

SMase enzymes are reportedly upregulated in certain cardiovascular and metabolic disorders, such as type 2 diabetes mellitus (T2DM) [13,26,27]. Sphingolipids have been implicated as important regulators of inflammatory processes in diabetes [28]. Stress conditions initiate changes in sphingolipid metabolism [29], and sphingolipids have emerged as key mediators of stress responses [30,31]. Extracellular stressors induce sphingolipid synthesis and turnover, thereby 'remodeling' sphingolipid profiles and their topological distribution within cells [32]. Emerging evidence not only demonstrates profound changes in sphingolipid pools and distribution under conditions of overnutrition [33–35], but also implicates sphingolipids in mediating cell-signaling responses that precipitate pathology associated with obesity [36]. In spite of the marked alterations in the metabolism and actions of sphingolipids in diabetes and recent observations indicating that ceramide may contribute to the development of diabetic endothelial dysfunction [37], relatively little is known about the effects of sphingolipids on vascular functions in T2DM. In the present study, we analyzed the effects of SMase on vascular tone under diabetic conditions in order to elucidate the signaling mechanisms involved.

2. Results

First, we verified the general metabolic and vascular phenotypes of the T2DM mice tested in the present study. Db/db mice reportedly develop obesity with elevated blood glucose levels and insulin resistance [38–40]. Accordingly, the body weight increased almost 2-fold (Figure 1A), whereas blood glucose levels increased 3-fold (Figure 1B) in db/db mice as compared to non-diabetic control littermates. Furthermore, the serum phosphorylcholine level was also significantly increased in the diabetic group (Figure 1C), which is consistent with the reported enhancement of SMase activity in type 2 diabetes [13,26,27]. According to literature data, acetylcholine (ACh)-evoked vasorelaxation of the aorta prepared from db/db mice is completely NOS-dependent [41], thus ACh was used to characterize the endothelial function. The vessels of db/db animals showed marked endothelial dysfunction, as indicated by the impairment of the dose-response relationship of ACh-induced vasorelaxation after precontraction with 10 μmol/L PE (Figure 1D). The E_{max} value decreased to 50.8 ± 2.0% in diabetic vessels as compared to controls (65.8 ± 3.9%). However, there was no significant difference in the EC_{50} values (34.7 ± 16.0 nM vs. 55.7 ± 15.7 nM), indicating unchanged potency in spite of the reduced efficacy of endogenous NO upon stimulation of endothelial NOS (eNOS) by ACh. In contrast, reactivity of the vascular smooth muscle to NO remained unaltered, as neither the E_{max} (105.2 ± 1.8% vs. 103.3 ± 2.2%) nor the EC_{50} (10.7 ± 1.3 nM vs. 14.1 ± 2.0 nM) values of sodium nitroprusside (SNP)-induced vasorelaxation differed in vessels of db/db animals as compared to controls (Figure 1E). Taken together, these results confirm the T2DM-like metabolic and vascular phenotypes in db/db mice and suggest the in vivo enhancement of SMase activity as well.

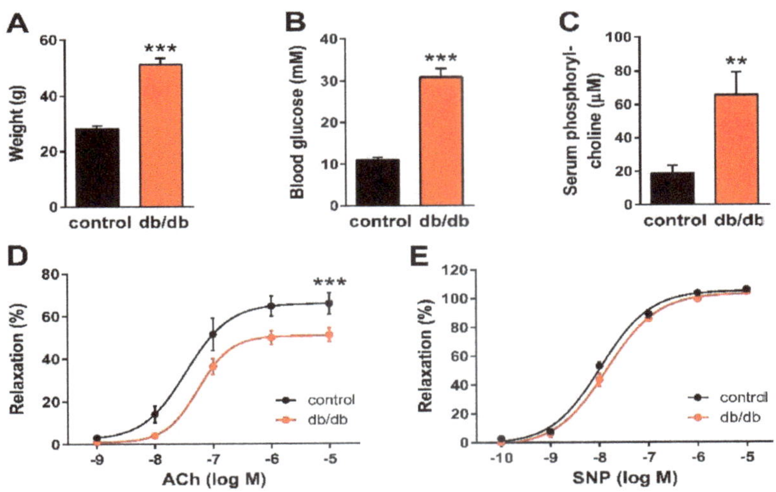

Figure 1. Manifestation of the metabolic and vascular phenotypes of T2DM in db/db mice. Body weight (**A**), as well as non-fasting blood glucose (**B**) and serum phosphorylcholine levels (**C**), increased in db/db mice as compared to controls (** $p < 0.01$, *** $p < 0.001$ vs. control group; Student's unpaired t-test, n = 13–22). ACh-induced relaxation diminished (**D**), while the reactivity of the vascular smooth muscle to sodium nitroprusside (SNP) remained unaltered (**E**) in vessels of db/db mice as compared to controls (mean ± SEM, *** $p < 0.001$ vs. control; dose-response curve fitted to n = 12–24).

Next, we determined the effect of nSMase on the active tone of control and db/db vessels (Figure 2A). After 10 μmol/L phenylephrine (PE)-induced precontraction, 0.2 U/mL nSMase elicited additional contraction in control vessels that reached its maximum at

7.2 min before relaxing back to the pre-SMase level by the end of the 20-min observation period. In contrast, nSMase in db/db vessels elicited completely different responses. After a marked initial relaxation elicited by 0.2 U/mL nSMase during the first 5 min, the tone of the db/db vessels remained in a relaxed state below the level of the initial tension. From the shape of the tension curve, it appeared that in addition to the overriding relaxation there was a delayed and transient constriction response with a time course similar to that observed in control vessels, but it was unable to overcome the robust dilatation. Evaluation of the AUC (Figure 2B) and the maximal changes in the vascular tone (Figure 2C) also supported the conclusion that there is a marked difference in the vascular effects of nSMase between control and db/db mice: contraction dominates in the former, whereas the latter is characterized by reduction of the vascular tone.

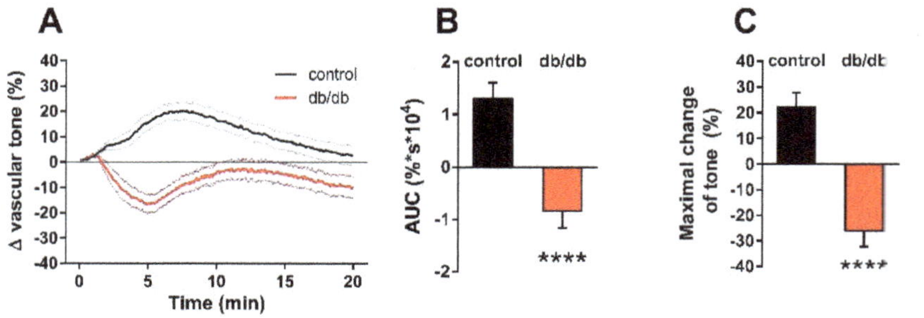

Figure 2. Effects of nSMase on the vascular tone. Application of 0.2 U/mL nSMase evoked a complex vascular effect with dominant contraction in control vessels and a more pronounced relaxation in vessels of db/db mice. Black and red lines on panel (**A**) represent average changes in tension of PE-precontracted vessels in control and db/db mice, respectively (dotted lines represent SEM). Both area under curve values (**B**) and maximal tension changes (**C**) were significantly different in vessels from db/db animals as compared to controls (mean ± SEM, Student's unpaired *t*-test, **** $p < 0.0001$ vs. control; n = 51–49).

Our next aim was to differentiate the constrictor and relaxant components of the vascular tension changes in response to nSMase. In porcine coronary arteries [14] and in the carotid arteries of spontaneously hypertensive rats [23–25], prostanoids acting on TP receptors have been implicated in mediating the vasoconstrictor effect of SMase. Therefore, we hypothesized that thromboxane prostanoid (TP) receptors also mediate the nSMase-induced vasoconstriction in our murine aorta model. To test this hypothesis, the TP receptor antagonist SQ 29,548 was administered to the organ chambers 30 min prior to administration of nSMase. Blockade of TP receptors not only abolished the vasoconstriction but also converted it to a transient vasorelaxation in control vessels (Figure 3A). The maximum relaxation was reached at 5.5 min after the administration of nSMase, and the vascular tone returned to baseline after 10 min. TP receptor inhibition also markedly changed the vascular response to nSMase in the db/db group: the vasorelaxation was enhanced to more than 70% and reached its maximum at 6.5 min. After its peak, the relaxation decreased, but the vascular tone failed to return to the pre-SMase level even after 20 min. Both the AUC (Figure 3B) and the peak vasorelaxation (Figure 3C) values showed marked differences between the two experimental groups, indicating that the strongly enhanced and prolonged vasorelaxant capacity is responsible for the differences between the vasoactive effects of nSMase in db/db and control vessels. This finding was very surprising in light of the diminished ACh-induced vasorelaxation that we had observed in db/db animals (Figure 1D) and was not consistent with the large body of literature indicating diminished endothelium-dependent vasorelaxation in T2DM.

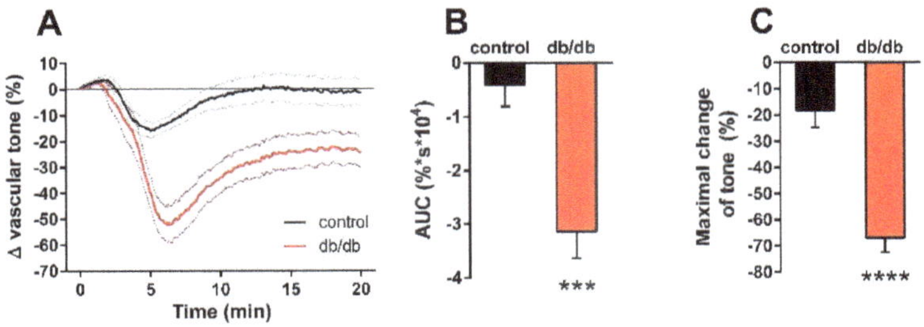

Figure 3. Effects of TP receptor blockade on nSMase-induced changes in the vascular tone. After inhibition of the TP receptor by 1 μM SQ 29,548, 0.2 U/mL nSMase relaxed both db/db and control vessels, with a significantly higher relaxation in the db/db group (**A**). Black and red lines in panel A represent average tension changes in PE-precontracted vessels of control and db/db mice, respectively, whereas dotted lines represent SEM. Both area under curve values (**B**) and maximal tension changes (**C**) were significantly different in vessels from db/db animals as compared to controls (mean ± SEM, Student's unpaired *t*-test, *** $p < 0.001$ vs. control; **** $p < 0.0001$ vs. control; $n = 20$).

Next, we aimed to analyze the mechanism of the enhanced nSMase-induced vasorelaxation in the vessels of db/db mice. Theoretically, it could be due to the enhancement of eNOS-mediated vasorelaxation or to the onset of an NO-independent mechanism. To clarify this question, the vessels were incubated with the NOS inhibitor L-NAME (100 μM) in addition to the TP receptor blocker SQ 29,548 (1 μM) for 30 min prior to 0.2 U/mL nSMase administration. L-NAME at a concentration of 100 μM abolished the vasorelaxation observed in the presence of 1 μM SQ 29,548 both in control and in db/db vessels (Figure 4A). There were no significant differences between the two groups either in the AUC (Figure 4B) or in the maximal change of tension values (Figure 4C). These results indicate that the same secondary signaling pathways—namely TP receptors and eNOS—mediate the vasoactive effects of nSMase in health and in T2DM.

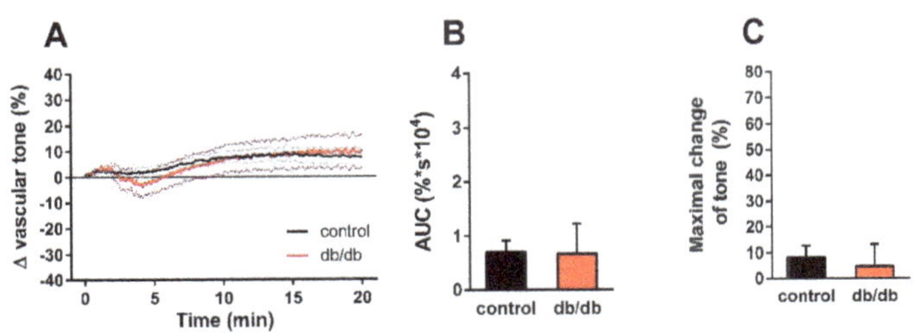

Figure 4. Effects of combined TP receptor and NOS blockade on nSMase-induced changes in vascular tone. After incubation of the vessels with 1 μM SQ 29,548 and 100 μM L-NAME for 30 min, 0.2 U/mL nSMase could no longer evoke a tension change in the thoracic aorta of control or db/db mice (**A**). Black and red lines in panel A represent average tension changes in PE-precontracted vessels of control and db/db mice, respectively (dotted lines represent SEM). Area under curve values (**B**) and maximal tension changes (**C**) were not different in vessels from db/db animals as compared to controls (mean ± SEM, $n = 9$–17).

Finally, we aimed to investigate the possible involvement of downstream sphingolipid metabolites in the vasorelaxing effect of nSMase observed in db/db mice. Thus, aortic segments isolated from db/db were treated with either the ceramidase inhibitor D-erythro MAPP (10 μM) or SKI-II (1 μM), a sphingosine kinase inhibitor. In order to examine the vasorelaxant effects exclusively, vessels were pre-treated with 1 μM SQ 29,568. Neither MAPP nor SKI-II could affect the vasorelaxation evoked by 0.2 U/mL nSMase (Figure 5A). In addition, evaluation of AUC (Figure 5B) and maximal vasorelaxing responses (Figure 5C) showed no significant differences in MAPP or SKI-II-treated vessels compared to vehicle treated ones. These results suggest that the effect of nSMase is not mediated by sphingosine or sphingosine-1-phosphate.

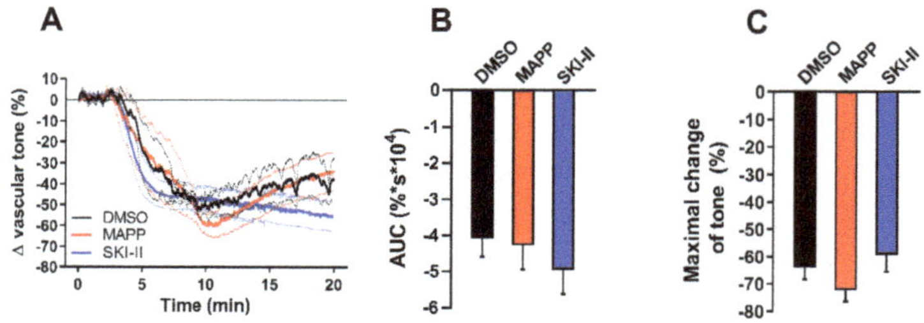

Figure 5. Effects of ceramidase and sphingosine-kinase inhibitors on nSMase-induced vascular tone changes in db/db mice. After incubating with 1 μM SQ 29,548 for 30 min, db/db vessels were treated with 10 μM D-erythro MAPP or 1 μM SKI-II. Neither MAPP nor SKI-II could affect the vasorelaxation induced by 0.2 U/mL nSMase (**A**). Black, red, and blue lines in panel A represent average tension changes in PE-precontracted vessels of vehicles, MAPP, and SKI-II treated vessels, respectively (dotted lines represent SEM). Area under curve values (**B**) and maximal tension changes (**C**) were not different in MAPP- or SKI-II-treated vessels as compared to vehicle controls (mean ± SEM, n = 4–5).

3. Discussion

Findings of the present study indicate that nSMase-induced changes in vascular tension involve both vasoconstriction and vasorelaxation in murine vessels. Our results suggest that the former is mediated by the release of prostanoids and activation of TP receptors, whereas the latter is mediated by eNOS. Surprisingly, nSMase-induced eNOS-mediated vasorelaxation is markedly enhanced in the vessels of db/db mice in spite of the endothelial dysfunction indicated by the diminished vasorelaxation evoked by ACh Therefore, nSMase appears to be able to induce enhanced NO release from endothelial cells in T2DM.

Vasoconstriction in response to SMase has been reported in a number of studies, although the mechanisms mediating this effect appear to be highly variable depending on the experimental conditions, including species, vascular region, and integrity of the endothelium (see Table 1). Release of prostanoids and consequent activation of TP receptors have been proposed in porcine coronary arteries [14] as well as in the carotid arteries of spontaneously hypertensive rats [23–25]. In our study, nSMase-induced contraction was found to be TP receptor-dependent in both control and db/db mice, indicating that nSMase stimulates the release of TXA_2 from the aortic rings.

There might be at least three different sources for the SMase-induced arachidonic acid formation necessary for TXA_2 production [42]. One such possibility is that diacylglycerol (DAG) would accumulate while sphingomyelin synthase converted the newly generated ceramide back to sphingomyelin, and DAG lipases would provide arachidonic acid for the production of TXA_2 [43]. Another mechanism might relate to the observation that C1P can allosterically activate phospholipase A_2 (PLA_2) [44], which leads to arachidonic acid formation [45]. It might be important in this context that the gene encoding ceramide kinase (*CERK*) is upregulated in T2DM [46]. Finally, S1P has been reported recently to regulate prostanoid production in a S1P receptor-dependent manner [47]. It should be noted that mechanisms other than prostanoid release might also mediate the vasoconstrictor effect of SMase, i.e., modulation of different ion channels or initiating the Rho signaling pathway [18,19,22,48].

Vasorelaxation in response to nSMase appears to be endothelial NO-dependent, as L-NAME completely abolished the decrease in vascular tone in both control and db/db vessels. Without L-NAME, relaxation was dramatically increased in db/db-derived vascular rings. This is unexpected because endothelial dysfunction with consequential decreased vasorelaxant capacity is considered to be a hallmark of T2DM-like conditions. A potential explanation may be related to the altered structure of the plasma membrane in T2DM [49]. Normally, sphingomyelin (SM) represents about 10–20% of the lipids in the plasma membrane, mostly residing in the outer leaflet. However, most of these are found in the caveolae, and SMase is thought to be a regulator of lipid microdomains [50,51]. Pilarczyk and colleagues provided evidence that in db/db mice, the endothelial lining of the aorta contains 10-fold larger lipid raft areas enriched in SM as compared to controls [49]. This arrangement might be related to the decreased NO-release in T2DM, as eNOS is inhibited by caveolin-1 [52], which is considered to be an important regulator of eNOS [53–55]. In our experimental setting, nSMase-induced degradation of sphingomyelin could interfere with this caveolar structure and induce the detachment of eNOS from caveolin-1, leading to high amounts of NO released from the endothelium of db/db vascular rings. This hypothesis is supported by the observations of Mogami et al. [21], indicating that SMase causes endothelium-dependent vasorelaxation through Ca^{2+}-independent endothelial NO production in bovine aortic valves and coronary arteries. They also reported SMase-induced translocation of endothelial NOS from plasma membrane caveolae to the intracellular region. Furthermore, protein expression levels of caveolin-1 were reported to be significantly higher in the aorta of db/db mice, and this was thought to be related to the impaired aortic relaxation of C57BL/KsJ mice [56]. In order to find out if downstream sphingolipid mediators such as sphingosine or sphingosine-1-phosphate contribute to the enhanced vasorelaxation seen in db/db aortic segments, we also investigated whether inhibition of ceramidase or sphingosine-kinase could interfere with the vasorelaxation. Neither D-erythro-MAPP nor SKI-II could diminish the nSMase-induced vasorelaxation in db/db vessels; therefore, it is unlikely that the release of these mediators participates in the effect. This finding strengthens our hypothesis that the enhanced relaxation in db/db vessels is mediated by a direct effect of nSMase on the membrane structure. On the other hand, we cannot rule out that the ceramide-related pathway might be involved in the SMase-induced contractions as well [18,23], as we did not address this question in our experiments. Finally, the potentially increased NO-sensitivity of guanylate cyclase (sGC) [57], which could be related to the dysfunctional NO-release observed in T2DM, should also be considered, as this would sensitize sGC to NO and result in enhanced NO-mediated vasorelaxation. However, this mechanism can be excluded in our present experiments, as the SNP dose-response curve remained unchanged in db/db vessels (Figure 1E), indicating that the sensitivity of the vascular smooth muscle to NO was not upregulated.

Sphingolipid metabolism is markedly altered in T2DM and related conditions [58–62], and the observed changes in endothelial lipid rafts [49] might be a consequence of the disrupted plasma membrane lipid metabolism. On the other hand, T2DM has several characteristics that resemble a chronic inflammatory disease [63]. Cytokines that accumulate in chronic inflammation, such as tumor necrosis factor alpha (TNF-α) and interleukin 1 beta (IL-1β), can also induce marked changes in sphingolipid metabolism [6,64,65]. Our observation that serum phosphorylcholine levels were increased in the db/db group might be a strong indicator of the altered in vivo sphingolipid metabolism in our animal model and agrees with the literature; however, it cannot be excluded that enzymes other than nSMase also contribute to this increase (e.g., other SMases, choline-kinase, phospholipase C, ENPP, etc.).

As a limitation of our study, it has to be mentioned that the characteristics of the pathophysiological conditions in the db/db mouse model differ from those of human T2DM in some aspects [66]. For example, db/db mice do not necessarily develop hypertension and may have high levels of high-density lipoprotein and a reduced tendency toward atherosclerosis [67]. Therefore, due to the more severe endothelial dysfunction, the enhancement of nSMase-induced eNOS-mediated vasorelaxation may be limited in humans with T2DM. A further limitation of our study is that we tested only one single dose of bacterial nSMase. This 0.2 U/mL dose of exogenously administered bacterial nSMase represents the upper range used in the literature [14–18,20–23], as our aim was to evaluate the consequences of a robust activation of sphingomyelinase degradation. Further studies may aim to elucidate the exact dose-response relationship for SMase-induced vasorelaxation and vasoconstriction in db/db mice or other T2DM-related conditions, which may also help to clarify the exact molecular mechanisms involved. Though bacterial and different mammalian nSMase orthologues share low overall homology, the 'catalytic core' residues are strongly conserved [68]. Therefore, it is likely that the addition of bacterial nSMase mimics the overactivation of mammalian nSMase and leads to similar cellular effects—in our case, the release of thromboxane and NO [69,70].

4. Materials and Methods

All procedures were carried out according to the guidelines of the Hungarian Law of Animal Protection (40/2013). The procedures were approved by the National Scientific Ethical Committee of Animal Experimentation (PE/EA/924-7/2021, accepted: 7 September 2021).

4.1. Animals and General Procedures

The BKS db diabetic mouse strain (JAX stock #000642) was obtained from the Jackson Laboratory (Bar Harbor, ME, USA) and has been maintained in our animal facility by mating repulsion double heterozygotes (*Dock7*^{m} +/+ *Lepr*^{db}). Littermate adult male diabetic (*Lepr*^{db}/*Lepr*^{db}, referred to as db/db) and misty (*Dock7*^{m}/*Dock7*^{m}, referred to as control) mice were selected for experiments. All mice investigated in this research were male and aged between 90 and 180 days. Animals were weighed, and blood samples were collected by cardiac puncture followed by transcardial perfusion with 10 mL of heparinized (10 IU/mL) Krebs solution under deep ether anesthesia, as described previously [71]. Nonfasting blood glucose was measured by a Dcont IDEÁL biosensor-type blood glucose meter (77 Elektronika Kft.; Budapest, Hungary). In some experiments, additional blood samples were collected, allowed to clot for 30 min at room temperature, and centrifuged at 2000× *g* for 15 min at 4 °C. Serum was snap frozen for a later phosphorylcholine assay, which was based on the method described by Hojjati and Jiang [72] using a commercially available kit (item No. 10009928, Cayman Chemical; Ann Arbor, MI, USA).

4.2. Myography

The thoracic aorta was removed and cleaned of fat and connective tissue under a dissection microscope (M3Z, Wild Heerbrugg AG; Gais, Switzerland) and immersed in a Krebs solution of the following composition (mmol/L): 119 NaCl, 4.7 KCl, 1.2 KH_2PO_4, 2.5 $CaCl_2 \cdot 2\ H_2O$, 1.2 $MgSO_4 \cdot 7\ H_2O$, 20 $NaHCO_3$, 0.03 EDTA, and 10 glucose at room temperature and pH 7.4. Vessels were cut into ~3 mm-long segments and mounted on stainless steel vessel holders (200 μm in diameter) in a conventional myograph setup (610 M multiwire myograph system; Danish Myo Technology A/S; Aarhus, Denmark). Special care was taken to preserve the endothelium.

The wells of the myographs were filled with an 8 mL Krebs solution aerated with carbogen. The vessels were allowed a 30-min resting period, during which the bath solution was warmed to 37 °C and the passive tension was adjusted to 15 mN, which was determined to be optimal in a previous study [71]. Subsequently, the tissues were exposed to a 124 mmol/L K^+ Krebs solution (made by isomolar replacement of Na^+ by K^+) for 1 min, followed by several washes with normal Krebs solution. Reactivity of the smooth muscle was tested by a contraction evoked by 10 μmol/L PE, and reactivity of the endothelium was tested by following the PE-evoked contraction with administration of 0.1 μmol/L ACh. After repeated washing, during which the vascular tension returned to the resting level, the segments were exposed to a 124 mmol/L K^+ Krebs solution for 3 min in order to elicit a reference maximal contraction. Subsequently, after a 30-min washout, increasing concentrations of PE (0.1 nmol/L to 10 μmol/L) and ACh (1 nmol/L to 10 μmol/L) were administered to determine the reactivity of the smooth muscle and the endothelium, respectively. Following a 30-min resting period, the vessels were precontracted to 70–90% of the reference contraction by an appropriate concentration of PE, and after contraction had stabilized, the effects of 0.2 U/mL nSMase (SMase from *B. cereus*, Sigma-Aldrich; St. Louis, MO, USA) were investigated for 20 min. Bacterial SMase functions at neutral pH and is reportedly a useful tool for mimicking the biological effects of activation of cellular SMase [73,74]. In some experiments, the selective TP receptor antagonist SQ 29,548 (1 μM) with or without the nitric oxide synthase (NOS) inhibitor L-NAME (100 μM) was applied to the baths 30 min prior to administration of nSMase. In order to block ceramidase or sphingosine-kinase enzymes, D-erythro MAPP (10 μM) or SKI-II (1 μM) were used, respectively. Finally, to test the sensitivity of the smooth muscle to NO, SNP (0.1 nmol/L to 10 μmol/L) was administered after a stable precontraction elicited by 1 μmol/L PE.

4.3. Data Analysis

An MP100 system and AcqKnowledge 3.72 software from Biopac System Inc. (Goleta, CA, USA) were used to record and analyze changes in the vascular tone. All data are presented as mean ± SE, and "*n*" indicates the number of vascular segments tested in myography experiments or the number of animals tested in the case of body weight, blood glucose, and serum phosphorylcholine levels. Maximal changes in the vascular tone were calculated as a percentage of precontraction. To evaluate the temporal pattern of nSMase-induced vasoactive responses, individual curves were constructed and averaged, showing the changes in vascular tone for 20 min after the application of nSMase. Area under the curve (AUC) values were calculated from individual experiments for quantification of the overall vasoactive effect. The statistical analysis was performed using the GraphPad Prism software v.6.07 from GraphPad Software Inc. (La Jolla, CA, USA). Student's unpaired *t*-test was applied when comparing two variables, and a p value of less than 0.05 was considered to be statistically significant. The effects of cumulative doses of PE and ACh were evaluated by dose-response curve fitting for the determination of E_{max} and EC_{50} values. All data presented in this study are available in the Supplementary Material File (Figures S1–S5).

4.4. Reagents

All reagents in this study, including nSMase, were purchased from Sigma-Aldrich (St. Louis, MO, USA), except SQ 29,548, which was from Santa Cruz Biotechnology (Dallas, TX, USA).

5. Conclusions

Administration of nSMase induces TP receptor-mediated vasoconstriction and eNOS-mediated vasorelaxation in murine vessels. In spite of endothelial dysfunction in db/db mice, the vasorelaxant effect of nSMase is markedly augmented. SMase-mediated disruption of SM in endothelial lipid rafts might represent a possible mechanism responsible for enhanced NO generation in T2DM. An intriguing interpretation of our finding is that the retraction of eNOS in sphingomyelin-rich microdomains of the endothelial plasma membrane could contribute significantly to the development of vascular dysfunction in T2DM.

Supplementary Materials: The following supporting information can be downloaded at: https://www.mdpi.com/article/10.3390/ijms24098375/s1.

Author Contributions: Conceptualization, É.R., L.K., G.T. and Z.B.; methodology, É.R. and Z.B.; validation, K.L., G.T. and Z.B.; investigation, É.R., A.J., R.C.P., A.P., Z.S. and D.K.; resources, Z.B.; data curation, É.R., R.C.P., A.J., A.P., Z.S. and D.K.; writing—original draft preparation, É.R., A.J., L.K., K.L., G.T. and Z.B.; visualization, É.R. and A.J.; supervision, L.K., K.L., G.T. and Z.B.; project administration, Z.B.; funding acquisition, É.R. and Z.B. All authors have read and agreed to the published version of the manuscript.

Funding: This research was funded by the Hungarian National Research, Development, and Innovation Office (OTKA K-112964, K-125174, K-139230, and PD-132851) and by the Ministry of Innovation and Technology of Hungary from the NRDI Fund (2020-1.1.6-JÖVŐ-2021-00010, 2020-1.1.6-JÖVŐ-2021-00013 and TKP2021-EGA-25).

Institutional Review Board Statement: The animal experiments were carried out according to the guidelines of the Hungarian Law of Animal Protection (40/2013) and were approved by the National Scientific Ethical Committee of Animal Experimentation (PE_EA_924-7_2021, Approval date 7 September 2021).

Informed Consent Statement: Not applicable.

Data Availability Statement: The data presented in this study are available in the Supplemental File.

Acknowledgments: The authors are grateful to Margit Kerék for expert technical assistance and to Erzsébet Fejes for critically reading the manuscript.

Conflicts of Interest: The authors declare no conflict of interest. The funders had no role in the design of the study, in the collection, analysis, or interpretation of data, in the writing of the manuscript, or in the decision to publish the results.

References

1. Peters, S.L.; Alewijnse, A.E. Sphingosine-1-phosphate signaling in the cardiovascular system. *Curr. Opin. Pharmacol.* **2007**, *7*, 186–192. [CrossRef] [PubMed]
2. Igarashi, J.; Michel, T. Sphingosine-1-phosphate and modulation of vascular tone. *Cardiovasc. Res.* **2009**, *82*, 212–220. [CrossRef] [PubMed]
3. Kerage, D.; Brindley, D.N.; Hemmings, D.G. Review: Novel insights into the regulation of vascular tone by sphingosine 1-phosphate. *Placenta* **2014**, *35*, S86–S92. [CrossRef] [PubMed]
4. Proia, R.L.; Hla, T. Emerging biology of sphingosine-1-phosphate: Its role in pathogenesis and therapy. *J. Clin. Investig.* **2015**, *125*, 1379–1387. [CrossRef]
5. Hemmings, D.G. Signal transduction underlying the vascular effects of sphingosine 1-phosphate and sphingosylphosphorylcholine. *Naunyn. Schmiedebergs Arch. Pharmacol.* **2006**, *373*, 18–29. [CrossRef] [PubMed]
6. De Palma, C.; Meacci, E.; Perrotta, C.; Bruni, P.; Clementi, E. Endothelial nitric oxide synthase activation by tumor necrosis factor alpha through neutral sphingomyelinase 2, sphingosine kinase 1, and sphingosine 1 phosphate receptors: A novel pathway relevant to the pathophysiology of endothelium. *Arterioscler. Thromb. Vasc. Biol.* **2006**, *26*, 99–105. [CrossRef] [PubMed]
7. Fyrst, H.; Saba, J.D. An update on sphingosine-1-phosphate and other sphingolipid mediators. *Nat. Chem. Biol.* **2010**, *6*, 489–497. [CrossRef] [PubMed]
8. Meyer zu Heringdorf, D.; Jakobs, K.H. Lysophospholipid receptors: Signalling, pharmacology and regulation by lysophospholipid metabolism. *Biochim. Biophys. Acta* **2007**, *1768*, 923–940. [CrossRef]
9. Strub, G.M.; Maceyka, M.; Hait, N.C.; Milstien, S.; Spiegel, S. Extracellular and intracellular actions of sphingosine-1-phosphate. *Adv. Exp. Med. Biol.* **2010**, *688*, 141–155.
10. Hla, T.; Dannenberg, A.J. Sphingolipid signaling in metabolic disorders. *Cell Metab.* **2012**, *16*, 420–434. [CrossRef]
11. Ernst, A.M.; Brugger, B. Sphingolipids as modulators of membrane proteins. *Biochim. Biophys. Acta* **2014**, *1841*, 665–670. [CrossRef] [PubMed]
12. Adada, M.; Luberto, C.; Canals, D. Inhibitors of the sphingomyelin cycle: Sphingomyelin synthases and sphingomyelinases. *Chem. Phys. Lipids* **2016**, *197*, 45–59. [CrossRef]
13. Pavoine, C.; Pecker, F. Sphingomyelinases: Their regulation and roles in cardiovascular pathophysiology. *Cardiovasc. Res.* **2009**, *82*, 175–183. [CrossRef] [PubMed]
14. Murohara, T.; Kugiyama, K.; Ohgushi, M.; Sugiyama, S.; Ohta, Y.; Yasue, H. Effects of sphingomyelinase and sphingosine on arterial vasomotor regulation. *J. Lipid Res.* **1996**, *37*, 1601–1608. [CrossRef] [PubMed]
15. Johns, D.G.; Jin, J.S.; Wilde, D.W.; Webb, R.C. Ceramide-induced vasorelaxation: An inhibitory action on protein kinase C. *Gen. Pharmacol.* **1999**, *33*, 415–421. [CrossRef] [PubMed]
16. Johns, D.G.; Osborn, H.; Webb, R.C. Ceramide: A novel cell signaling mechanism for vasodilation. *Biochem. Biophys. Res. Commun.* **1997**, *237*, 95–97. [CrossRef]
17. Zheng, T.; Li, W.; Wang, J.; Altura, B.T.; Altura, B.M. Effects of neutral sphingomyelinase on phenylephrine-induced vasoconstriction and Ca(2+) mobilization in rat aortic smooth muscle. *Eur. J. Pharmacol.* **2000**, *391*, 127–135. [CrossRef]
18. Zheng, T.; Li, W.; Wang, J.; Altura, B.T.; Altura, B.M. Sphingomyelinase and ceramide analogs induce contraction and rises in [Ca(2+)](i) in canine cerebral vascular muscle. *Am. J. Physiol. Heart Circ. Physiol.* **2000**, *278*, H1421–H1428. [CrossRef] [PubMed]
19. Altura, B.M.; Gebrewold, A.; Zheng, T.; Altura, B.T. Sphingomyelinase and ceramide analogs induce vasoconstriction and leukocyte-endothelial interactions in cerebral venules in the intact rat brain: Insight into mechanisms and possible relation to brain injury and stroke. *Brain Res. Bull.* **2002**, *58*, 271–278. [CrossRef]
20. Jang, G.J.; Ahn, D.S.; Cho, Y.E.; Morgan, K.G.; Lee, Y.H. C_2-ceramide induces vasodilation in phenylephrine-induced pre-contracted rat thoracic aorta: Role of RhoA/Rho-kinase and intracellular Ca^{2+} concentration. *Naunyn. Schmiedebergs Arch. Pharmacol.* **2005**, *372*, 242–250. [CrossRef] [PubMed]
21. Mogami, K.; Kishi, H.; Kobayashi, S. Sphingomyelinase causes endothelium-dependent vasorelaxation through endothelial nitric oxide production without cytosolic Ca^{2+} elevation. *FEBS Lett.* **2005**, *579*, 393–397. [CrossRef] [PubMed]

22. Cogolludo, A.; Moreno, L.; Frazziano, G.; Moral-Sanz, J.; Menendez, C.; Castaneda, J.; Gonzalez, C.; Villamor, E.; Perez-Vizcaino, F. Activation of neutral sphingomyelinase is involved in acute hypoxic pulmonary vasoconstriction. *Cardiovasc. Res.* **2009**, *82*, 296–302. [CrossRef] [PubMed]
23. Spijkers, L.J.; van den Akker, R.F.; Janssen, B.J.; Debets, J.J.; De Mey, J.G.; Stroes, E.S.; van den Born, B.J.; Wijesinghe, D.S.; Chalfant, C.E.; MacAleese, L.; et al. Hypertension is associated with marked alterations in sphingolipid biology: A potential role for ceramide. *PLoS ONE* **2011**, *6*, e21817. [CrossRef] [PubMed]
24. Spijkers, L.J.; Janssen, B.J.; Nelissen, J.; Meens, M.J.; Wijesinghe, D.; Chalfant, C.E.; De Mey, J.G.; Alewijnse, A.E.; Peters, S.L. Antihypertensive treatment differentially affects vascular sphingolipid biology in spontaneously hypertensive rats. *PLoS ONE* **2011**, *6*, e29222. [CrossRef] [PubMed]
25. van den Elsen, L.W.; Spijkers, L.J.; van den Akker, R.F.; van Winssen, A.M.; Balvers, M.; Wijesinghe, D.S.; Chalfant, C.E.; Garssen, J.; Willemsen, L.E.; Alewijnse, A.E.; et al. Dietary fish oil improves endothelial function and lowers blood pressure via suppression of sphingolipid-mediated contractions in spontaneously hypertensive rats. *J. Hypertens.* **2014**, *32*, 1050–1058; discussion 1058. [CrossRef] [PubMed]
26. Shamseddine, A.A.; Airola, M.V.; Hannun, Y.A. Roles and regulation of neutral sphingomyelinase-2 in cellular and pathological processes. *Adv. Biol. Regul.* **2015**, *57*, 24–41. [CrossRef]
27. Russo, S.B.; Ross, J.S.; Cowart, L.A. Sphingolipids in obesity, type 2 diabetes, and metabolic disease. *Handb. Exp. Pharmacol.* **2013**, *216*, 373–401. [CrossRef]
28. Cowart, L.A. Sphingolipids: Players in the pathology of metabolic disease. *Trends Endocrinol. Metab.* **2009**, *20*, 34–42. [CrossRef]
29. Hannun, Y.A.; Obeid, L.M. The ceramide-centric universe of lipid-mediated cell regulation: Stress encounters of the lipid kind. *J. Biol. Chem.* **2002**, *277*, 25847–25850. [CrossRef] [PubMed]
30. Sawai, H.; Hannun, Y.A. Ceramide and sphingomyelinases in the regulation of stress responses. *Chem. Phys. Lipids* **1999**, *102*, 141–147. [CrossRef] [PubMed]
31. Hannun, Y.A.; Luberto, C. Ceramide in the eukaryotic stress response. *Trends Cell Biol.* **2000**, *10*, 73–80. [CrossRef] [PubMed]
32. van Meer, G.; Holthuis, J.C. Sphingolipid transport in eukaryotic cells. *Biochim. Biophys. Acta* **2000**, *1486*, 145–170. [CrossRef] [PubMed]
33. Holland, W.L.; Summers, S.A. Sphingolipids, insulin resistance, and metabolic disease: New insights from in vivo manipulation of sphingolipid metabolism. *Endocr. Rev.* **2008**, *29*, 381–402. [CrossRef] [PubMed]
34. Unger, R.H.; Orci, L. Lipoapoptosis: Its mechanism and its diseases. *Biochim. Biophys. Acta* **2002**, *1585*, 202–212. [CrossRef]
35. Boden, G. Pathogenesis of type 2 diabetes. Insulin resistance. *Endocrinol. Metab. Clin. N. Am.* **2001**, *30*, 801–815, v. [CrossRef] [PubMed]
36. Samad, F. Contribution of sphingolipids to the pathogenesis of obesity. *Future Lipidol.* **2007**, *2*, 625–639. [CrossRef]
37. Symons, J.D.; Abel, E.D. Lipotoxicity contributes to endothelial dysfunction: A focus on the contribution from ceramide. *Rev. Endocr. Metab. Disord.* **2013**, *14*, 59–68. [CrossRef]
38. Aasum, E.; Hafstad, A.D.; Severson, D.L.; Larsen, T.S. Age-dependent changes in metabolism, contractile function, and ischemic sensitivity in hearts from db/db mice. *Diabetes* **2003**, *52*, 434–441. [CrossRef] [PubMed]
39. Coleman, D.L. Obese and diabetes: Two mutant genes causing diabetes-obesity syndromes in mice. *Diabetologia* **1978**, *14*, 141–148. [CrossRef] [PubMed]
40. Do, O.H.; Low, J.T.; Gaisano, H.Y.; Thorn, P. The secretory deficit in islets from db/db mice is mainly due to a loss of responding beta cells. *Diabetologia* **2014**, *57*, 1400–1409. [CrossRef] [PubMed]
41. Sallam, N.A.; Laher, I. Redox Signaling and Regional Heterogeneity of Endothelial Dysfunction in db/db Mice. *Int J. Mol. Sci.* **2020**, *21*, 6147. [CrossRef]
42. Ramadan, F.M.; Upchurch, G.R., Jr.; Keagy, B.A.; Johnson, G., Jr. Endothelial cell thromboxane production and its inhibition by a calcium-channel blocker. *Ann. Thorac. Surg.* **1990**, *49*, 916–919. [CrossRef]
43. Epand, R.M.; So, V.; Jennings, W.; Khadka, B.; Gupta, R.S.; Lemaire, M. Diacylglycerol Kinase-epsilon: Properties and Biological Roles. *Front. Cell. Dev. Biol.* **2016**, *4*, 112. [CrossRef] [PubMed]
44. Subramanian, P.; Vora, M.; Gentile, L.B.; Stahelin, R.V.; Chalfant, C.E. Anionic lipids activate group IVA cytosolic phospholipase A2 via distinct and separate mechanisms. *J. Lipid Res.* **2007**, *48*, 2701–2708. [CrossRef] [PubMed]
45. Pettus, B.J.; Bielawska, A.; Spiegel, S.; Roddy, P.; Hannun, Y.A.; Chalfant, C.E. Ceramide kinase mediates cytokine- and calcium ionophore-induced arachidonic acid release. *J. Biol. Chem.* **2003**, *278*, 38206–38213. [CrossRef] [PubMed]
46. Mitsutake, S.; Date, T.; Yokota, H.; Sugiura, M.; Kohama, T.; Igarashi, Y. Ceramide kinase deficiency improves diet-induced obesity and insulin resistance. *FEBS Lett.* **2012**, *586*, 1300–1305. [CrossRef] [PubMed]
47. Machida, T.; Matamura, R.; Iizuka, K.; Hirafuji, M. Cellular function and signaling pathways of vascular smooth muscle cells modulated by sphingosine 1-phosphate. *J. Pharmacol. Sci.* **2016**, *132*, 211–217. [CrossRef] [PubMed]
48. Goto, K.; Kitazono, T. Endothelium-dependent hyperpolarization (EDH) in diet-induced obesity. *Endocr. Metab. Sci.* **2020**, *1*, 100062. [CrossRef]
49. Pilarczyk, M.; Mateuszuk, L.; Rygula, A.; Kepczynski, M.; Chlopicki, S.; Baranska, M.; Kaczor, A. Endothelium in spots—high-content imaging of lipid rafts clusters in db/db mice. *PLoS ONE* **2014**, *9*, e106065. [CrossRef]

50. Mitsutake, S.; Zama, K.; Yokota, H.; Yoshida, T.; Tanaka, M.; Mitsui, M.; Ikawa, M.; Okabe, M.; Tanaka, Y.; Yamashita, T.; et al. Dynamic modification of sphingomyelin in lipid microdomains controls development of obesity, fatty liver, and type 2 diabetes. *J. Biol. Chem.* **2011**, *286*, 28544–28555. [CrossRef] [PubMed]
51. Romiti, E.; Meacci, E.; Tanzi, G.; Becciolini, L.; Mitsutake, S.; Farnararo, M.; Ito, M.; Bruni, P. Localization of neutral ceramidase in caveolin-enriched light membranes of murine endothelial cells. *FEBS Lett.* **2001**, *506*, 163–168. [CrossRef] [PubMed]
52. Jasmin, J.F.; Frank, P.G.; Lisanti, M.P. Caveolins and caveolae: Roles in signaling and disease mechanism. In *Advances in Experimental Medicine and Biology*; Springer Science + Business Media, LCC.: New York, NY, USA, 2012; pp. 3–13.
53. Garcia-Cardena, G.; Martasek, P.; Masters, B.S.; Skidd, P.M.; Couet, J.; Li, S.; Lisanti, M.P.; Sessa, W.C. Dissecting the interaction between nitric oxide synthase (NOS) and caveolin. Functional significance of the nos caveolin binding domain in vivo. *J Biol. Chem.* **1997**, *272*, 25437–25440. [CrossRef] [PubMed]
54. Frank, P.G.; Woodman, S.E.; Park, D.S.; Lisanti, M.P. Caveolin, caveolae, and endothelial cell function. *Arterioscler. Thromb. Vasc. Biol.* **2003**, *23*, 1161–1168. [CrossRef] [PubMed]
55. Shaul, P.W. Regulation of endothelial nitric oxide synthase: Location, location, location. *Annu. Rev. Physiol.* **2002**, *64*, 749–774. [CrossRef]
56. Lam, T.Y.; Seto, S.W.; Lau, Y.M.; Au, L.S.; Kwan, Y.W.; Ngai, S.M.; Tsui, K.W. Impairment of the vascular relaxation and differential expression of caveolin-1 of the aorta of diabetic +db/+db mice. *Eur. J. Pharmacol.* **2006**, *546*, 134–141. [CrossRef]
57. Miller, M.A.; Morgan, R.J.; Thompson, C.S.; Mikhailidis, D.P.; Jeremy, J.Y. Adenylate and guanylate cyclase activity in the penis and aorta of the diabetic rat: An in vitro study. *Br. J. Urol.* **1994**, *74*, 106–111. [CrossRef]
58. Samad, F.; Hester, K.D.; Yang, G.; Hannun, Y.A.; Bielawski, J. Altered adipose and plasma sphingolipid metabolism in obesity: A potential mechanism for cardiovascular and metabolic risk. *Diabetes* **2006**, *55*, 2579–2587. [CrossRef]
59. Arora, T.; Velagapudi, V.; Pournaras, D.J.; Welbourn, R.; le Roux, C.W.; Oresic, M.; Backhed, F. Roux-en-Y gastric bypass surgery induces early plasma metabolomic and lipidomic alterations in humans associated with diabetes remission. *PLoS ONE* **2015**, *10*, e0126401. [CrossRef]
60. Fox, T.E.; Bewley, M.C.; Unrath, K.A.; Pedersen, M.M.; Anderson, R.E.; Jung, D.Y.; Jefferson, L.S.; Kim, J.K.; Bronson, S.K.; Flanagan, J.M.; et al. Circulating sphingolipid biomarkers in models of type 1 diabetes. *J. Lipid Res.* **2011**, *52*, 509–517. [CrossRef]
61. Gorska, M.; Baranczuk, E.; Dobrzyn, A. Secretory Zn^{2+}-dependent sphingomyelinase activity in the serum of patients with type 2 diabetes is elevated. *Horm. Metab. Res.* **2003**, *35*, 506–507. [CrossRef]
62. Gorska, M.; Dobrzyn, A.; Baranowski, M. Concentrations of sphingosine and sphinganine in plasma of patients with type 2 diabetes. *Med. Sci. Monit.* **2005**, *11*, CR35–CR38. [PubMed]
63. Donath, M.Y.; Shoelson, S.E. Type 2 diabetes as an inflammatory disease. *Nat. Rev. Immunol.* **2011**, *11*, 98–107. [CrossRef] [PubMed]
64. Dressler, K.A.; Mathias, S.; Kolesnick, R.N. Tumor necrosis factor-alpha activates the sphingomyelin signal transduction pathway in a cell-free system. *Science* **1992**, *255*, 1715–1718. [CrossRef] [PubMed]
65. Wiegmann, K.; Schutze, S.; Machleidt, T.; Witte, D.; Kronke, M. Functional dichotomy of neutral and acidic sphingomyelinases in tumor necrosis factor signaling. *Cell* **1994**, *78*, 1005–1015. [CrossRef] [PubMed]
66. Wang, B.; Chandrasekera, P.C.; Pippin, J.J. Leptin- and leptin receptor-deficient rodent models: Relevance for human type 2 diabetes. *Curr. Diabetes Rev.* **2014**, *10*, 131–145. [CrossRef]
67. Cohen, M.P.; Clements, R.S.; Hud, E.; Cohen, J.A.; Ziyadeh, F.N. Evolution of renal function abnormalities in the db/db mouse that parallels the development of human diabetic nephropathy. *Exp. Nephrol.* **1996**, *4*, 166–171.
68. Clarke, C.J.; Snook, C.F.; Tani, M.; Matmati, N.; Marchesini, N.; Hannun, Y.A. The extended family of neutral sphingomyelinases. *Biochemistry* **2006**, *45*, 11247–11256. [CrossRef]
69. Schulz, M.K.; Gutterman, D.K.; Freed, J.K. Activation of Neutral Sphingomyelinase Contributes to Nitric Oxide-Mediated Flow-Induced Dilation in the Human Microvasculature. *FASEB J.* **2020**, *34*, 1. [CrossRef]
70. Bautista-Perez, R.; del Valle-Mondragon, L.; Cano-Martinez, A.; Perez-Mendez, O.; Escalante, B.; Franco, M. Involvement of neutral sphingomyelinase in the angiotensin II signaling pathway. *Am. J. Physiol. Renal. Physiol.* **2015**, *308*, F1178–F1187. [CrossRef]
71. Horvath, B.; Orsy, P.; Benyo, Z. Endothelial NOS-mediated relaxations of isolated thoracic aorta of the C57BL/6J mouse: A methodological study. *J. Cardiovasc. Pharmacol.* **2005**, *45*, 225–231. [CrossRef]
72. Hojjati, M.R.; Jiang, X.C. Rapid, specific, and sensitive measurements of plasma sphingomyelin and phosphatidylcholine. *J. Lipid Res.* **2006**, *47*, 673–676. [CrossRef] [PubMed]
73. Raines, M.A.; Kolesnick, R.N.; Golde, D.W. Sphingomyelinase and ceramide activate mitogen-activated protein kinase in myeloid HL-60 cells. *J. Biol. Chem.* **1993**, *268*, 14572–14575. [CrossRef] [PubMed]
74. Linardic, C.M.; Hannun, Y.A. Identification of a distinct pool of sphingomyelin involved in the sphingomyelin cycle. *J. Biol. Chem.* **1994**, *269*, 23530–23537. [CrossRef] [PubMed]

www.ingramcontent.com/pod-product-compliance
Lightning Source LLC
Chambersburg PA
CBHW040717270926
51850CB00023B/2678
9783036576855

MDPI
St. Alban-Anlage 66
4052 Basel
Switzerland
Tel. +41 61 683 77 34
Fax +41 61 302 89 18
www.mdpi.com

International Journal of Molecular Sciences Editorial Office
E-mail: ijms@mdpi.com
www.mdpi.com/journal/ijms